FeFET Devices, Trends, Technology and Applications

Scrivener Publishing
100 Cummings Center, Suite 541J
Beverly, MA 01915-6106

Semiconductor Devices and Their Applications

Series Editor: Balwinder Raj, PhD
Shingo Yamaguchi, PhD
Shiromani Balmukund Rahi, PhD

Scope:
Nanoelectronic devices such as double gate MOSFET, FinFET, Nanowire, CNTFET, TFET, FeFET, Gate All Around (GAA), HEMT, and spin-based devices are used for chip design. The study of these devices is important for sectors whose applications in chip design have a huge impact for bringing revolutionary advances in nanoelectronic devices, circuits and systems due to improved electronic properties of the nanoscale semiconductor devices. The knowledge and resources for the fabrication of these nanoelectronic devices is also available, as its fabrication flow is similar to conventional MOSFET. It is predicted by the International Technology Road Map for Semiconductors (ITRS) that, with advancement of technology the conventional fabrication flow can be used for nanoelectronic devices. Artificial intelligence (AI), machine learning (ML), and advanced electronic circuits involve learning from every data input and using those inputs to generate new rules for future business analytics. AI and machine learning are now giving us new opportunities to use big data, as well as unleash new use cases with new data types. With the increasing use of AI dealing with highly sensitive information such as healthcare, adequate security measures are required to securely store and transmit this information. This book series will provides a broader coverage of the basic aspects of advanced circuits design and applications in AI/ML.

This Wiley-Scrivener book series presents concise summaries of cutting-edge research and practical applications across a wide spectrum in the field advanced devices, circuits and systems for AI applications. The series covers a range of content from professional to academic. Typical topics might include timely report of state-of-the art analytical techniques, a bridge between new research results, as published in journal articles, and a contextual literature review, a snapshot of a hot or emerging topic, an in-depth case study or clinical example and a presentation of core concepts that students must understand in order to make independent contributions.

Publishers at Scrivener
Martin Scrivener (martin@scrivenerpublishing.com)
Phillip Carmical (pcarmical@scrivenerpublishing.com)

FeFET Devices, Trends, Technology and Applications

Edited by

Balwinder Raj

Shiromani Balmukund Rahi

and

Nandakishor Yadav

Scrivener Publishing

WILEY

This edition first published 2025 by John Wiley & Sons, Inc., 111 River Street, Hoboken, NJ 07030, USA and Scrivener Publishing LLC, 100 Cummings Center, Suite 541J, Beverly, MA 01915, USA

For more information about Scrivener publications please visit www.scrivenerpublishing.com.

Wiley Global Headquarters
111 River Street, Hoboken, NJ 07030, USA

For details of our global editorial offices, customer services, and more information about Wiley products visit us at www.wiley.com.

Library of Congress Cataloging-in-Publication Data

ISBN 9781394287277

Front cover images supplied by Adobe Firefly
Cover design by Russell Richardson

Set in size of 11pt and Minion Pro by Manila Typesetting Company, Makati, Philippines

Printed in the USA

10 9 8 7 6 5 4 3 2 1

Contents

Preface

Ferroelectric materials and devices are essential components of nonvolatile memory systems like FeRAM, wherein their polarization changes are employed for encoding binary information. These are additionally used in a variety of sensors, such as temperature and piezoelectric meters, to take advantage of their outstanding piezoelectric capabilities. Ferroelectric nanowires may be made using processes like chemical vapor deposition and sol-gel procedures, which allow for precise oversight of their diameters. This book explores the characteristics, novel materials used, modification in device structure, and advancement of model FET devices. Even though many devices, following Moore's law and more than Moore, are proposed and designed, a complete history of the existing and proposed semiconductor devices is not available to the readers. The book focuses on developments and research that is undergone in emerging ferroelectric FET devices and their applications. It also provides unique coverage of topics covering recent advancements and novel concepts in the field of miniaturized ferroelectric devices to motivate young researchers.

The scope of the book is generic and provides an easy-to-understand approach, making it excellent for those who are new to the subject, and hence, it would be for scientists, researchers, and postgraduate students who are interested in learning the latest emerging ferroelectric materials, nanoscale devices, and its properties. The reader of the book is able to get an insight analysis of all the recent developments and developed ferroelectric device structures along with their applications. The properties, characterization, and their relative applications are proposed in a different manner. This book provides the state-of-the-art research techniques available in the field of ferroelectric and its devices research connected to the applications. The students also find the applications of nanoscale semiconductor devices in this book which makes them feel interesting. A brief summary of the chapters is given below:

Chapter 1: The evolution of silicon-based complementary metal-oxide semiconductor (CMOS) technology has been essential in shaping modern semiconductor electronics. In this evolution, the steady scaling of transistors has contributed a lot, and these steady developments of transistors enabled the recent high-performance semiconductor chips. In this book chapter, we are going to explore the historical scaling advancements in CMOS technology, highlighting the challenges faced in the past and present. Past challenges encompassed issues such as short-channel effects and leakage currents, while current challenges revolve around power dissipation and quantum effects at nanoscale dimensions. Looking forward, the chapter also discusses potential future innovations including novel materials, device architectures, and computing paradigms. By addressing these challenges and embracing new technologies, Si-based CMOS is poised to continue its transformative journey in the semiconductor industry. Even though there are still a couple of issues to be overcome, it seems like the CMOS technology is able to keep on improving, with the help of material development and fabrication-equipment development. This chapter has been written to understand the basic principles of semiconductors and the demands for improving scaling and subthreshold swing, which are essential for designing FeFET.

Chapter 2: Ferroelectric polymers in organic field-effect transistors (OFETs) have garnered significant attention as they offer cost-effective means and simplified manufacturing processes based on solution processability and flexible memory modules. They can be fabricated by various solution processing techniques such as spin coating, Langmuir-Blodgett, ink-jet printing, and roll-to-roll printing. Organic ferroelectric field-effect transistors (FeFETs) can be used in large electronic applications: flexible displays, sensors, radio frequency identification (RFID) tags, etc. An additional benefit of PVDF copolymers like poly(vinylidene fluoride-trifluoroethylene) [P(VDF-TrFE)] is their ability to rapidly crystallize into the β phase. The β phase shows the maximum dipole moment within various reported polar phases. Ferroelectric polymer material–based devices have shown good memory functionalities which can have great significance for various applications. Particularly, OFET having ferroelectric polymer as a gate dielectric layer has attracted huge interest because of their large potentiality aimed at the growth of nonvolatile memory devices at low cost. This work highlights the diverse applications of FeFETs in advancing nonvolatile memory, artificial synapses, and wearable technologies, leveraging ferroelectric polymers.

Chapter 3: Advancement in next-generation technologies demands multifunctional devices with ultrafast speed, low cost, and smaller size. Therefore, immense efforts are being made to understand the complex and cooperative behavior of current, charge, electric dipole moments, and/or spin degrees of freedom of materials. For example, ferroelectrics (a non-centrosymmetric material with switchable spontaneous polarization) near or below the size of 3.2 nm are appealing to explore their potential application for next-generation devices. Ferroelectric materials are generally used as a tuneable capacitor and data storage devices in various electronic systems such as mobiles, computers, space centers, and satellites. They can also be utilized as crucial electronic components such as a phase shifter, electrocaloric effect-based cooler, memory, and energy harvester in satellites and/or space flight-based advance devices. To improve the functionality of these materials further, extensive studies are required. We briefly present here the basic concept of ferroelectricity, necessities, facile and advanced synthesis techniques, and potential applications of ferroelectric materials in novel devices.

Chapter 4: This chapter introduces a hetero-buried oxide (HBOX)–ferroelectric tunnel FET. Numerous electrical parameters are tuned concerning the silicon and BOX layer thickness adjustment, including ON/OFF-state currents and I_{ON}/I_{OFF} ratio. A negative capacitance–based ferroelectric gate stack is also introduced for obtaining the steep subthreshold slope (SS). To ensure compatibility with the typical FET production approach, a ferroelectric (Fe) layer comprised of silicon-doped hafnium oxide is employed at the bottom of the gate. The Fe layer thickness is carefully optimized to improve Analog and DC parameters. A doped pocket (n+) has been imparted within the source in conjunction with the gate overlapping and drain underlapping. In addition to the design features, a pocket improves the electric field around the tunnel junction, making it more likely for the carriers to tunnel toward the channel. The hetero-buried oxide is integrated into the silicon-on-insulator material. A high k-BOX remains at the source and a low k-BOX is exploited at the drain sides to minimize the ambipolar properties of the TFET and increase the ON-state current. The proposed structure's characteristics have been optimized to accommodate changes in its physical dimensions. This arrangement provides a 38.5 mV/dec subthreshold slope and an I_{ON}/I_{OFF} ratio of around 10^9. The Fe HBOX TFET is known to operate at an impressive speed, depending on the capacitances recovered from it. It is suitable for analog/RF applications as it exhibits superior transconductance and a reduced amount of output conductance. The steep subthreshold slope makes it ideal for low-power-demanding applications. The proposed ferroelectric TFET device evaluations are performed with the Sentaurus TCAD 2D software.

Chapter 5: A thorough analysis of ferroelectric materials in cutting-edge nanoelectronics devices is presented here. Ferroelectric material–based FETs describe how mitigating the problem arises in conventional CMOS. A ferroelectric polarization field is injected into a field-effect transistor (FET) to control semiconductor carriers, resulting in a ferroelectric field-effect transistor (FeFET). This article compares several ferroelectric materials' characteristics, such as HfO_2, PZT, and other 2D materials, and shows how ferroelectricity varies with temperature. After outlining the distinct electrical and physical characteristics of ferroelectric materials, we concentrate on different structures of field-effect transistors with ferroelectric materials based on 2D and 3D structures. FeFET's different architectures, such as MFS and MFIS, and applications based on these structures are also given. The various FeFET structures are compared to each other in terms of performances and limitations to the real world. The other structures of field-effect transistors are proposed based on ferroelectric materials for various applications such as memory and sensors. To pique the attention of scientists interested in the continued development of this rapidly developing field of study, this article thoroughly analyzes the current and upcoming challenges in ferroelectric-based nanoelectronics. FeFETs are appealing for cutting-edge electrical and optoelectronic applications, such as developing memory, artificial neural networks, high-performance photodetectors, and intelligent sensors due to their ferroelectric and semiconductor flexible interaction. In this review, we also covered the potential applications of FeFET in the future for better device operation and which ferroelectric materials have the best FeFET structure and can retain ferroelectricity up to 1-nm technology.

Chapter 6: This chapter investigates how negative capacitance (NC) from ferroelectric materials can be incorporated into existing double-gate tunnel field-effect transistors (DG-TFETs). This approach aims to overcome the limitations of conventional FETs (bulk MOSFETs, FinFETs, GAAFETs, and nanosheet FETs) in ultralow power applications. This chapter explores how ferroelectric gate materials can improve performance in double-gate tunnel field-effect transistors (DG-TFETs). This approach allows for lower power supply voltages without sacrificing key device characteristics like on-current (I_{ON}), off-current (I_{OFF}), and their ratio (I_{ON}/I_{OFF}). Experiments show impressive results, achieving very low leakage currents (I_{OFF} around 10^{-20} A/μm) and ultralow on-currents (I_{ON} around 10^{-5} A/μm while significantly reducing subthreshold swing (SS). The chapter also addresses the challenges of integrating TFETs with existing CMOS technology. Solutions like using materials with smaller bandgaps and implementing type II staggered heterojunctions are proposed to improve tunneling efficiency and boost on-current. Overall, this study highlights the potential of ferroelectric materials to revolutionize low-power electronics. By enhancing DG-TFET performance, these materials pave the way for significantly lower power supply voltages in future semiconductor devices.

Chapter 7: Internal heating in portable devices is one of the most challenging tasks in front of researchers owing to excessive utilization of it in the daily lives of people worldwide. So, these devices are an ingenuous part of our lives owing to the capability to have an excess amount of information and its utilization. These devices contain smaller-size (nanosize) constituents that have better quality in almost all prospectives than their older counterparts. In the present scenario, these devices must consume less power to improve their applicability and utilization for a longer duration. The negative capacitance (NC) features developed in the field-effect transistor (FET) owing to the presence of the ferroelectric materials in the gate stack reduces the subthreshold limit and leakage current which make NCFET an important part of present portable as well as advanced devices. This chapter describes pieces of evidence of NC in the ferroelectric material sandwiched within the gate stack and its performance on NCFET transistor applicability. It also covers the present status of evidence of NC in hafnia, AlScN, and fluorite-based advanced and futuristic ferroelectric transistors and its futuristic perspective role in NCFET devices.

Chapter 8: Tunnel field-effect transistor (TFET)–based biosensors represent a cutting-edge technology with immense potential for revolutionizing the landscape of sensing and diagnostics. Their versatility, sensitivity, and compatibility with emerging nanotechnologies position TFET biosensors at the forefront of innovation, opening new avenues for applications in healthcare, environmental monitoring, and beyond. As research in this field continues to progress, TFET biosensors are poised to play a pivotal role in shaping the future of analytical and diagnostic sciences. TFET biosensors operate on the principle of detecting changes in the electrical characteristics of a semiconductor channel induced by biomolecular interactions at the sensor surface. The intrinsic sensitivity of TFETs to charge variations makes them ideal candidates for the real-time, label-free detection of various biomolecules, such as proteins, nucleic acids, and small molecules.

Chapter 9: This chapter gives an elaboration of the distinct class of semiconductor materials, known as transparent conducting oxides (TCO). These materials have novel characteristics on account of exhibiting the unusual combination of two contemporary properties: high optical transparency and high electrical conductivity. These unique optoelectronic properties of these materials depend upon different parameters which are also deeply discussed in the chapter. For the optical properties, contributing factors like optical gap, hybridization of the orbitals, defects, and doping are discussed in detail. On a similar platform, effective masses and electron distribution laws are discussed in order to justify the electrical properties of the TCOs. Furthermore, different types of TCO materials are also discussed in the chapter. The materials are elaborated with their structures that are responsible for imparting unique properties to the specific material. Transparent conducting oxides form the backbone of modern technology owing to the unique combination of transparency and conductivity. The chapter also describes different routes to deposit the transparent conducting oxides in the form of nanostructures like thin films and nanoparticles. Deposition techniques play a significant role in imparting specific properties to the deposited nanostructures. The two types of deposition techniques, i.e., physical deposition techniques and chemical deposition techniques, along with the specific requirements are explored in the chapter. Among the chemical deposition techniques, the chapter lays special emphasis on the sol-gel–based spin-coating technique on account of it being the most inexpensive and easy-to-handle deposition technique. Furthermore, the importance of transparent conducting oxides from the application point of view is discussed. The application of transparent conducting oxides in photovoltaics, display devices, sensors, and multifunctional coatings is also elaborated in detail.

Chapter 10: Owing to the versatile ability to interact with various types of external disturbances such as light, force, heat, and electricity, ferroelectric plays a special role in several devices like traducers, ferroelectric random-access memory, FeFET, ferroelectric tunnel junction, ferroelectric diodes, and nanobiosensors. The characteristic feature of ferroelectrics enhances performance in FeFET biosensors compared to conventional FET biosensors. This is due to improvements in the subthreshold swing slope, better control over the channel, and lower power consumption. Additionally, ferroelectric biosensors outperform traditional biosensors because of the induced spontaneous polarization resulting from interactions with biomolecule energy and/or physical stimuli. Due to this property of ferroelectrics, the output characteristics of ferroelectric biosensors and FeFET biosensors are improved. This is because the excess availability of carriers at the interface surface results from the additional generation of spontaneous polarization, which induces electron-hole pairs. Consequently, this leads to enhanced sensitivity, selectivity, and quicker response times.

To further understand the functionality of ferroelectric, extensive studies are required. Herein, we summarize the ferroelectric-based biosensor design, working principle, complex interaction mechanisms, and monitoring efficiency toward the specific targeted biomolecule (or analyte) in the presence of interfering contents. In addition, a detailed discussion of its applications in human health, environmental, and agricultural monitoring, supported by recent reports, is provided in the chapter. It also gives insights into the future perspective of ferroelectricity in biosensor devices.

Chapter 11: Recently, ferroelectric materials have garnered considerable attention in the field of material and electronic engineering due to their potential uses in nanoscale devices, spurred by the ongoing scaling of semiconductor technology as outlined in the International Technology Roadmap for Semiconductors (ITRS) and other engineering organization. This chapter is going to give some introduction to ferroelectric applications, elucidating the characteristics of ferroelectric materials and discussing the fundamental scaling principles outlined in the ITRS roadmap. This chapter then explores the realm of nanoscale devices, examining various configurations such as ultrathin body MOSFETs, gate-all-around MOSFETs, and advanced engineering techniques such as local high doping. Furthermore, this chapter investigates the incorporation of ferroelectric materials into various modern nanoscale devices, emphasizing their possible advantages and potential applications. Additionally, this chapter is also going to examine the placement of ferroelectric technologies within the ITRS roadmap framework and provide insights into the future prospects of ferroelectric applications in the semiconductor industry as well. All in all, it seems like with the help of the guideline provided by the ITRS roadmap and advancement of ferroelectric materials, the semiconductor devices (including memory devices and memory devices) would be able to further able to achieve more improvement in terms of low power consumption and fast switching operation.

Chapter 12: Electron mobility is one of the most important parameters in electronic applications and transistor design. With the advent of the recent ferroelectric field-effect transistor (FeFET), the importance of understanding electron mobility has become more and more significant. In this regard, this chapter is going to address recent electron mobility models for FeFET. From the classical electron mobility models to quantum mechanical electron mobility models, various electron models are explained so that the readers can understand the electron mobility concept in modern semiconductor transistors with more ease. Specifically, as the dimension of the device has been decreased over and over again, various complex quantum mechanisms made it difficult to interpret the movement of electrons and calculate electron mobility. In this chapter, we would like to look into the recent electron mobility models which have been developed for modern transistors and aimed to have high accuracy, compared to the conventional electron mobility models.

Acknowledgements

This book is based on research work carried out by a number of authors. Many people have contributed greatly to this book on FeFET devices, trends, technology and applications. We would like to thank and acknowledge the efforts and support of all contributors for completing the book. We are also grateful for a number of friends and colleagues in encouraging us to start work on this book. We as the editors would like to acknowledge all of them for their valuable help and generous ideas in improving the quality of this book. With our feelings of gratitude, we would like to introduce them in turn. The first mention is the authors and reviewers of each chapter of this book. Without their outstanding expertise, constructive reviews, and devoted effort, this comprehensive book would become something without content. The second mention is the Wiley and Scrivener staff for their constant encouragement, continuous assistance, and untiring support. Without their technical support, this book would not be completed. The third mention is the editor's family for being the source of continuous love, unconditional support, and prayers not only for this work but also throughout our lives. Last but far from least, we express our heartfelt thanks to the Almighty for bestowing over us the courage to face the complexities of life and complete this work.

Balwinder Raj
Shiromani Balmukund Rahi
Nandakishor Yadav
July 2024

1

Scaling and Challenge of Si-Based CMOS: Past, Present, and Future

Shiromani Balmukund Rahi[1] and Young Suh Song[2]*

[1]University School of Information and Communication Technology, Gautam Buddha University, Greater Noida, Uttar Pradesh, India
[2]Department of Computer Science, Korea Military Academy, Seoul, Republic of Korea

Abstract

The evolution of silicon-based Complementary Metal-Oxide-Semiconductor (CMOS) technology has been essential in shaping modern semiconductor electronics. The steady scaling of transistors has contributed significantly to this evolution, and the steady development of transistors has enabled the development of high-performance semiconductor chips. In this chapter, we explore the historical scaling advancements in CMOS technology, highlighting the challenges faced in the past and present. Past challenges encompass issues such as short-channel effects and leakage currents, whereas current challenges revolve around power dissipation and quantum effects at nanoscale dimensions. This chapter also discusses potential future innovations, including novel materials, device architectures, and computing paradigms. By addressing these challenges and embracing new technologies, Si-based CMOS has been poised to continue its transformative journey in the semiconductor industry. Even though there are still a couple of issues to be overcome, it seems that CMOS technology is able to continue improving, with the help of material development and fabrication-equipment development. This chapter has been written to understand the basic principles of semiconductors and the demands for improving scaling and subthreshold swing, which are essential for designing FeFETs.

***Keywords*:** Quantum computing, digital twins, technological development, semiconductor industry, semiconductor

**Corresponding author*: songyeongseo84@gmail.com

Balwinder Raj, Shiromani Balmukund Rahi and Nandakishor Yadav (eds.) FeFET Devices, Trends, Technology and Applications, (1–26) © 2025 Scrivener Publishing LLC

1.1 Introduction to Si-Based CMOS Technology

Silicon-based Complementary Metal-Oxide-Semiconductor (CMOS) technology is a fundamental element in modern semiconductor electronics [1]. Its core principles and ability to scale have significantly advanced the semiconductor industry, leading to unprecedented performance and integration [2]. Gaining a solid understanding of the fundamentals of Si-based CMOS is crucial for recognizing its historical significance and evolutionary trajectory [3].

CMOS technology relies on silicon as its semiconductor material, leveraging its abundance, stability, and well-understood properties to create electronic components [4]. The CMOS architecture comprises complementary pairs of Metal-Oxide-Semiconductor Field-Effect Transistors (MOSFETs), specifically NMOS (N-type MOS) and PMOS (P-type MOS). These complementary transistor pairs work together to facilitate efficient switching and logic operation.

A major strength of the Si-based CMOS technology is its scalability. Engineers and researchers have continually pushed the boundaries of miniaturization by reducing transistor dimensions to enhance performance while lowering power consumption [5]. This scaling trend, often associated with Moore's law, has resulted in an exponential growth in transistor density per unit area and improved computational capabilities in electronic devices.

The evolution of Si-based CMOS technology has seen successive generations marked by advancements in fabrication techniques, material innovation, and architectural improvements. From the early days of planar CMOS technology to the development of FinFETs and beyond, the industry has adapted to meet the demands for faster, smaller, and more energy-efficient devices.

Furthermore, Si-based CMOS technology has widespread application in various electronic devices such as microprocessors, memory chips, sensors, and integrated circuits used in communication and computing systems [6–8]. Its versatility and scalability have been instrumental in driving the digital revolution and fostering innovations in areas such as artificial intelligence, the Internet of Things (IoT), and mobile computing.

1.2 Basic Concept of Transistor Scaling

Transistor scaling is a fundamental concept in semiconductor technology, which involves reducing the size of transistors while improving their performance and energy efficiency [9, 10]. This concept is essential for meeting the demands of modern electronic devices that require smaller, faster, and more power-efficient components.

At its core, transistor scaling refers to shrinking the dimensions of transistors, such as the gate length and channel width, to enhance their switching speed and reduce power consumption [11, 12]. The basic idea behind scaling is to pack more transistors onto a semiconductor chip, thereby increasing its computational power and functionality.

One of the key metrics used to quantify transistor scaling is feature size, typically measured in nanometers (nm). As technology advances, feature sizes decrease, allowing more transistors to be integrated into smaller areas [13–18]. This trend, often described by Moore's law, has driven rapid advancements in semiconductor technology over the past few decades. Specifically, as the transistor size decreases, the time constant (T_d) of the transistor can also be decreased (Equations 1.1–1.5). Equations 1.1–1.5 and Figure 1.1 can explain the earlier sentence (I_D, C, V, W, L, and μ represent the drain current, capacitance, voltage applied to MOSFET (V_{DS}), width of MOSFET, length of MOSFE, and electron mobility, respectively).

$$I_D = \frac{C\Delta V}{\Delta t} \tag{1.1}$$

$$T_d = \Delta t = \frac{C\Delta V}{I_D} \tag{1.2}$$

when $V_G - V_T = V_{DD}$,

$$C = C_o * W * L \tag{1.3}$$

$$I_D = \frac{\mu W C_o V_{DD}^2}{2L} \tag{1.4}$$

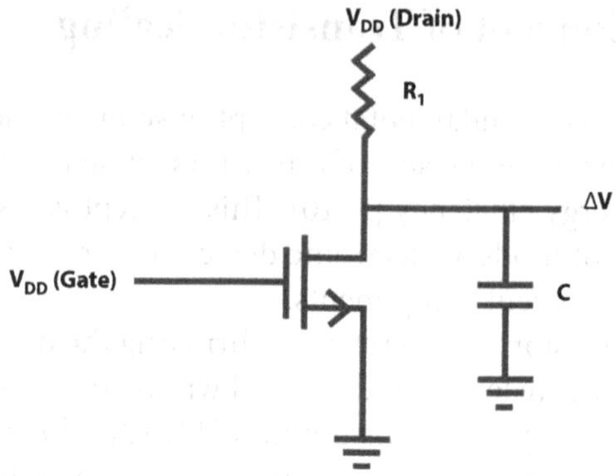

Figure 1.1 Illustration shows the example-circuit for explaining the importance of scaling.

$$T_d = \frac{2L^2}{\mu V_{DD}} \tag{1.5}$$

As shown in Equation 1.5, the length (L) of the transistor decreases and the time constant (T_d) decreases [19–27]. It is important to note that the time constant decreases 'quadratically' as the transistor length decreases. This implies that when the length of the transistor becomes half (1/2), the time constant can be 0.25 times smaller, and therefore the frequency can become four times larger. This implies that the transistor can operate four times faster than before (for example, from 1 GHz to 4 GHz) [28–35].

Transistor scaling provides several advantages for electronic devices. First, smaller transistors have shorter gate lengths, which reduce the time required for signals to propagate through the transistor, leading to faster switching speeds [36–42]. This speed improvement is crucial for high-performance applications, such as microprocessors and memory chips.

Second, scaled transistors typically consume less power during their operation [13–16]. This is due to the reduced leakage currents and improved control over the flow of electrons through the transistor channels [13–15]. Lower power consumption is desirable for extending the battery life in portable devices and reducing the overall energy consumption in large-scale computing systems. However, transistor scaling also poses challenges as the dimensions reach nanoscale. Issues such as leakage currents, short-channel effects, and variability have become more prominent, requiring innovative solutions in transistor design, materials, and fabrication processes [43–49].

In the above paragraph, I explained the importance of scaling by showing the improvement in the speed of the transistor (frequency) and low power consumption [50–56]. However, the recent research shows that it is difficult to improve 'the speed of transistor.' The reason for this can include several self-heating effects in transistors, the generation time of electrons and holes, and power consumption issues [16–18]. Therefore, it seems that many semiconductor companies have recently fabricated multi-core CPU [57–64].

1.3 Past Challenges in Scaling Si-Based CMOS

The scaling of silicon-based Complementary Metal-Oxide-Semiconductor (CMOS) technology has been a remarkable journey, filled with achievements and challenges that must be overcome [65–72]. As engineers and researchers have pushed the limits of miniaturization to improve the performance and integration density, several significant challenges have emerged in the past [73–79].

One of the primary challenges in scaling the Si-based CMOS technology is the onset of short-channel effects. As the transistor dimensions were scaled down, the channel length of MOSFETs became shorter, leading to increased leakage currents and reduced control over the transistor's behavior [80–87]. Short-channel effects limit the performance gains achieved through scaling and require innovative solutions for transistor design and material engineering [19–21].

Another notable challenge is the escalation of power consumption and heat dissipation as transistors become smaller and more densely packed on semiconductor chips [22–26]. This increase in power density poses thermal management challenges, affecting device reliability and overall system performance [27, 28]. Engineers have to develop new thermal management techniques and materials to effectively address these issues [29–34].

Furthermore, as feature sizes approach the nanoscale, variability in transistor characteristics has become a significant concern [35–42]. Process variations during fabrication, inherent material fluctuations, and random dopant fluctuations lead to variations in the transistor performance across a chip and between manufacturing batches [43–47]. Managing and mitigating these variability challenges require advanced modeling, process control, and statistical techniques [48–57].

For further understanding, Figure 1.2 and Equations 1.6–1.9 can be used to explain the effect of scaling on the threshold voltage. As the channel length (L) decreases because of scaling, the threshold voltage usually

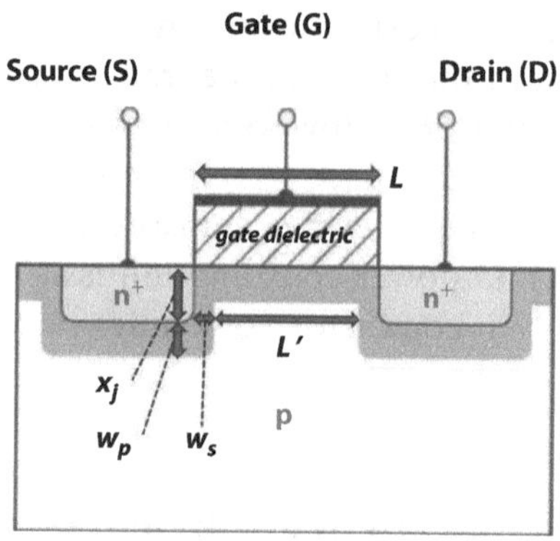

Figure 1.2 Schematic illustration describes the effect of scaling on decrease of threshold voltage.

increases. This phenomenon is often called 'threshold voltage roll-off' [88–93]. In Equations 1.6–1.9, V_T, V_{FB}, Q_d, and C_{ox} represent threshold voltage, flat-band voltage, depletion charge, and oxide-capacitance.

$$V_T = V_{FB} + 2\varphi_F - \frac{Q'_d}{C_{ox}} \tag{1.6}$$

$$Q'_d L = Q_d \frac{L + L'}{2} \tag{1.7}$$

$$(x_j + x_D)^2 = (x_j + w_s)^2 + x_D^2 \tag{1.8}$$

$$V_T = V_{FB} + 2\varphi_F - \frac{Q_d}{C_{ox}}[1 - (\frac{x_j}{L})(\sqrt{1 + 2x_D / x_j} - 1)] \tag{1.9}$$

Equations 1.6–1.9 explains the decrease of threshold voltage, according to 'charge-sharing model.' According to the charge sharing model, as the channel length decreases, the depletion charges on the source and drain sides are shared [94–99]. Namely, under a long-channel device, it is possible to neglect the depletion region because of the source and drain; however, under a short-channel device, 'the depletion region because of the source and drain' cannot be neglected. Therefore, in short-channel devices, the threshold voltage decreases (Equation 1.9) and the off current (I_{off})

increases. Therefore, short-channel devices consume a significant amount of power [100–106].

Moreover, achieving manufacturability and yield at smaller technology nodes is a challenge. The fabrication of complex integrated circuits with billions of transistors requires highly precise and reliable manufacturing processes. Defects, lithography limitations, and yield issues have become more pronounced as feature sizes decrease, necessitating continuous advancements in manufacturing technology and process optimization [107–112].

Despite these challenges, the semiconductor industry has successfully navigated past hurdles in scaling the Si-based CMOS technology. Innovative solutions such as new transistor architectures (e.g., FinFETs), advanced materials (e.g., high-k dielectrics), process enhancements, and design optimizations play pivotal roles in overcoming past scaling challenges and enabling continued progress in semiconductor technology [113–120].

1.4 Present Challenges and Limitations of Si-Based CMOS

Despite significant advancements in silicon-based Complementary Metal-Oxide-Semiconductor (CMOS) technology over time, there are various challenges and limitations in the current landscape, especially as transistor sizes continue to decrease recently (under 20 nm) and new performance requirements have emerged [121–126].

A primary present challenge for Si-based CMOS technology relates to managing power dissipation and heat. Shrinking transistors and their denser integration on semiconductor chips lead to increased power density and elevated heat generation. This rise in power dissipation not only affects device reliability but also calls for effective thermal management solutions to prevent overheating and maintain optimal operational performance [127–132].

Furthermore, as transistor dimensions approach the atomic scale (under 20 nm), quantum effects become more prominent, presenting significant hurdles for Si-based CMOS devices. Quantum phenomena, such as tunneling, limitations in carrier mobility, and confinement effects, can negatively impact transistor behavior, resulting in performance decline and heightened variability. Tackling these quantum-related challenges requires innovative device designs, novel materials, and advanced fabrication techniques [133–138].

In addition, further scaling in Si-based CMOS technology encounters limitations due to physical constraints. The conventional planar transistor structure reaches its scaling thresholds, leading to issues such as short-channel effects, gate leakage, and subthreshold currents. These limitations impede the continuous reduction of transistor dimensions without compromising the functionality or reliability [139–141].

Another critical limitation is the growing complexity of manufacturing processes at advanced technology nodes. Smaller feature sizes and intricate device structures pose challenges, such as process variability, higher defect density, and the need for enhanced yield optimization. Overcoming these manufacturing hurdles requires substantial investments in research, development, and production technologies [142–145].

1.5 Representative Methods for Scaling MOSFET

Three scaling methods are widely accepted for scaling technologies (Figure 1.3). Constant voltage (CV) scaling, quasi-constant voltage (QCV) scaling, and constant electric field (CE) scaling are representative scaling methods [146, 147]. Each has its own strengths and weaknesses, and their characteristics are shown in Tables 1.1 and 1.2. It is worth noting that CE, CV, and QCV scaling methods have their own strengths and weaknesses (Table 1.2).

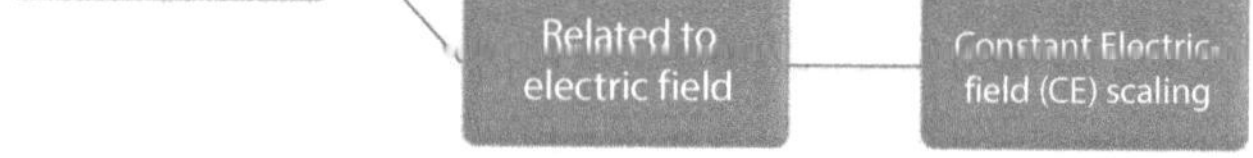

Figure 1.3 Representative methods for MOSFET scaling.

Table 1.1 Explanation of three scaling methods.

Type	Description
CV scaling	Constant voltage scaling is a strategy in which the supply voltage (V_{dd}) is decreased in proportion to the reduction in the transistor feature size. This method aids in lowering the power consumption because the power dissipation in CMOS circuits is directly proportional to the square of the supply voltage. Decreasing the voltage reduces the dynamic power consumption. However, lowering the supply voltage also presents challenges such as heightened sensitivity to process variations and diminished noise margins. Techniques such as adaptive voltage scaling (AVS) and dynamic voltage and frequency scaling (DVFS) are utilized to dynamically adjust the supply voltage based on workload demands, while ensuring performance and energy efficiency.
QCV scaling	Quasi-constant-voltage scaling is a variant of constant-voltage scaling, where the supply voltage is decreased at a slower rate compared to the reduction in feature sizes. This method helps balance the trade-off between power consumption and performance degradation. This allows for some level of voltage scaling, while maintaining acceptable performance levels. Quasi-constant voltage scaling techniques aim to alleviate the impact of reduced voltage on circuit performance by optimizing the circuit design, enhancing transistor characteristics, and employing adaptive control mechanisms.
CE scaling	Constant electric-field scaling is an approach that aims to maintain a constant electric field across the gate oxides of transistors as their dimensions are scaled down. The electric field across the gate oxide influences transistor behavior, including the threshold voltage and subthreshold slope. By preserving a constant electric-field, transistor performance can be maintained or even enhanced despite scaling. This technique requires adjustments to the gate oxide thickness, doping profiles, and device geometries to achieve the desired electric field characteristics while scaling down the transistor dimensions.

Table 1.2 Comparison between three scaling methods.

	CE	CV	QCV
Linear dimension	$1/\lambda$	$1/\lambda$	$1/\lambda$
Potential	$1/\lambda$	1	$1/\lambda^{0.5}$
Electric field	1	λ	$\lambda^{0.5}$
Current	$1/\lambda$	λ	1
Power density	1	λ^3	$\lambda^{0.5}$
Capacitance	$1/\lambda$	$1/\lambda$	$1/\lambda$
Impurity concentration	λ	λ	Λ
Gate delay	$1/\lambda$	$1/\lambda^2$	$1/\lambda^{1.5}$

1.6 Future Prospects and Innovations in Si-Based CMOS Technology

Several technologies have contributed to potential future advancements and innovations in the Si-based CMOS technology. As these technologies progress simultaneously, the Si-based CMOS technology can continue to evolve and innovate. The key technologies used are summarized in Table 1.3.

1.7 Navigating the Evolution of Si-Based CMOS Technology

The progression of silicon-based Complementary Metal-Oxide-Semiconductor (CMOS) technology has been a dynamic journey characterized by notable advancements and transformative innovations. As we explore the historical, current, and future aspects of Si-based CMOS technology, several fundamental insights have emerged that influence its ongoing development.

Si-based CMOS technology has served as a fundamental pillar in modern semiconductor electronics, driving exceptional levels of performance, integration density, and energy efficiency. The persistent focus on scalability and innovation has propelled the semiconductor industry, facilitating the creation of robust computing systems, communication devices, and sensor technologies.

Table 1.3 Possible methods to realize further scaling in the future.

Categories	**Explanation**
Novel Materials	The incorporation of new semiconductor materials holds significant promise for overcoming the limitations of Si-based CMOS technology. Materials such as high-k dielectrics can enhance charge transport properties, power consumption, and overall performance.
Negative Capacitance (NC)	Achieving negative capacitance through continuous deposition of extremely thin films is a groundbreaking approach. When applied to gate dielectric materials, this technology enables devices to operate at significantly lower gate voltages, leading to a substantial reduction in the power consumption. Further elaboration on this topic can be found in subsequent chapters of this book.
3D Integration	Advancements in 3D chip design and integration techniques have enabled the integration of more transistors into smaller spaces. Implementing structures such as FinFETs within 3D transistor designs can lower the power consumption and boost the performance.
Quantum Computing	Quantum computing technologies offer a complementary approach to the traditional CMOS technology. Quantum computing demonstrates a superior performance for specific computational tasks, making it an attractive alternative for future high-speed computing requirements.
2D Materials	Innovations in nanodevices and the use of 2D materials, such as graphene and transition metal dichalcogenides, can revolutionize CMOS technology. These materials exhibit exceptional electronic properties in confined spaces, paving the way for future advancements in CMOS technology.
Energy-Efficient Design	Emphasizing energy efficiency in circuit and algorithm designs is crucial for the future of the CMOS technology. Energy-efficient designs play a pivotal role across various sectors, from mobile devices to datacenters.

Nevertheless, this journey encountered a number of challenges. Previous obstacles in scaling Si-based CMOS, such as short-channel effects and power dissipation, have demanded inventive solutions and advancements in materials. Current challenges, including quantum effects and physical constraints, require interdisciplinary approaches and innovative technologies to be surmounted.

In the future, the Si-based CMOS technology holds substantial promise and potential. Emerging technologies, such as negative capacitance materials, 3D integration, quantum computing, and nanodevices, present new avenues for further innovation and enhanced performance. When combined with energy-efficient designs and materials, these technologies can reshape the landscape of CMOS technology and facilitate groundbreaking applications in diverse industries.

Successfully navigating the evolution of Si-based CMOS technology requires a strategic mindset that fosters collaboration, extensive research, and continuous development. It is crucial to address technical hurdles while also considering broader implications such as environmental sustainability and societal impacts.

In summary, Si-based CMOS technology is poised to continuously advance, driving technological progress, economic prosperity, and societal advantages. By remaining at the forefront of innovation and embracing emerging technologies, we can navigate the evolving realm of CMOS technology and unlock new opportunities in the future.

1.8 The Future of Transistors: 2D FET

The future of transistors lies in the 2D Field-Effect Transistor (FET) technology, which represents a significant and remarkable paradigm shift in semiconductor design. This paradigm shift from silicon technology to 2D materials would enable more scaling because 2D materials are known to have few dangling bonds [Figure 1.4]. Unlike traditional transistors made from bulk materials, such as silicon, 2D FETs are constructed using atomically thin materials, such as graphene, transition metal dichalcogenides (TMDs), and black phosphorus. Graphene is known for its excellent electrical and thermal conductivity, and it could act as semi-metal [Figure 1.4]. In contrast, hexagonal boron nitride (h-BN) and MoS_2 act as insulators and semiconductors, respectively. 2D materials include semi-metallic 2D materials, semiconductor 2D materials, and insulator 2D materials.

First, semi-metallic 2D materials possess characteristics similar to those of metals and semiconductors. For example, graphene, a widely recognized

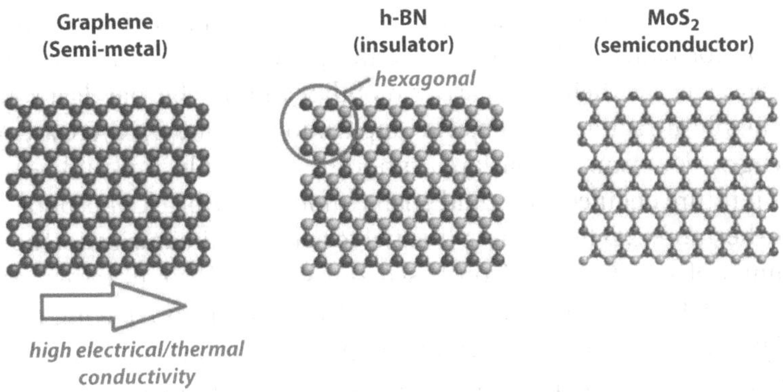

Figure 1.4 Example of 2D materials: Graphene (semi-metal, hexagonal boron nitride (h-BN)), and MoS_2 (semiconductor). The characteristics of 2D materials could depend on fabrication method, fabrication-temperature, and their width (W).

semi-metallic material, demonstrates this duality. In graphene, the conduction and valence bands intersect at a singular point called the Dirac point, which results in distinctive electronic properties. Because of its semi-metallic attributes, graphene displays notable electrical conductivity and remarkable mechanical strength, rendering it applicable across a spectrum of fields, including electronics, photonics, and materials science.

Second, semiconductor 2D materials, on the other hand, feature a bandgap between their valence and conduction bands, facilitating semiconducting behavior. Transition metal dichalcogenides (TMDs) are notable examples of semiconductor 2D materials. Unlike graphene, TMDs possess a finite bandgap, which allows for precise control over the electron flow, thus enabling the development of electronic and optoelectronic devices. Semiconductor 2D materials are instrumental in the fabrication of transistors, light-emitting diodes (LEDs), and photovoltaic devices.

Finally, insulating 2D materials are distinguished by their wide bandgap, which impedes the flow of electrons under normal circumstances. Hexagonal boron nitride (h-BN) is one of the most recognized insulator 2D materials. H-BN exhibits excellent insulating properties, high thermal stability, and chemical inertness, rendering it a promising candidate for encapsulating other 2D materials and safeguarding against their degradation. Insulator 2D materials play a vital role in device manufacturing by furnishing dielectric layers to isolate the electronic components and mitigate leakage currents.

One of the key advantages of 2D FETs is their atomic-scale thickness, which allows precise control over the flow of electrons. This enables better electrostatic control and reduces leakage current, leading to improved transistor performance. Additionally, the unique electronic properties of 2D materials, such as their high carrier mobility and tunable bandgaps, make them highly attractive for transistor applications. However, there are some challenges. For example, black phosphorous is unstable. It would be much better if future engineering could enable more stable black phosphorous.

Furthermore, 2D FETs offer potential for device miniaturization beyond the limits of traditional silicon-based transistors. Their ultrathin nature allows the fabrication of densely packed transistor arrays, paving the way for higher transistor densities and more powerful integrated circuits.

Another area of interest is the potential for novel functionalities in 2D FETs. Researchers are exploring phenomena such as quantum tunneling and spintronics in these devices that could enable new types of logic operations and information processing.

Overall, 2D FETs hold great promise for advancing transistor technology and driving innovation in various fields, including computing and telecommunications. As research in this area continues, we expect to see further improvements in transistor performance, energy efficiency, and device functionality, ushering in a new era of electronics.

1.9 Conclusion

Overall, although there have been some challenges in achieving continuous scaling of CMOS transistors, it seems that several emerging techniques that adopt new materials could further enable more scaling. Engineers should select the proper scaling method or a new scaling method to effectively design next-generation transistors. By suppressing several short-channel effects using emerging technologies, it seems like it is possible to design and fabricate more scaled CMOS transistors for integrated semiconductor chips.

Acknowledgment

In this chapter, the knowledge in previous research has been widely utilized, and the source of information is listed in a reference format (e.g., [1], [2], ...). The author would like to sincerely thank the authors who have made their sincere efforts to publish many articles and research related to this chapter.

References

1. Hoentschel, J. and Wei, A., From the present to the future: Scaling of planar VLSI-CMOS devices towards 3D-FinFETs and beyond 10nm CMOS technologies; manufacturing challenges and future technology concepts. *2015 China Semiconductor Technology International Conference*, Shanghai, China, pp. 1–4, 2015, doi: 10.1109/CSTIC.2015.7153333.
2. Shahidi, G.G., Chip Power Scaling in Recent CMOS Technology Nodes. *IEEE Access*, 7, 851–856, 2019, doi: 10.1109/ACCESS.2018.2885895.
3. Bohr, M.T. and Young, I.A., CMOS Scaling Trends and Beyond. *IEEE Micro*, 37, 6, 20–29, November/December 2017, doi: 10.1109/MM.2017.4241347.
4. Prakash, A. and Jain, P., Optimization of Leakage Power Dissipation in CMOS Inverter using Self biased and W/L Scaling Techniques. *2023 2nd International Conference on Futuristic Technologies (INCOFT)*, Belagavi, Karnataka, India, pp. 1–5, 2023, doi: 10.1109/INCOFT60753.2023.
5. Chiang, T.K., A Short-Channel-Effect-Degraded Noise Margin Model for Junctionless Double-Gate MOSFET Working on Subthreshold CMOS Logic Gates. *IEEE Trans. Electron Devices*, 63, 8, 3354–3359, Aug. 2016, doi: 10.1109/TED.2016.2581826.
6. Chang, V.S., Enabling Multiple-Vt Device Scaling for CMOS Technology beyond 7nm Node. *2020 IEEE Symposium on VLSI Technology*, Honolulu, HI, USA, pp. 1–2, 2020, doi: 10.1109/VLSITechnology18217.2020.9265050.
7. Tanaka, M. and Omura, I., Scaling rule for very shallow trench IGBT toward CMOS process compatibility. *2012 24th International Symposium on Power Semiconductor Devices and ICs*, Bruges, Belgium, pp. 177–180, 2012, doi: 10.1109/ISPSD.2012.6229052.
8. Kerber, A., Cimino, S., Guarin, F., Nigam, T., Assessing device reliability margin in scaled CMOS technologies using ring oscillator circuits. *2017 IEEE Electron Devices Technology and Manufacturing Conference (EDTM)*, Toyama, Japan, pp. 28–30, 2017, doi: 10.1109/EDTM.2017.7947495.
9. Carusone, A.C., Yasotharan, H., Kao, T., CMOS Technology Scaling Considerations for Multi-Gbps Optical Receivers With Integrated Photodetectors. *IEEE J. Solid-State Circuits*, 46, 8, 1832–1842, Aug. 2011, doi: 10.1109/JSSC.2011.2157254.
10. Chung, C.-C., Su, W.-S., Lo, XXXC.-K., A 0.52/1 V Fast Lock-in ADPLL for Supporting Dynamic Voltage and Frequency Scaling. *IEEE Trans. Very Large Scale Integr. VLSI Syst.*, 24, 1, 408–412, Jan. 2016, doi: 10.1109/TVLSI.2015.2407370.
11. Venkatesan, S., A Wafer Scale Hybrid Integration Platform for Co-packaged Photonics using a CMOS based Optical InterposerTM. *2022 IEEE Symposium on VLSI Technology and Circuits (VLSI Technology and Circuits)*, Honolulu, HI, USA, pp. 381–382, 2022.
12. Niitsu, K., Analysis and Techniques for Mitigating Interference From Power/Signal Lines and to SRAM Circuits in CMOS Inductive-Coupling Link for

Low-Power 3-D System Integration. *IEEE Trans. Very Large Scale Integr. VLSI Syst.*, 19, 10, 1902–1907, Oct. 2011, doi: 10.1109/TVLSI.2010.2056711.
13. Rathore, S., Self-Heating Aware Threshold Voltage Modulation Conforming to Process and Ambient Temperature Variation for Reliable Nanosheet FET. *2023 IEEE International Reliability Physics Symposium (IRPS)*, Monterey, CA, USA, pp. 1–5, 2023, doi: 10.1109/IRPS48203.2023.10117918.
14. Tayal, S., Incorporating Bottom-Up Approach Into Device/Circuit Co-Design for SRAM-Based Cache Memory Applications in. *IEEE Trans. Electron Devices*, 69, 11, 6127–6132, Nov. 2022, doi: 10.1109/TED.2022.3210070.
15. Song, Y.S., Kim, J.H., Kim, G., Kim, H.-M., Kim, S., Park, XXXB.-G., Improvement in Self-Heating Characteristic by Incorporating Hetero-Gate-Dielectric in Gate-All-Around MOSFETs. *IEEE J. Electron Devices Soc.*, 9, 36–41, 2021, doi: 10.1109/JEDS.2020.3038391.
16. Greenwood, B., Gate oxide yield improvement for 0.18µm power semiconductor devices with deep trenches: DP: Discrete and power devices. *2017 28th Annual SEMI Advanced Semiconductor Manufacturing Conference (ASMC)*, Saratoga Springs, NY, USA, pp. 346–351, 2017, doi: 10.1109/ASMC.2017.7969258.
17. Song, Y.S., Kim, S., Kim, J.H., Kim, G., Lee, J.-H., Choi, W.Y., Enhancement of Thermal Characteristics and On-Current in GAA MOSFET by Utilizing Al2O3-Based Dual-κ Spacer Structure. *IEEE Trans. Electron Devices*, 70, 1, 343–348, Jan. 2023, doi: 10.1109/TED.2022.3223321.
18. Jaisawal, R.K., Self-Heating and Interface Traps Assisted Early Aging Revelation and Reliability Analysis of Negative Capacitance FinFET. *2023 7th IEEE Electron Devices Technology & Manufacturing Conference (EDTM)*, Seoul, Korea, Republic of, pp. 1–3, 2023, doi: 10.1109/EDTM55494.2023.10103127.
19. Adly, F., Yoo, P.D., Muhaidat, S., Al-Hammadi, Y., Lee, U., Ismail, M., Randomized General Regression Network for Identification of Defect Patterns in Semiconductor Wafer Maps. *IEEE Trans. Semicond. Manuf.*, 28, 2, 145–152, May 2015, doi: 10.1109/TSM.2015.2405252.
20. Kang, S.J., Kim, J.H., Song, Y.S., Go, S., Kim, S., Investigation of Self-Heating Effects in Vertically Stacked GAA MOSFET With Wrap-Around Contact. *IEEE Trans. Electron Devices*, 69, 3, 910–914, March 2022, doi: 10.1109/TED.2022.3140283.
21. Song, Y.S., Tayal, S., Rahi, S.B., Kim, J.H., Upadhyay, A.K., Park, XXXB.-G., Thermal-Aware IC Chip Design by Combining High Thermal Conductivity Materials and GAA MOSFET. *2022 5th International Conference on Circuits, Systems and Simulation (ICCSS)*, Nanjing, China, pp. 135–140, 2022, doi: 10.1109/ICCSS55260.2022.
22. Chang, C.W., Liu, S.E., Lin, B.L., Chiu, C.C., Lee, Y.-H., Wu, K., Thermal behavior of self-heating effect in FinFET devices acting on back-end interconnects. *2015 IEEE International Reliability Physics Symposium*, Monterey, CA, USA, pp. 2F.6.1–2F.6.5, 2015, doi: 10.1109/IRPS.2015.7112696.

23. Lee, S.-W., Jeon, P.J., Choi, K., Min, S.-W., Kwon, H., Im, S., Analysis of Self-Heating Effect on Short Channel Amorphous InGaZnO Thin-Film Transistors. *IEEE Electron Device Lett.*, 36, 5, 472–474, May 2015, doi: 10.1109/LED.2015.2411742.
24. Kumar, N., Kaushik, P.K., Gupta, A., Singh, P., Self-Heating Effects (SHEs) in Gate-All-Around FETs with Horizontally Stacked Multiple Junctionless Nanowires. *2022 IEEE Delhi Section Conference (DELCON)*, New Delhi, India, pp. 1–4, 2022, doi: 10.1109/DELCON54057.2022.9753115.
25. Qian, C., Investigation of self-heating effect in SOI tunnel field-effect transistor. *2015 IEEE 11th International Conference on ASIC (ASICON)*, Chengdu, China, pp. 1–4, 2015, doi: 10.1109/ASICON.2015.7517159.
26. Aoki, H., Sakairi, H., Kuroda, N., Nakamura, Y., Chikamatsu, K., Nakahara, K., AlGaN/GaN MIS HEMT Modeling of Frequency Dispersion and Self-Heating Effects. *2018 IEEE International Symposium on Radio-Frequency Integration Technology (RFIT)*, Melbourne, VIC, Australia, pp. 1–3, 2018, doi: 10.1109/RFIT.2018.8524087.
27. Su, Y., Lai, J., Sun, L., Investigation of Self-Heating Effects in Vacuum Gate Dielectric Gate-all-Around Vertically Stacked Silicon Nanowire Field Effect Transistors. *IEEE Trans. Electron Devices*, 67, 10, 4085–4091, Oct. 2020, doi: 10.1109/TED.2020.3017452.
28. Xu, C., Guo, X., Jiang, H., Zhang, Z., Liu, S., Modeling and simulation of self-heating effect with temperature difference air flow sensor. *2014 15th International Conference on Electronic Packaging Technology*, Chengdu, China, pp. 655–659, 2014, doi: 10.1109/ICEPT.2014.6922740.
29. Kumar, N., Ashok Bhinge, K., Gupta, A., Singh, P., Electro-Thermal Properties and Self-Heating Effect in Multi-Nanosheet FETs: Junctionless Mode Versus Inversion Mode. *2023 7th IEEE Electron Devices Technology & Manufacturing Conference (EDTM)*, Seoul, Korea, Republic of, pp. 1–3, 2023, doi: 10.1109/EDTM55494.2023.10103095.
30. Wu, Q., Xu, Y., Wang, Z., Xia, L., Yan, B., Xu, R., Implementation of self-heating and trapping effects in surface potential model of AlGaN/GaN HEMTs. *2017 IEEE MTT-S International Microwave Symposium (IMS)*, Honololu, HI, USA, pp. 236–239, 2017, doi: 10.1109/MWSYM.2017.8059084.
31. Liu, L., Lu, Y., Guo, J., Coupled Electro–Thermal Simulation for Self-Heating Effects in Graphene Transistors. *IEEE Trans. Electron Devices*, 60, 8, 2598–2603, Aug. 2013, doi: 10.1109/TED.2013.2268458.
32. Jiang, H., Liu, X., Xu, N., He, Y., Du, G., Zhang, X., Investigation of Self-Heating Effect on Hot Carrier Degradation in Multiple-Fin SOI FinFETs. *IEEE Electron Device Lett.*, 36, 12, 1258–1260, Dec. 2015, doi: 10.1109/LED.2015.2487045.
33. Liao, M.-H., Hsieh, C.-P., Lee, XXXC.-C., Systematic Investigation of Self-Heating Effect on CMOS Logic Transistors From 20 to 5 nm Technology Nodes by Experimental Thermoelectric Measurements and Finite Element

Modeling. *IEEE Trans. Electron Devices*, 64, 2, 646–648, Feb. 2017, doi: 10.1109/TED.2016.2642404.

34. Park, C. and Yun, I., Degradation of Off-Phase Leakage Current of FinFETs and Gate-All-Around FETs Induced by the Self-Heating Effect in the High-Frequency Operation Regime. *IEEE Trans. Nanotechnol.*, 19, 308–314, 2020, doi: 10.1109/TNANO.2020.2986540.
35. Alias, N.E., Hassan, R., Johari, Z., Statistical variability in n-channel SOI FinFET in the presence of random discrete dopant. *2016 IEEE International Conference on Semiconductor Electronics (ICSE)*, Kuala Lumpur, Malaysia, pp. 232–235, 2016, doi: 10.1109/SMELEC.2016.7573634.
36. Hirai, T. and Kano, M., Adaptive Virtual Metrology Design for Semiconductor Dry Etching Process Through Locally Weighted Partial Least Squares. *IEEE Trans. Semicond. Manuf.*, 28, 2, 137–144, May 2015, doi: 10.1109/TSM.2015.2409299.
37. Tsunomura, T., Nishida, A., Hiramoto, T., Effect of Channel Dopant Profile on Difference in Threshold Voltage Variability Between NFETs and PFETs. *IEEE Trans. Electron Devices*, 58, 2, 364–369, Feb. 2011, doi: 10.1109/TED.2010.2091452.
38. Race, S., Ziemann, T., Kovacevic-Badstuebner, I., Stark, R., Tiwari, S., Grossner, U., Design for Reliability of SiC Multichip Power Modules: The Effect of Variability. *2021 33rd International Symposium on Power Semiconductor Devices and ICs (ISPSD)*, Nagoya, Japan, pp. 399–402, 2021, doi: 10.23919/ISPSD50666.2021.9452198.
39. Magnone, P., Impact of Hot Carriers on nMOSFET Variability in 45- and 65-nm CMOS Technologies. *IEEE Trans. Electron Devices*, 58, 8, 2347–2353, Aug. 2011, doi: 10.1109/TED.2011.2156414.
40. Bazizi, E.M., Impact of backplane configuration on the statistical variability in 22nm FDSOI CMOS. *2015 International Conference on Simulation of Semiconductor Processes and Devices (SISPAD)*, Washington, DC, USA, pp. 345–348, 2015, doi: 10.1109/SISPAD.2015.7292330.
41. Hiramoto, T., Direct Measurement of Correlation Between SRAM Noise Margin and Individual Cell Transistor Variability by Using Device Matrix Array. *IEEE Trans. Electron Devices*, 58, 8, 2249–2256, Aug. 2011, doi: 10.1109/TED.2011.2138142.
42. Martinez Brito, J.P., Lubaszewski, M., Bampi, S., Within-die and die-to-die variability on 65nm CMOS : oscillators experimental results. *2015 International Workshop on CMOS Variability (VARI)*, Salvador, Brazil, pp. 27–32, 2015, doi: 10.1109/VARI.2015.7456559.
43. Lorenz, J., Simultaneous simulation of systematic and stochastic process variations. *2014 International Conference on Simulation of Semiconductor Processes and Devices (SISPAD)*, Yokohama, Japan, pp. 289–292, 2014, doi: 10.1109/SISPAD.2014.6931620.
44. Yang, W.-B., Lin, Y.-Y., Lo, XXXY.-L., Analysis and design considerations of static CMOS logics under process, voltage and temperature variation

in 90nm CMOS process. *Sapporo Jpn.*, 1653–1656, 2014, doi: 10.1109/InfoSEEE.2014.6946202.
45. Chen, S., Signal inpainting on graphs via total variation minimization. *2014 IEEE International Conference on Acoustics, Speech and Signal Processing (ICASSP)*, Florence, Italy, pp. 8267–8271, 2014, doi: 10.1109/ICASSP.2014.6855213.
46. Baer, E., Burenkov, A., Evanschitzky, P., Lorenz, J., Simulation of process variations in FinFET transistor patterning. *2016 International Conference on Simulation of Semiconductor Processes and Devices (SISPAD)*, Nuremberg, Germany, pp. 299–302, 2016, doi: 10.1109/SISPAD.2016.7605206.
47. Freijedo, J., Semião, J., Rodríguez-Andina, J.J., Vargas, F., Teixeira, I.C., Teixeira, J.P., Modeling the effect of process variations on the timing response of nanometer digital circuits. *2011 12th Latin American Test Workshop (LATW)*, Beach of Porto de Galinhas, Brazil, pp. 1–5, 2011, doi: 10.1109/LATW.2011.5985927.
48. Butler, D. and Zhang, H., Intelligent software sensors and process prediction for glass container forming processes based on multivariate statistical process control techniques. *Proceedings of 2012 UKACC International Conference on Control*, Cardiff, UK, pp. 281–285, 2012, doi: 10.1109/CONTROL.2012.6334643.
49. Terpák, J., Laciak, M., Kacur, J., Durdán, M., Flegner, P., Tréfa, G., The Mathematical Model for Indirect Measurement of Carbon Concentration in the Steelmaking Process and its Utilization in Process Control. *2021 22nd International Carpathian Control Conference (ICCC)*, Velké Karlovice, Czech Republic, pp. 1–4, 2021, doi: 10.1109/ICCC51557.2021.9454605.
50. Ding, W., Xie, L., Wu, Z., PWM digital control of aeration in wastewater treatment process. *2013 25th Chinese Control and Decision Conference (CCDC)*, Guiyang, China, pp. 64–68, 2013, doi: 10.1109/CCDC.2013.6560895.
51. Wu, S. and Zhang, R., Improved LQ Tracking Control Design for Industrial Processes Under Uncertainty: The Extended Nonminimal State Space Approach. *IEEE Trans. Syst. Man Cybern.: Syst.*, 52, 2, 1356–1360, Feb. 2022, doi: 10.1109/TSMC.2020.3014839.
52. Abdullah, M. and Rossiter, J.A., Alternative Method for Predictive Functional Control to Handle an Integrating Process. *2018 UKACC 12th International Conference on Control (CONTROL)*, Sheffield, UK, pp. 26–31, 2018, doi: 10.1109/CONTROL.2018.8516787.
53. Xu, X., Xie, H., Shi, J., Iterative Learning Control (ILC) Guided Reinforcement Learning Control (RLC) Scheme for Batch Processes. *2020 IEEE 9th Data Driven Control and Learning Systems Conference (DDCLS)*, Liuzhou, China, pp. 241–246, 2020, doi: 10.1109/DDCLS49620.2020.9275065.
54. Jun, Z. and Liu, XXXG.-P., Design and implementation of networked real-time control system with image processing capability. *2014 UKACC International Conference on Control (CONTROL)*, Loughborough, UK, pp. 39–43, 2014, doi: 10.1109/CONTROL.2014.6915112.

55. Ren, J.-C., Liu, D., Wan, Y., Model-Free Adaptive Iterative Learning Control Method for the Czochralski Silicon Monocrystalline Batch Process. *IEEE Trans. Semicond. Manuf.*, 34, 3, 398–407, Aug. 2021, doi: 10.1109/TSM.2021.3074625.
56. Li, K., Wang, F., He, D., Zhang, S., A knowledge based intelligent control method for dehydration and mixing process. *2017 29th Chinese Control And Decision Conference (CCDC)*, Chongqing, China, pp. 477–482, 2017, doi: 10.1109/CCDC.2017.7978141.
57. Young, S.S., Laxman, R.T., Shubam, T., Shiromani, B.R., Arun Samuel, T.S., *Handbook of Emerging Materials for Semiconductor Industry*, Springer Nature, Singapore, 2024.
58. Bhardwaj, A., Kumar, P., Raj, B., Anand, S., Design and Performance Optimization of doping-less Vertical Nanowire TFET using Gate Stack Technique. *J. Electron. Mater. (JEMS)*, Springer, 41, 7, 4005–4013, 2022.
59. Singh, J. and Raj, B., Tunnel Current Model of Asymmetric MIM Structure Levying Various Image Forces to Analyze the Characteristics of Filamentary Memristor. *Appl. Phys. A*, Springer, 125, 3, 203.1 to 203.11, Feb 2019.
60. Goyal, C., Singh Ubhi, J., Raj, B., Low Leakage Zero Ground Noise Nanoscale Full Adder using Source Biasing Technique. *J. Nanoelectron. Optoelectronics*, American Scientific Publishers, 14, 360–370, March 2019.
61. Singh, G., Sarin, R.K., Raj, B., A Novel Robust Exclusive-OR Function Implementation in QCA Nanotechnology with Energy Dissipation Analysis. *J. Comput. Electronics*, Springer, 15, 2, 455–465, June 2016.
62. Wadhwa, G., Kamboj, P., Raj, B., Design optimisation of junctionless TFET biosensor for high sensitivity. *Adv. Nat. Sci.: Nanosci. Nanotechnol.*, 10, 045001, 7, 2019.
63. Bansal, P. and Raj, B., MEMRISTOR MODELING AND ANALYSIS FOR LINEAR DOPANT DRIFT KINETICS. *J. Nanoeng. Nanomanufacturing*, American Scientific Publishers, 6, 1–7, 2016.
64. Ashima, Vaithiyanathan, D., Raj, B., Performance Analysis of Charge Plasma induced Graded Channel Si nanotube. *J. Eng. Res. (JER)*, EMSME Special Issue, 146–154, Aug 2021.
65. Tomar, A.S., Magraiya, V.K., Raj, B., Scaling of Access and Data Transistor for High Performance DRAM Cell Design. *Quantum Matter*, 2, 412–416, Oct. 2013.
66. Jain, N. and Raj, B., Parasitic Capacitance and Resistance Model Development and Optimization of Raised Source/Drain SOI FinFET Structure for Analog Circuit Applications. *J. Nanoelectron. Optoelectronins*, ASP USA, 13, 531–539, Apl 2018.
67. Singh, S., Vishvakarma, S.K., Raj, B., Analytical Modeling of Split-Gate Junction-Less Transistor for a Biosensor Application. *Sens. Bio-sensing*, Elsevier, 18, 31–36, April 2018.
68. Gopal, M. and Raj, B., Low Power 8T SRAM Cell Design for High Stability Video Applications. *ITSI Transaction Electr. Electron. Eng.*, 1, 5, 91–97, 2013.

69. Raj, B., Mitra, J., D.K. Bihani, Rangharajan V., A.K. Saxena, Dasgupta, S., Analysis of Noise Margin, Power and Process Variation for 32 nm FinFET Based 6T SRAM Cell. *J. Comput. (JCP)*, Academic Publisher FINLAND, 5, 6, 1–8, 2010.
70. Sharma, D., Mehra, R., Raj, B., Comparative Analysis of Photovoltaic Technologies for High Efficiency Solar Cell Design. *Superlattices Microstructures*, Elsevier, 153, 106861, May 2021.
71. Kaur, P., Buttar, A.S., Raj, B., A comprehensive Analysis of Nanoscale Transistor based Biosensor: A Review. *Indian J. Pure Appl. Phys.*, 59, 304–318, April 2021.
72. Yadav, D., Raj, B., Raj, B., Design and Simulation of Low Power Microcontroller for IoT Applications. *J. Sens. Letters*, ASP, 18, 401–409, May 2020.
73. Singh, S. and Raj, B., A 2-D Analytical Surface Potential and Drain current Modeling of Double-Gate Vertical t-shaped Tunnel FET. *J. Comput. Electron.*, Springer, 19, 1154–1163, Apl 2020.
74. Singh, J. and Raj, B., An Accurate and Generic Window function for Non-linear Memristor Model. *J. Comput. Electron.*, Springer, 18, 2, 640–647, June 2019.
75. Kaur, M., Gupta, N., Kumar, S., Raj, B., Singh, A.K., Comparative RF and Crosstalk Analysis of Carbon Based Nano Interconnects. *IET Circuits Devices Syst.*, 15, 6, 493–503, Feb 2021.
76. Kandasamy, N., Ahmad, F., Ajitha, D., Raj, B., Telagam, N., Quantum Dot Cellular Automata based Scan Flip Flop and Boundary Scan Register. *IETE J. Res.*, 66, 535-548, 2020.
77. Singh, J. and Raj, B., Design and Investigation of 7T2M NVSARM with Enhanced Stability and Temperature Impact on Store/Restore Energy. *IEEE Trans. Very Large Scale Integr. Syst.*, 27, 6, 1322–1328, June 2019.
78. Bhardwaj, A.K., Gupta, S., Raj, B., Singh, A., Impact of Double Gate Geometry on the Performance of Carbon Nanotube Field Effect Transistor Structures for Low Power Digital Design. *Comput. Theor. Nanoscience*, ASP, 16, 1813–1820, 2019.
79. Jain, N. and Raj, B., Thermal Stability Analysis and Performance Exploration of Asymmetrical Dual-k underlap Spacer (ADKUS) SOI FinFET for Security and Privacy Applications. *Indian J. Pure Appl. Phys. (IJPAP)*, 57, 352–360, May 2019.
80. Jain, N. and Raj, B., Dual-k Spacer Region Variation at the Drain Side of Asymmetric SOI FinFET Structure: Performance Analysis towards the Analog/RF Design Applications. *J. Nanoelectron. Optoelectronics*, American Scientific Publishers, 14, 349–359, March 2019.
81. Singh, J., Sharma, S., Raj, B., Khosla, M., Analysis of barrier layer thickness on performance of In1-xGaxAs based Gate Stack Cylindrical Gate Nanowire MOSFET. *JNO ASP*, 13, 1473–1477, Oct 2018.

82. Jain, N. and Raj, B., Analysis and Performance Exploration of High-k SOI FinFETs Over the Conventional Low-k SOI FinFET toward Analog/RF Design. *J. Semicond. (JoS)*, IOP Science, 39, 12, 124002-1-7, Dec 2018.
83. Goyal, C., Ubhi, J.S., Raj, B., A reliable leakage reduction technique for approximate full adder with reduced ground bounce noise. *J. Math. Probl. Engineering*, Hindawi, 2018, Article ID 3501041, 16, 15 Oct 2018.
84. Anuradha, Singh, J., Raj, B., Mamta Khosla, Design and Performance Analysis of Nano-scale Memristor-based Nonvolatile SRAM. *J. Sens. Letter*, American Scientific Publishers, 16, 798–805, Oct 2018.
85. Wadhwa, G. and Raj, B., Parametric Variation Analysis of Charge-Plasma-based Dielectric Modulated JLTFET for Biosensor Application. *IEEE Sens. J.*, 18, 15, AUGUST 1, 6070 - 6077, 2018 (SCI.
86. Singh, J. and Raj, B., Comparative Analysis of Memristor Models for Memories Design. *JoS IoP*, 39, 7, 074006-1-12, July 2018.
87. Yadav, D., Chouhan, S.S., Vishvakarma, S.K., Raj, B., Application Specific Microcontroller Design for IoT based WSN. *Sens. Letter*, ASP, 16, 374–385, May 2018.
88. Singh, G., Sarin, R.K., Raj, B., Fault-Tolerant Design and Analysis of Quantum-Dot Cellular Automata Based Circuits. *IEEE/IET Circuits Devices Syst.*, 12, 638–64, 2018.
89. Singh, J. and Raj, B., Modeling of Mean Barrier Height Levying Various Image Forces of Metal Insulator Metal Structure to Enhance the Performance of Conductive Filament Based Memristor Model. *IEEE Nanotechnol.*, 17, 2, 268–267, March 2018.
90. Jain, A., Sharma, S., Raj, B., Analysis of Triple Metal Surrounding Gate (TM-SG) III-V Nanowire MOSFET for Photosensing Application. *Opto-electronics J.*, Elsevier, 26, 2, 141–148, May 2018.
91. Jain, A., Sharma, S., Raj, B., Design and Analysis of High Sensitivity Photosensor Using Cylindrical Surrounding Gate MOSFET For Low Power Sensor Applications. *Eng. Sci. Technology Int. J.*, Elsevier's, 19, 4, 1864–1870, December 2016.
92. Saiphani Kumar, G., Singh, A., Raj, B., Design and Analysis of Gate All Around CNTFET based SRAM cell Design. *J. Comput. Electronics*, Springer, 17, 1, 138–145, March 2018.
93. Singh, G., Sohi, B.S., Raj, B., Material Properties Analysis of Graphene Base Transistor (GBT) for VLSI Analog Circuits. *Indian J. Pure Appl. Phys. (IJPAP)*, 55, 896–902, Dec-2017.
94. Kumar, S. and Raj, B., Estimation of Stability and Performance metric for Inward Access Transistor based 6T SRAM Cell Design using n-type/p-type DMDG-GDOV TFET. *IEEE VLSI Circuits Syst. Letter*, 3, 2, 25–39, June 2017.
95. Sharma, S., Kumar, A., Pattanaik, M., Raj, B., Body Biased Multimode Multi-Threshold CMOS Technique for Ground Bounce Noise Reduction in Static CMOS Adders. *Int. J. Inf. Electron. Eng.*, 3, 3, 567–572, 2013.

96. Singh, H., Kumar, P., Raj, B., Performance Analysis of Majority Gate SET Based 1-bit Full Adder. *Int. J. Comput. Commun. Eng. (IJCCE)*, IACSIT Press Singapore, 2, 4, 146–154, 2013. ISSN: 2010-3743.
97. Bhardwaj, A.K., Gupta, S., Raj, B., Investigation of Parameters for Schottky Barrier (SB) Height for Schottky Barrier Based Carbon Nanotube Field Effect Transistor Device. *J. Nanoelectron. Optoelectronics*, ASP, 15, 783–791, July 2020.
98. Bansal, P. and Raj, B., MEMRISTOR: A VERSATILE NONLINEAR MODEL FOR DOPANT DRIFT AND BOUNDARY ISSUES. *JCTN*, American Scientific Publishers, 14, 5, 2319–2325, May 2017.
99. Jain, N. and Raj, B., An Analog and Digital Design Perspective Comprehensive Approach on Fin-FET (Fin-Field Effect transistor) Technology - A Review. *Rev. Adv. Sci. Eng.*, (RASE) ASP, 5, 1–14, 2016.
100. Raj, B., Saxena, A.K., Dasgupta, S., Analytical Modeling for the Estimation of Leakage Current and Subthreshold Swing Factor of Nanoscale Double Gate FinFET Device, in: *Microelectronics International*, vol. 26, pp. 53–63, 2009.
101. Soniya, Singh, S., Wadhwa, G., Raj, B., An Analytical Modeling for Dual Source Vertical Tunnel Field Effect Transistor. *Int. J. Recent Technol. Eng. (IJRTE)*, 8, 2, 603–608, July 2019.
102. Singh, S. and Raj, B., Design and analysis of hetrojunction Vertical T-shaped Tunnel Field Effect Transistor. *J. Electron. Mater.*, Springer, 48, 10, 6253–6260, October 2019.
103. Goyal, C., Ubhi, J.S., Raj, B., A Low Leakage CNTFET based Inexact Full Adder for Low Power Image Processing Applications. *Int. J. Circuit Theory Applications*, Wiley, 47, 9, 1446–1458, September 2019.
104. Raj, B., Saxena, A.K., Dasgupta, S., A Compact Drain Current and Threshold Voltage Quantum Mechanical Analytical Modeling for FinFETs. *J. Nanoelectron. Optoelectron. (JNO)*, USA, 3, 2, 163–170, 2008.
105. Wadhwa, G. and Raj, B., An Analytical Modeling of Charge Plasma based Tunnel Field Effect Transistor with Impacts of Gate underlap Region. *Superlattices Microstructures*, Elsevier, 142, 106512, June 2020.
106. Singh, S. and Raj, B., Modeling and Simulation analysis of SiGe hetrojunction Double GateVertical t-shaped Tunnel FET. *Superlattices Microstructures*, Elsevier, 142, 106496, June 2020.
107. Jain, N. and Raj, B., Device and Circuit Co-Design Perspective Comprehensive Approach on FinFET Technology - A Review. *J. Electron Devices*, 23, 1, 1890–1901, 2016.
108. Kumar, S. and Raj, B., Analysis of I_{ON} and Ambipolar Current for Dual-Material Gate-drain Overlapped DG-TFET. *J. Nanoelectron. Optoelectronics*, American Scientific Publishers USA, 11, 323–333, June 2016.
109. Anjum, N., Bali, T., Raj, B., Design And Simulation Of Handwritten Multiscript Character Recognition. *Int. J. Adv. Res. Comput. Commun. Eng.*, 2, 7, 2544–2549, July 2013.

110. Singh, K. and Raj, B., Performance and analysis of temperature dependent Multi-walled Carbon Nanotubes as Global Interconnects at different technology nodes. *J. Comput. Electronics*, Springer, 14, 2, 469–476, June 2015.
111. Kumar, S. and Raj, B., Compact channel potential analytical Modeling of DG-TFET based on Evanescent-mode Approach. *J. Comput. Electronics*, Springer, 14, 2, 820–827, July 2015.
112. Singh, K. and Raj, B., Temperature Dependent Modeling and Performance Evaluation of Multi-Walled CNT and Single-Walled CNT as Global Interconnects. *J. Electron. Materials*, Springer, 44, 12, 4825–4835, 2015.
113. Sharma, V.K., Pattanaik, M., Raj, B., INDEP approach for leakage reduction in nanoscale CMOS circuits. *Int. J. Electronics*, Taylor & Francis, 102, 2, 200–215, 2014.
114. Singh, K. and Raj, B., Influence of Temperature on MWCNT bundle, SWCNT bundle and Copper interconnects for Nanoscaled Technology nodes. *J. Mater. Science Mater. Electronics*, Springer, 26, 8, 6134–6142, 2015.
115. Anjum, N., Bali, T., Raj, B., Design and Simulation of Handwritten Gurumukhi and Devanagri Numerical Recognition. *Int. J. Comput. Applications*, Foundation of Computer Science New York USA, 73, 12, 16–21, 2013.
116. Khandelwal, S., Gupta, V., Raj, B., Gupta, R.D., Process Variability Aware Low Leakage Reliable Nano Scale DG-FinFET SRAM Cell Design Technique. *J. Nanoelectron. Optoelectron.*, 106, 810–817, December 2015.
117. Sharma, V.K., Pattanaik, M., Raj, B., ONOFIC Approach: Low Power High Speed Nanoscale VLSI Circuits Design. *Int. J. Electronics*, Taylor & Francis, 101, 1, 61–73, 2014.
118. Khandelwal, S., Raj, B., Gupta, R.D., FinFET based 6T SRAM Cell Design: Analysis of Performance Metric, Process Variation and Temperature Effect. *J. Comput. Theor. Nanoscience*, ASP USA, 12, 2500–2506, 2015.
119. Singh, S., Yadav, S., Rahul, J., Srivastava, A., Raj, B., Impact of HfO_2 in Graded Channel Dual Insulator Double Gate MOSFET. *J. Comput. Theor. Nanoscience*, American Scientific Publishers, 12, 6, 950–953, April 2015.
120. Sharma, V.K., Pattanaik, M., Raj, B., PVT variations aware low leakage INDEP approach for nanoscale CMOS Circuits. *Microelectron. Reliability*, Elsevier, 54, 90–99, 2014.
121. Raj, B., Saxena, A.K., Dasgupta, S., Quantum Mechanical Analytical Modeling of Nanoscale DG FinFET: Evaluation of Potential, Threshold Voltage and Source/Drain Resistance. *Elsevier's J. Mater. Sci. Semicond. Processing*, Elsevier, 16, 4, 1131–1137, 2013.
122. Gopal, M., Prasad, S.S.D., B.R., 8T SRAM Cell Design for Dynamic and Leakage Power Reduction. *Int. J. Comput. Applications*, Foundation of Computer Science, New York, USA, 71, 9, 43–48, June 2013.
123. Pattanaik, M., Raj, B., Sharma, S., Kumar, A., Diode Based Trimode Multi-Threshold CMOS Technique for Ground Bounce Noise Reduction in Static CMOS Adders. *Adv. Mater. Res.*, Trans Tech Publications, Switzerland, 548, 885–889, 2012.

124. Raj, B., Saxena, A.K., Dasgupta, S., Nanoscale FinFET Based SRAM Cell Design: Analysis of Performance metric, Process variation, Underlapped FinFET and Temperature effect. *IEEE Circuits Syst. Mag.*, 11, 2, 38–50, 2011.
125. Sharma, V.K., Pattanaik, M., Balwinder, R., Leakage Current ONOFIC Approach for Deep Submicron VLSI Circuit Design. *Int. J. Electrical Computer Electron. Commun. Eng. World Acad. Sciences Eng. Technol.*, 7, 4, 239–244, 2013.
126. Kaur, P., Gill, S.S., Raj, B., Comparative Analysis of OFETs Materials and Devices for Sensor Applications. *J. Silicon*, Springer, 14, 4463–4471, 2022.
127. Sharma, S.K., Kumar, P., Raj, B., Raj, B., $In_{1-x}Ga_xAs$ Double Metal Gate-Stacking Cylindrical Nanowire MOSFET for highly sensitive Photo detector. *J. Silicon*, Springer, 14, 3535–3541, 2022.
128. Raj, B., Saxena, A.K., Dasgupta, S., Analytical Modeling of Quasi Planar Nanoscale Double Gate FinFET with Source/Drain Resistance and field dependent carrier mobility: A quantum mechanical study. *J. Comput. (JCP)*, Academic Publisher FINLAND, 4, 9, 1–8, 2009.
129. Bhushan, S., Khandelwal, S., Raj, B., Analyzing Different Mode FinFET based Memory Cell at Different Power Supply for Leakage Reduction. *Seventh International Conference on Bio-Inspired Computing: Theories and Application, (BIC-TA 2012) Advances in Intelligent Systems and Computing*, vol. 20289-100, 20122013.
130. Singh, J. and Raj, B., Temperature Dependent Analytical Modeling and Simulations of Nanoscale Memristor. *J. Eng. Sci. Technology Int. J.*, Elsevier's, 21, 862–868, Oct 2018.
131. Singh, S., Bala, S., Raj, B., Raj, B., Improved Sensitivity of Dielectric Modulated Junctionless Transistor for Nanoscale Biosensor Design. *Sens. Letter*, ASP, 18, 328–333, Apl 2020.
132. Kumar, V., Vishvakarma, S.K., Raj, B., Design and Performance Analysis of ASIC for IoT Applications. *Sens. Letter*, ASP, 18, 31–38, Jan 2020.
133. Akanksha Jaiswal, Sarin, R.K., Raj, B., Sukhija, S., A Novel Circular Slotted Microstrip-fed Patch Antenna with Three Triangle Shape Defected Ground Structure for Multiband Applications. *Adv. Electromagn. (AEM)*, 7, 3, 56–63, Aug 2018.
134. Wadhwa, G. and Raj, B., Label Free Detection of Biomolecules using Charge-Plasma-Based Gate Underlap Dielectric Modulated Junctionless TFET. *J. Electron. Mater. (JEMS)*, Springer, 47, 8, 4683–4693, August 2018.
135. Singh, G., Sarin, R.K., Raj, B., Design and Performance Analysis of a New Efficient Coplanar Quantum-Dot Cellular Automata Adder. *Indian J. Pure Appl. Phys. (IJPAP)*, 55, 97–103, February 2017.
136. Kaur, P., Pandey, V., Raj, B., Comparative Study of Efficient Design, Control and Monitoring of Solar Power using IoT. *Sens. Letter*, ASP, 18, 419–426, May 2020.

137. Bhardwaj, A.K., Gupta, S., Raj, B., Development & Analysis of compact model for Double Gate Schottky Barrier CNTFET. *J. Nanoelectron. Optoelectronics*, ASP, 15, 1199–1208, Aug 2020.
138. Wadhwa, G., Kamboj, P., Singh, J., Raj, B., Design and Investigation of Junctionless DGTFET for Biological Molecule Recognition. *Trans. Electr. Electron. Materials*, Springer, 22, 282–289, 2021.
139. Verma, S.K., Singh, S., Wadhwa, G., Raj, B., Detection of Biomolecules using Charge-Plasma based Gate Underlap Dielectric Modulated Dopingless TFET. *Trans. Electr. Electron. Mater. (TEEM)*, Springer, 21, 528–535, June 2020.
140. Jain, N. and Raj, B., Impact of underlap spacer region variation on Electrostatic and analog/RF performance of symmetrical high-k SOI FinFET at 20 nm channel length. *J. Semicond. (JoS)*, IOP Science, 38, 12, 122002, Dec 2017.
141. Singh, S. and Raj, B., Analytical Modeling and Simulation analysis of T-shaped III-V heterojunction Vertical T-FET. *Superlattices Microstructures*, Elsevier, 147, 106717, Nov 2020.
142. Singh, G., Sarin, R.K., Raj, B., Design and Analysis of Area Efficient QCA Based Reversible Logic Gates. *J. Microprocess. Microsystems*, Elsevier, 52, 59–68, May 2017.
143. Singh, S. and Raj, B., Parametric variation analysis on hetero-junction Vertical t-shape TFET for Supressing ambipolar conduction. *Indian J. Pure Appl. Phys.*, 58, 478–485, June 2020.
144. Soniya, Singh, S., Wadhwa, G., Raj, B., Design and Analysis of Dual Source Vertical Tunnel Field Effect Transistor for high performance. *Trans. Electr. Electron. Materials*, Springer, 21, 74–82, Oct 2019.
145. Kaur, M., Gupta, N., Kumar, S., Raj, B., Singh, A.K., RF Performance Analysis of Intercalated Graphene Nanoribbon Based Global Level Interconnects. *J. Comput. Electronics*, Springer, 19, 1002–1013, June 2020.
146. Wadhwa, G. and Raj, B., Design and Performance Analysis of Junctionless TFET Biosensor for high sensitivity. *IEEE Nanotechnol.*, 18, 567–574, 2019.
147. Singh, J. and Raj, B., Enhanced Nonlinear Memristor Model Encapsulating Stochastic Dopant Drift. *JNO ASP*, 14, 958–963, 2019.

2

Ferroelectric Polymer-Based Field-Effect Transistor (FeFET) and its Applications

Dhrubojyoti Roy[1*], Mohua Chakraborty[1], Dipankar Bandyopadhyay[2] and Partho Sarathi Gooh Pattader[2]

[1]Department of Physics, C V Raman Global University, Bhubaneswar, Odisha, India
[2]Department of Chemical Engineering, Indian Institute of Technology Guwahati, Guwahati, Assam, India

Abstract
Ferroelectric polymers in organic field-effect transistors (OFETs) have garnered significant attention because they offer cost-effective means and simplified manufacturing processes based on solution processability and flexible memory modules. They can be fabricated by various solution-processing techniques, such as spin coating, Langmuir–Blodgett, ink-jet printing, and roll-to-roll printing. Organic ferroelectric field-effect transistors (FeFETs) can be used in large electronic applications such as flexible displays, sensors, and radio frequency identification (RFID) tags. An additional benefit of PVDF copolymers, such as poly(vinylidene fluoride-trifluoroethylene) (P(VDF-TrFE)), is their ability to rapidly crystallize into the β phase. The β phase showed the maximum dipole moment among the various reported polar phases. Ferroelectric polymer material-based devices have shown good memory functionality, which can be of great significance for various applications. In particular, OFET with ferroelectric polymers as gate dielectric layers have attracted significant interest because of their large potential for the growth of nonvolatile memory devices at low cost. This work highlights the diverse applications of FeFETs in advancing nonvolatile memory, artificial synapses, and wearable technologies, leveraging ferroelectric polymers.

***Keywords*:** Ferroelectric polymer, ferroelectric field effect transistors (FeFETs), nonvolatile memory, synaptic device

**Corresponding author*: dhrubojyoti.roy@cgu-odisha.ac.in

Balwinder Raj, Shiromani Balmukund Rahi and Nandakishor Yadav (eds.) FeFET Devices, Trends, Technology and Applications, (27–56) © 2025 Scrivener Publishing LLC

2.1 Introduction

Ferroelectric polymers have drawn considerable attention in the last few decades because of their phase transition features. The unveiling of piezoelectric properties in stretched and poled polyvinylidene fluoride (PVDF) films by Kawai in 1969 [1] has attracted significant attention for PVDF and its copolymer P(VDF-TrFE). PVDF polymer materials are widely recognized for their unique ability to convert mechanical energy into electrical signals, making them pivotal in the development of advanced technologies such as sensors, actuators, and energy harvesters. The various PVDF crystalline phases are α-, β-, γ-, and δ- phases [2–5]. The phases are correlated with various polymer chain conformations. The β phase is a highly ordered crystalline structure with a *TTT* (all-trans) planar zigzag conformation with fully extended polymer chains, α and δ phases have a *TGTG'* (trans-gauche-trans-gauche) conformation often occurring in semi-crystalline polymers, where chains can adopt both ordered (crystalline) and disordered (amorphous) regions, and the γ phase has a T_3GT_3G' chain conformation involving three trans segments followed by a gauge segment occurring in certain types of polymers [6, 7]. Mostly α, β and γ phases of PVDF used for applications and investigated [1–5, 7]. The ferroelectric properties of PVDF and P(VDF-TrFE) crystalline phases were mainly obtained from molecular chain packing in an all-trans conformation. In ferroelectric polymers, such as PVDF, the arrangement of substituted fluorine atoms on one side and hydrogen atoms on the other side of the molecular chain supports the formation of a net dipole moment perpendicular to the chain axis. When crystallized, such polymer chains with aligned dipoles develop a collective net dipole moment, which is characteristic of the β phase. In the α phase crystal, the polymer chains are packed such that the dipoles are antiparallel and the the net dipole moment is negligible [8]. Ferroelectric polymers exhibit phase transitions, i.e., trans to gauche conformational transitions at the Curie transition. Above the Curie temperature, in ferroelectric crystals, segmental rotation of the polymer chain readily occurs, which reduces steric hindrance, leading to the formation of a gauche conformation [9]. PVDF and its copolymers are thermoplastic fluoropolymers with good mechanical flexibility and low thermal conductivity, making them ideal for commercial applications as electrical insulators. In PVDF, the dipole moment arising from the $-CH_2CF_2-$ molecular unit contributed to a dielectric constant of approximately 8 at 1 kHz. The crystallization of these PVDF polymers in solution-processed films results in rough films (σRMS >3 nm). In contrast, copolymers of P(VDF-TrFE) have better film morphology than PVDF and therefore have wider applicability in electronic devices [10–15]. In P(VDF-TrFE) crystals, the presence of strong local molecular polarization, in contrast to PVDF, results in a slightly elevated dielectric constant of approximately 10 at 1 kHz.

Following the rise in the demand for electronic memory devices that can be embedded in organic integrated circuits [16], the P(VDF–TrFE) polymer provides better opportunities for application as a flexible low-cost memory device owing to its prominent ferroelectric behavior [17–19]. Several studies have focused on investigating the ferroelectric properties of PVDF-based polymers and copolymers, including saturation polarization [20, 21], dipole switching time [22], coercive field [23], thin-film crystallinity [24], modulation of the interfacial layer [25], and the behavior of electroactive polymers under electron beam irradiation [26, 27].

Ferroelectric polymer film-based memory devices, when integrated into a miniaturized electronic device, must possess films in the nanoscale or microscale range. Ferroelectric memory devices are characterized by a range of structural designs, such as capacitors with a metal-ferroelectric-metal (MFM) configuration [28–30], diodes with a metal-ferroelectric-insulator-semiconductor (MFIS) configuration [17, 29, 31–34], and ferroelectric field-effect transistors (FeFETs) [17, 33–36]. Submicron-thick ferroelectric polymer film preparation can be made using spin-coating [36–39], electrospinning [40] or Langmuir Blodgett (LB) technique [17, 41]. In general, because of the different defect problems in the LB P(VDF-TrFE) films, they have limited applicability in the fabrication process of real devices. Spin coating is an effective and reliable process for fabricating high-quality P(VDF-TrFE) films on sizeable surfaces for various device applications at minimal cost. A properly poled and thermally annealed spin-coated film of P(VDF-TrFE) exhibited robust ferroelectricity.

There has been a steady rise in interest in solution-processed thin-film transistors for applications in portable, wearable, and miniaturized electronics, such as flexible displays [42, 43] and identification tags [44]. All such gadgets require nonvolatile memory elements where data can be stored and can simultaneously perform programming, reading, and erasing of data electrically. FeFETs are appealing as memory elements because they offer rapid and nondestructive data readout processes while consuming minimal power [45]. In the basic configuration of an FeFET depicted in Figure 2.1, the structure comprises a stack of metal–ferroelectric–semiconductor layers, where the ferroelectric layer serves as the gate dielectric. Over the past decade, numerous studies have showcased P(VDF-TrFE) polymer-based FeFETs, highlighting their exceptional features, such as memory window, data stability, retention characteristics, and the ability to

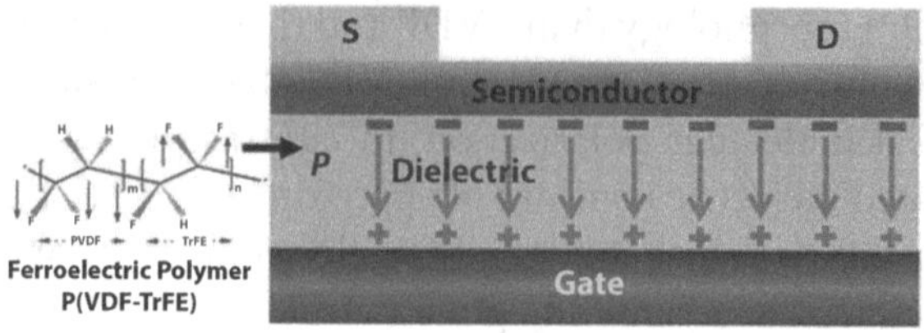

Figure 2.1 Diagram illustrating the basic top-source bottom-dielectric configuration of a FeFET structure, where the dielectric layer is composed of the ferroelectric polymer P(VDF-TrFE) in a downward polarization state under negative gate bias.

achieve multilevel nonvolatile conductance states [46–55]. These advancements aim to develop nondestructive readout devices that operate with minimal power consumption. Therefore, FeFETs based on polymer materials with decent data storage competence, apt to circuit integration and industrially scalable viability can be the future for next-generation organic flexible electronics.

2.2 Fabrication of Gate Dielectric Layer and FeFET

As stated earlier, among the various techniques for the preparation of ferroelectric polymer thin films, the spin-coating process is most widely used for thin film preparation because of its high reliability for fabricating quality films over large surface areas at a minimal cost.

The spin-coating process is described in detail in this study, as it is usually used for the fabrication of solution-processable flexible devices. Generally, three processing steps are used to prepare thin films of PVDF-based polymers.

1. Preparation of polymer solutions using various solvents.
2. Spin-coating of the solution in various kinds of substrates at different revolutions per minute (RPM).
3. Annealing under various conditions to improve the film crystallinity.

The preparation of the thin films using the spin-coating technique is shown in Figure 2.2. The PVDF-TrFE copolymer thin-film preparation process is described in detail. PVDF-TrFE copolymer with varied wt% of TrFE can be obtained from various suppliers such as PIZEOTECH, France; Solvay, USA; and MSI Sensors, PA, USA, and can be easily spin-coated

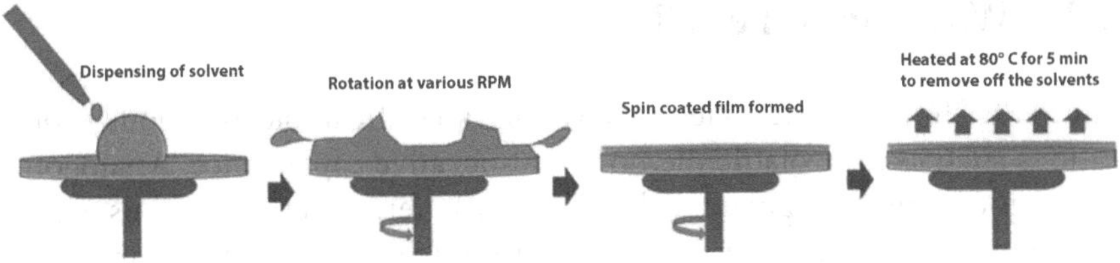

Figure 2.2 Schematic diagram of spin-coating process.

on the various types of substrates, such as PTFE, silicon (Si), and indium tin oxide (ITO), prepared by the friction-transfer method from its solution. The known melting (T_m) and Curie (T_c) temperatures of PVDF-TrFE (70/30 VDF/TrFE) were approximately 150 °C and 80 °C, respectively. Approximately 0.1–3 wt% PVDF-TrFE solution was prepared using methyl ethyl ketone (MEK), dimethylformamide (DMF), dimethyl sulfoxide (DMSO), or their mixer generally spin-coated at 2,000–3,000 rpm for 60 s for thin film formation. In thin films, P(VDF-TrFE) with a VDF content exceeding 70 mol% generally adopt an all-trans β-crystal structure, as presented in Figure 2.1, where the c-axis represents the polymer molecular chain alignment and CF_2 dipoles are directed along the b-axis. Annealing of the prepared films is usually performed at approximately 135 °C (glass transition temperature of PVDF-TrFE) for 1–2 h to obtain a higher β-crystal structure in PVDF-TrFE films and enhance their ferroelectric properties. Various processing steps are undertaken on the prepared ferroelectric thin films by obtaining preferential orientation of molecular chains to improve the degree of crystallinity of the β-phase, which enhances ferroelectric polarization and ease of switching of polarization under an ac field. Earlier studies have shown the regulation and improvement of both the β-phase crystallinity and orientation of ferroelectric polymer films using melt drawing [56], quenching [57–59], solvent evaporation, and temperature gradient [60–63]. Thermal processing at various temperatures has a marked effect on the crystalline structures of the deposited films and, therefore, on their ferroelectric properties [64].

For FeFET device fabrication, after preparation of PVDF-based ferroelectric polymer thin films as gate dielectric layers, various organic or inorganic semiconductor active layers are deposited by either solution processing or vacuum evaporation techniques. Thereafter, electrodes were deposited by physical deposition techniques using the shadow masking technique or patterning by either conventional photolithography or wet chemical etching processing. Metals with a suitable work function are used to have an ohmic contact with the semiconductor active layer.

2.3 Working of FeFET

The FeFET ferroelectric dielectric layer plays a critical role in various applications such as nonvolatile memory and negative capacitance. Analogous to magnetization-magnetic field hysteresis, ferroelectric crystals show electric polarization-electric field hysteresis with two stable polarization states in their ferroelectric domains, i.e., up polarization state and down polarization state. Polarization-electric field hysteresis is mainly described by zero field remnant polarization P_r, saturation polarization P_s at a given applied field, and coercive field E_c to switch the polarization state in the ferroelectric domains of the crystals.

Non-volatile memory structures based on FeFETs are more promising than resistive or capacitor-based data because of their ease of reading and writing non-destructive data. The ferroelectric dielectric layer in FeFET zero-field remnant polarization can have two dipole orientation states, either upward or downward. Based on the orientation of the polarization state in the ferroelectric dielectric layer, either positive or negative surface charges are induced in the semiconductor channel layer along the semiconductor-dielectric interface. These compensation surface charges in the semiconductor channel stabilize the polarization state of the ferroelectric dielectric layer. The orientation of the polarization state in the dielectric layer of the FeFET at zero gate bias can lead to the development of two conduction states in the semiconductor channel layer: either high or low. The on/off current ratio is the ratio of the high and low drain current values corresponding to the on- and off- states of the FeFET measured at zero gate bias at a very small drain bias. A distinct and high on/off ratio of the FeFET is a prerequisite for its integration in a memory chip. The high and low drain current states or the source-drain resistance can be designated as Boolean "0" and "1" states according to the known binary nonvolatile memory structure with nondestructive readout [18, 20]. The above features are presented schematically in Figures 2.3a,b. Therefore, in FeFET direction of polarization state in dielectric set by controlled gate biasing regulates the channel conductance at semiconductor-dielectric interface and hence establishes two logic states, "0" and "1" for non-volatile memory functionality. Other advantages of FeFETs include their ability to operate at low power, achieve high write/read speeds, provide random-access functionality, and exhibit high endurance. Electrical studies that are generally undertaken to ensure the non-volatile property of FeRAM and FeFETs are polarization versus electric field (*P*–*E*) and drain current (I_{DS}) versus gate voltage (V_G) transfer hysteresis curves shown in Figures 2.3a,b. The *P*–*E*

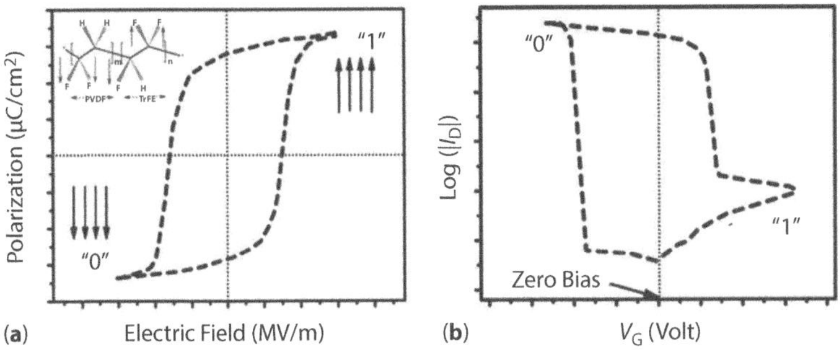

Figure 2.3 Schematic presentation of (a) Polarization–Electric Field (P–E) hysteresis plot of MIM capacitor based on ferroelectric materials. (b) I_D–V_G transfer hysteresis curve of ferroelectric material-based field effect transistor with p-type semiconductor.

curve of FeRAM has a hysteretic characteristic that arises from the ferroelectric property of the material, as shown in Figure 2.3a. Similarly, in the FeFET, the I_{DS} vs V_G transfer curve shows hysteretic characteristics, where the voltage width of the hysteresis provides information about the memory window for nonvolatile applications.

The idea of the FeFET is to include a high dielectric constant layer as a gate dielectric with ferroelectric properties in existing FET structures that induce nonvolatile properties as the ferroelectric layer remembers the electric field to which it had been exposed. In addition, the ferroelectric layer reduces the operating power of the device by exploiting negative capacitance property. There is a huge possibility that it can outperform current embedded flash memory. Thus, FeFETs are promising for future applications. Some of the recent application works related to Non-Volatile Memory (NVM) devices, photodetectors, synaptic devices, artificial skins, and wearable devices based on FeFETs are briefly presented in this study.

2.4 Applications of FeFET Device

i. Nonvolatile Memory Device

The classification of the memory device according to the present-day baseline, prototype, and emerging technologies is presented in Figure 2.4. Currently, there is a large focus on various emerging NVMs technologies, and a significant amount of research being conducted in this area on a regular basis. PCM, STTRAM, and RRAM are all back-end-of-line (BEOL) memories that use two-terminal memory elements combined between two metal films. Emerging NVM technologies each have their own unique

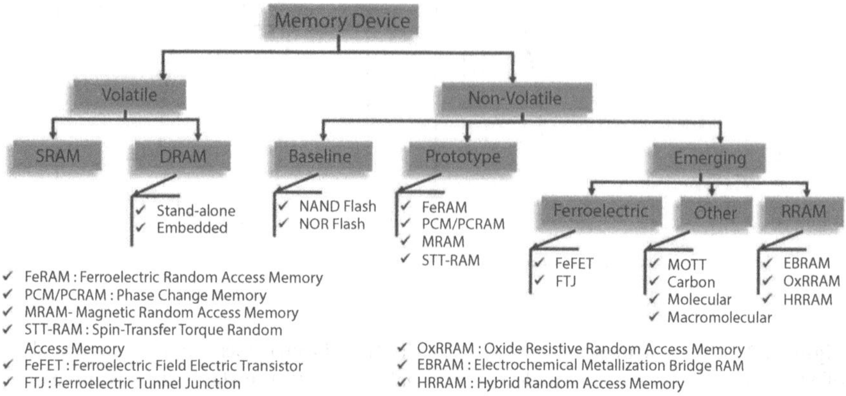

Figure 2.4 Schematic illustration of memory device classification.

strengths and weaknesses based on how they are made and how they perform. These NVM technologies have many possibilities other than their use in the main memory part of the memory hierarchy, such as cache or storage. NVMs have better scalability than DRAM and lower energy costs. A brief introduction to PCM, STTRAM, and RRAM is provided below.

The writing mechanism of MRAM is enhanced using the principle of spin-transfer torques, i.e., STTRAM, using which we can achieve higher-density NVMs at a lower cost. In the STTRAM device, the switching mechanism of electron spin takes place using spin-polarized current, i.e., a current having more electrons of either up or down spin. Among all NVMs, the highest known performance is for STTRAM, with a writing speed better than 10 ns and an endurance greater than 10^{12} cycles. PCM switching relies on reversible phase transitions induced by Joule heating. These transitions occur between a crystalline phase, characterized by low resistance, and an amorphous phase, characterized by high resistance, predominantly utilizing transition-metal chalcogenide materials. PCM shows the most promising performance among NVMs, with a switching speed of <100 ns and good data retention properties [65]. RRAM functions by modifying the internal resistance state within a solid-state dielectric material for data storage. RRAM can achieve a high endurance greater than 10^{12} cycles at a high switching speed of <1 ns and is attractive for realizing high storage density at a low cost.

Presently, FeFETs are a notable growing NVM technology, in which a ferroelectric film is employed as a dielectric layer. The direction of polarization switching in the ferroelectric layer results in modulation of the channel layer conductance of the FeFET device. The floating gate principle is used for data storage in flash memory; similarly, the ferroelectric polarization states of ferroelectric materials are used for data storage. FeFETs can exhibit memory and synaptic properties, which can play a vital role in integration circuits and neuromorphic computing [65, 66]. One transistor (1T) structural design based on FeFETs can assist in the development of both switch and memory combined with a 1T selector [67–69]. Diodes and nonlinear devices have limited compatibility with three-terminal FeFET-based devices as selector elements. Two-terminal switch elements, such as memristors, can work with one transistor-one memristor 1T1R or one diode-one memristor 1D1R platform [69, 70]. 1D1R and 1T1R can easily read data in an integrated memory array structure. The fast-switching speed and low power requirements make FeFETs a strong contender for next-gen NVM devices. However, there are some challenges that need to be standardized for FeFET technology application in future emerging NVMs devices, such as endurance and retention time of data, ease in reading, writing and erasing of data, and scalability and integration capability in ICs. Table 2.1 presents the advantages and challenges of the four emerging NVM technologies.

Since the first demonstration of an FeFET device based on PVDF and its copolymer [31, 71], significant efforts have been made to improve its functioning parameters such as flexibility, endurance, and reliability, as well as on material design optimization of the ferroelectric [50, 51] and semiconductor layers [18, 48, 50, 51]. Kang *et al.* demonstrated improved nonvolatile characteristics observed for FeFETs using nanoscopic confinement of the ferroelectric polymer PVDF-TrFE on nanopatterned organosilicate (OS) lamellae [72]. The nanoscale periodic lamellae of OS were patterned using block copolymer self-assembly (Figure 2.5(a)) and then hybridized with PVDF-TrFE (Figure 2.5(b)) to be applied as a gate dielectric in FeFETs [72]. The confined crystallization of PVDF-TrFE facilitates ferroelectric polarization switching because of the better crystal alignment and elimination of structural defects through nanoconfinement, as shown in Figure 2.5(c). The gate dielectric exhibited a low leakage current (Figure 2.5(d)). The fabricated FeFET device with a single-crystalline TIPS-PEN channel and PVDF-TrFE nanopatterned film as the gate dielectric displays a hysteretic I_{DS} transfer curve with a memory window, and the ON current increases with the gate voltage increment, as shown in Figure 2.5(e). The ON and OFF current alterations with changes in gate biasing are shown in Figure 2.5(f), where the ON/OFF current ratio saturated at ±8 V. Figure 2.5(g), shows the 2-hour data retention property of the device in the ON and OFF states, which was evaluated by measuring I_{DS} under 0 V gate bias

Table 2.1 Advantages and challenges of emerging nonvolatile memories.

NVM emerging technologies	Advantages	Challenges
PCM	• Stablished Technology • Better device scaling property	• Write speed slower ~10 ns • Large programming current • Limited endurance—Reduces after certain number of operations
STTRAM	• Physics well defined • High write speed • Better endurance	• Issue with Scalability-Thermal Stability • Read Disturb as current used for reading may change the memory state
RRAM	• Common material, structure, and behavior • Low power operation • Easier to adapt in production • Better device scalability	• Resistance variation is much greater • Endurance smaller • Reliability
FeFET	• Low power operation • 1T cell structure • Scalability	• Fabrication and Integration • Retention • Read/Write Disturb

and at a fixed −1 V source-to-drain (V_{DS}) biasing following switching of the polarization state.

Roy *et al.* presented the fundamental role of material processing in controlling the ferroelectric crystalline microstructures in PVDF-TrFE spin-coated films to obtain highly oriented polarized dipoles. Better-aligned dipoles support the minimization of polarization fluctuations along the semiconductor-dielectric interface, which ultimately minimizes energetic disorder and improves the mobility of the transport channel of the FeFET device [73]. The parallel orientation of the electric dipoles at the interface is obtained via modification of the microstructures of the ferroelectric film by preparing the film using a fast quenching process. The absorbance peak at 1,398 (Figure 2.6(c)) confirms that the fast-quenching process facilitated the alignment of the polymer chains along the surface of the substrate.

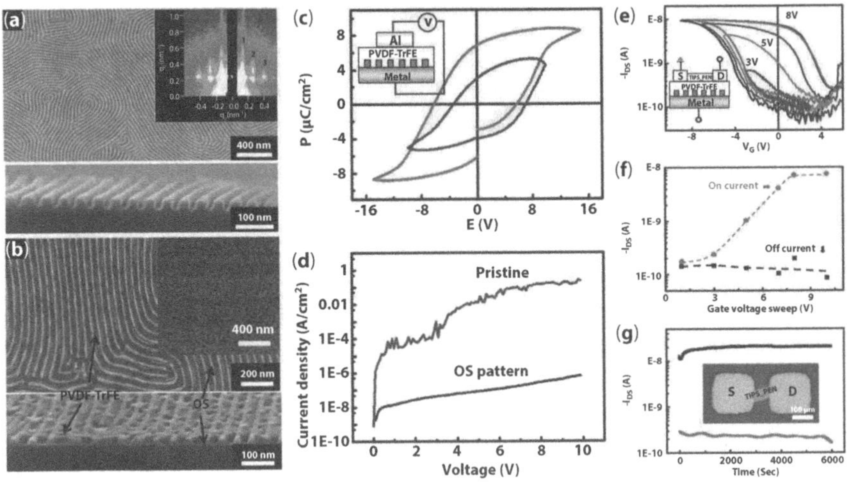

Figure 2.5 (a) SEM image of the morphology of periodic OS lamellae. The 2D GISAXS pattern of the nanopatterned is shown in the inset of (a). SEM micrograph side-view image is presented at the bottom of (a). (b) SEM micrograph of annealed spin-coated PVDF-TrFE film on periodic OS lamellae. AFM image in height contrast is shown in the inset of (b). A side view is presented at the bottom of figure (b). (c) Presents the *P* versus *E* hysteresis loop of the fabricated PVDF-TrFE capacitor. A schematic of the capacitor is presented in the inset. (d) I–V curve of PVDF-TrFE films. (e) Transfer characteristics (I_D-V_G) of FeFET device with hybrid gate dielectric layer. A schematic representation of the FeFET is shown in the inset. (f) Plot illustrating the ON and OFF currents versus gate voltage sweeps. (g) Data retention performance of the FeFET memory. Reproduced with permission. [72] Copyright 2020, ACS.

Ferroelectric electrical properties of fabricated MIM capacitors with fast-quenched films shows high remnant polarization of P_r ~5.5 ± 0.2 μC/cm², whereas normally-cooled film shows P_r ~2.7 ± 0.2 μC/cm², at electric field of 200 MV/m. Likewise, the fabricated FeFET device with a fast-quenched film as the implanted gate dielectric layer showed a 35% higher carrier mobility value than a normally cooled film. In addition, studies related to the buffering of the semiconductor/dielectric interface to engineer polarization fluctuation and contact resistance at the interface can improve the carrier mobility and power consumption of the FeFET device [74–77].

Fully mechanically flexible FeFET-based NVM electronics prepared using plastic and cellulose paper have been investigated by a number of researchers [78–82]. The flexible devices exhibited good memory properties, low operating voltage, current turn on/off ratio, and a distinct memory window. Kim *et al.* presented a highly flexible FeFET-based NVM device prepared using P(VDF-TrFE)/PVP as a bilayer dielectric layer and

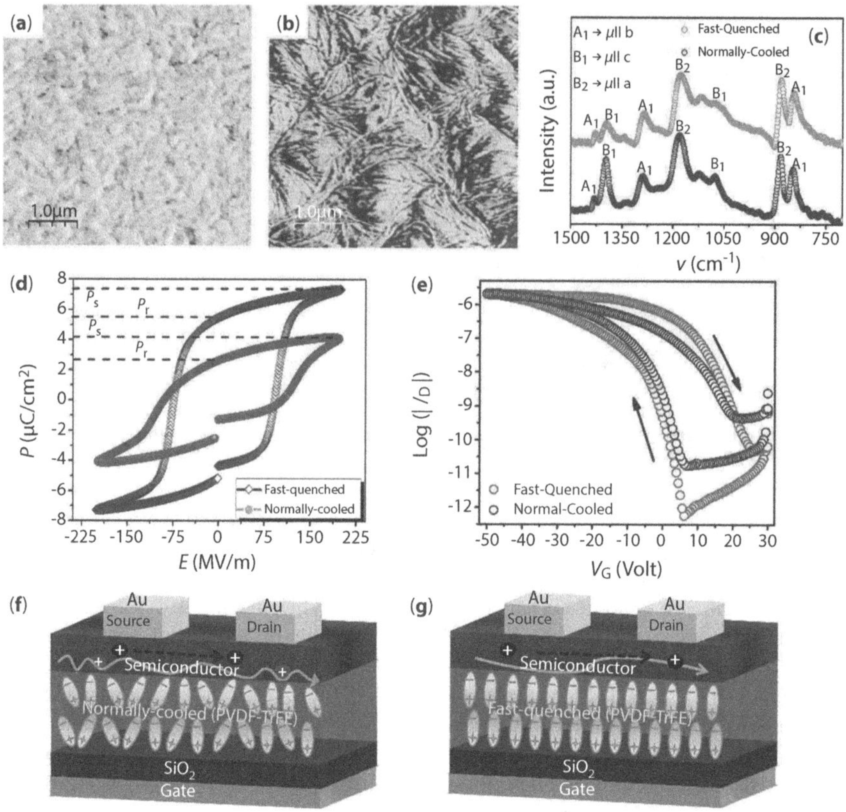

Figure 2.6 AFM phase-contrast image of (a) normally cooled and (b) fast-quenched P(VDF-TrFE) polymer film. (c) FTIR absorbance spectra obtained in GIRAS mode and (d) *P–E* hysteresis loop of fast-quenched and normally-cooled P(VDF-TrFE) thin film. (e) I_D–V_G transfer hysteresis curve of FeFET devices with CuPc as semiconductor channel layer. FeFET device with (f) higher polarization fluctuation and (g) lower polarization fluctuation in the channel layer at the semiconductor-dielectric interface [73] Copyright 2020, ACS.

a quinoidal oligo-thiophene derivative (QQT(CN)4) as the semiconductor channel layer [81]. The reliability of the prepared memory device was inspected by studying time-varying data retention capacity and endurance tests with manifold data write/erase capabilities. A high-performance air-stable FeFET-based NVM with outstanding mechanical elasticity that can function in p- and n-type dual modes is presented. In both switching modes, the prepared NVM device exhibited good mechanical properties. The device showed notable data retaining and enduring characteristics of >6,000 s and 1,000 folding cycles, respectively. The prepared FeFET devices showed high robustness without any extra covering layer for any mechanical action, such as bending or folding. The performance response of the device remained consistent even after 1,000 folding cycles, as shown in Figures 2.7(c,d). The data retention and endurance characteristics of the flexible FeFET devices with a bending radius of 4 mm are shown in Figures 2.7(e,f). The nanoindentation study confirmed the presence of elastic nature and solid bonding at the interface of the semiconductor channel layer and ferroelectric polymer film. This research marks a significant stride in the development of highly flexible organic memories, achieved by employing dual-mode flexible FeFET devices known for their superior mechanical elasticity.

Researchers have also conducted several studies on hybrid organic–inorganic FeFET memories to compare the memory and FeFET characteristics of fully organic-derived FeFET devices. Most commonly, among ferroelectric polymers, P(VDF–TrFE) is applied as the gate dielectric layer, and ZnO and IGZO films are utilized as inorganic semiconducting channel layers [10, 30, 52, 71, 82] in fabricated devices. The device carrier mobility, memory characteristics, and switching on/off ratio were further analyzed. Yoon *et al.* presented a completely transparent FeFET-based NVM device fabricated using a hybrid gate dielectric stack of a P(VDF-TrFE) ferroelectric polymer and Al_2O_3 oxide layers [83]. The ferroelectric polymer layer

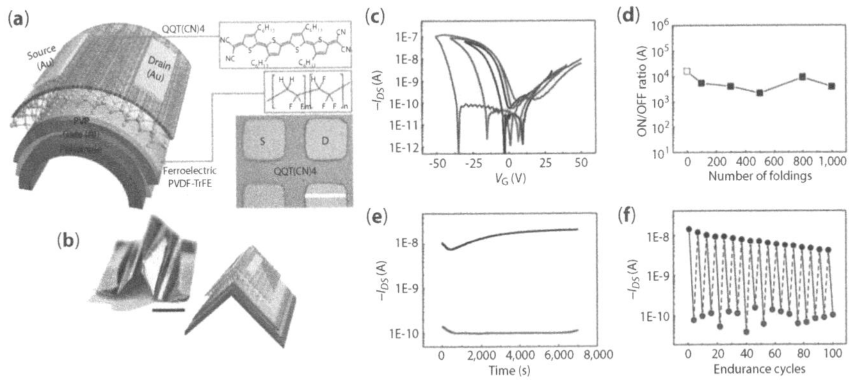

Figure 2.7 (a) Illustration of the fabricated FeFET device and molecular structure of QQT(CN)4 and PVDF-TrFE molecules applied as semiconducting and dielectric layers; (b) Picture of a prepared flexible device with an array of FeFETs on a PI substrate. (c) I_D–V_G transfer hysteresis curve of the fabricated FeFET device at V_{DS} = 5 V. (d) On/Off ratio and (e) data retention characteristics of the prepared FeFET device obtained for 1,000 folding cycles. (f) Endurance performance of device after 1,000 folding cycles. Reproduced with permission. [81] Copyright 2014, Nature Publishing Group.

exhibits the NVM property of the developed transistor, where Al–Zn–Sn–O (AZTO) is applied as a semiconducting channel.

The fabricated FeFET is fully transparent in the visible range. Figure 2.8b presents the variation in the I_D–V_G transfer hysteretic curves measured at different sweep ranges of gate voltage. These findings indicate that the fabricated FeFET device can function effectively at very low operating voltages. For NVM applications, the learning of the memory function of the device is essential using pulse signals of different time widths. The on/off memory margin increased with the pulse width. The initial on/off ratio was ~10^3, and the ratio reduced to ~10 after 1 h, as shown in Figure 2.8f, which shows that the retention behavior of the NVM FeFET device is receptive to the memory read-out voltage V_G. The application of two new strategies in fabricated FeFET devices, i.e., applying AZTO as a semiconducting channel and Al_2O_3 layer in the dielectric ensures good memory characteristics. A memory window of 7.5 V, mobility (μ_{lin}) of 32.2 cm^{2}V^{-1}s^{-1}, subthreshold swing (SS) of approximately 0.45 Vdecade^{-1}, an on/off ratio ~10^8, and low gate leakage current of 10^{-13} A is exhibited by fabricated FeFET device. Further, a work on flexible hybrid FeFET device was demonstrated by Petti *et al.*

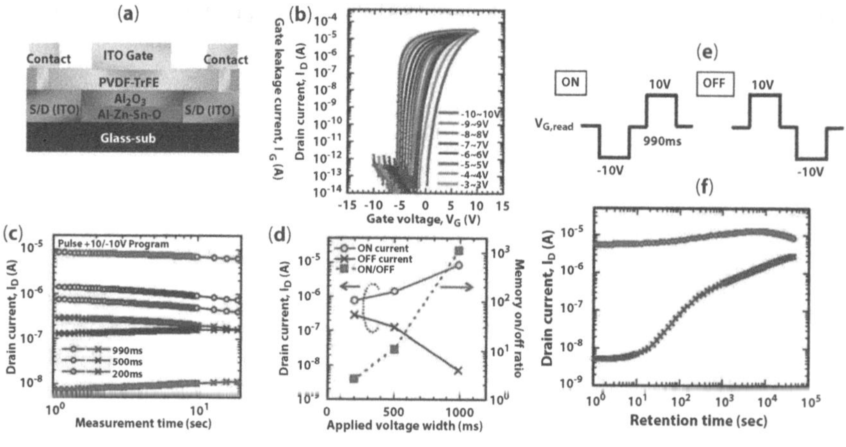

Figure 2.8 (a) Illustrative image of transparent NVM FeFET device. (b) I_D–V_G transfer hysteretic curves obtained as gate voltage sweep range successively increased from −3 to 3 V and from −10 to 10 V. (c) Variations of I_{DS} with time as the change in applied pulse width of signals from 990, 500, and 200 ms obtained for On and Off of the FeFET device. (d) Detailed measured I_{DS} values for the respective signals and equivalent on/off ratios observed under varying pulse widths. (e) Signal pulse progression for assessing the memory retention properties of the fabricated FeFET device. (f) Variation in I_{DS} measured in both the On and Off states with a delay in memory retention time. Figures (b–f) were reproduced with permission [83]. Copyright 2010, Wiley.

using a-IGZO semiconducting layer and P(VDF-TrFE) as gate dielectric layer [84]. NVM study presents that the memory window of the device is around 3.2 V with On/Off ratio ~10^6, at gate voltage of ±6 V and measured standard mobility of ~8 $cm^2V^{-1}s^{-1}$. The FeFET device showed threshold voltage shifts in the application of tensile and compressive strains.

ii. Photodetector

The exploration of plastic and paper substrates for flexible photodetectors has gained significant interest, as they have prospective applications in wearable optoelectronic devices. Recently, a few studies have been conducted on ferroelectric polymer-based FeFET devices using transition metal dichalcogenide (TMDC) semiconductors such as MoS_2, $MoTe_2$, and In_2Se_3 for photodetector applications [85, 86, 87–92]. Wu *et al.* a 2D In_2Se_3 photodetector having a board detection range from visible to near IR and possesses a quick response time [90]. Similarly, Zheng *et al.* demonstrated an indium phosphide (InP) single-nanowire ferroelectric polymer side-gated near-infrared (NIR) photodetector with dark current measured in picoamperes with a gain of approximately 10^5 [91]. Wang *et al.* presented an interesting study on MoS_2 based photodetector by fabricating an FeFET device using a ferroelectric dielectric layer. Reducing the dark current response of the device remnant polarization direction in the ferroelectric-semiconductor layer plays a significant role [92]. The photoswitching property at a light (λ) of 635 nm was largely obtained in the case of the P_{up} polarization state of the ferroelectric gate dielectric. Figure 2.9d illustrates how I_{ph} of the device varies with the laser power at V_{sd} = 5V. This study revealed that the prepared device has good photodetection properties. The MoS_2 device displayed exceptional photosensitivity compared with conventional MoS_2 transistor-based photodetectors. The measured high photosensitivity of the device is 2,570 A W^{-1} at a given laser power equal to 1 nW and 2.2×10^{12} Jones, respectively (Figure 2.9(f)), owing to the sharp electrostatic field presented by the stable remnant polarization near the semiconductor–dielectric interface, which resulted in full depletion of the MoS_2 channel. This supports the improved response of the fabricated FET-based detector, even without gate biasing. In addition to the excellent photoresponsivity, time-resolved photoresponse experiments at various frequencies confirmed the stability and reliability of the photodetectors. Infrared (0.85–1.55 μm) wave detection by a photodetector shows that it is an excellent device for application in optical communications. The present work highlights the prospects of 2D materials for advancing next-generation optoelectronic devices.

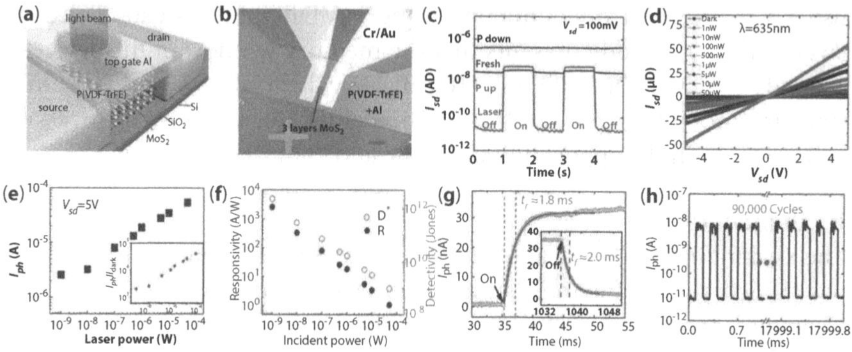

Figure 2.9 (a) Three-dimensional schematic and (b) optical images of the FET-based triple-layer MoS_2 photodetector with PVDF-TrFE as the ferroelectric dielectric layer. (c) Photoswitching behavior of three-layer MoS_2 photodetector in three states (λ = 635 nm, V_{SD} = 100 mV, P = 100 nW). (d) I_{DS} current of the photodetector in the dark and at various light-illuminated states of power from 1 nW to 50 μW. (e) Variation in photocurrent under different light lumen powers, while the inset presents the ratio of photoresponse current to dark current. (f) Photoresponsivity of MoS_2 based photodetector. Photocurrent response of FET device (g) with fitted exponential functions and (h) endurance testing for 90,000 illumination cycles of operation at V_{DS} = 100 mV and light power P = 100 nW in the P_{up} state of the ferroelectric layer. Reproduced with permission [92]. Copyright 2015, Wiley-VCH.

The photodetector application of ferroelectric polymer-based FETs on flexible substrates was further demonstrated by Zhao *et al.*. They integrated a P(VDF-TrFE) ferroelectric layer with a red light-responsive organic semiconductor, P(DPP-TzBT), as an organic FeFET-photodetector [93]. This study presents the detailed stability and reliability of flexible FETs measured undered various operating conditions and mechanical folding cycles. The device transfer curve shows hysteresis characteristics under both light and dark conditions, with ferroelectric polarization reversal occurring at approximately ±25 V (Figure 2.10(b)). The data retention characteristics of the device obtained at V_{read} = −10 V and V_{gate} = 0 V after a short V_{write} pulse (±100 V for 1–2 s) are presented in Figure 2.10c, showing that states with low and high resistance remain stable up to 10^3 s which proves the extended data retention ability. The NVM characteristics (Figure 2.10(d)) show good repeatable write and erase actions of the device at pulse gate biases of −60 V and +100 V, respectively. The transient photocurrent response with increasing intensity of the 730 nm light pulse is shown in Figure 2.10(e). Better light to dark current ratio attained at V_{gate} = 60 V state, compared to “0” and “1” states because positive biasing expedites the recombination process of generated holes and electrons using light.

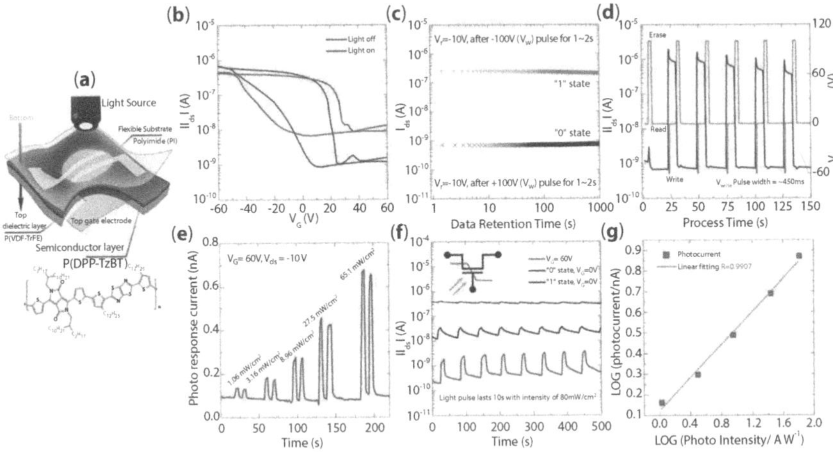

Figure 2.10 (a) Schematic representation of photosensing of FeFET device on polyimide (PI) substrate. (b) Hysteretic I_D–V_G transfer curve of FeFET device in the dark and under light of 730 nm wavelength with power 80 mW/cm². (c) Data retention behavior, (d) erase and write endurance test and (e) photoresponse behavior with variation in light lumen power of the fabricated FeFET device (f) Photoresponse performance of the device at three different resistance states with a light pulse array of 80 mW/cm² and width 10 s. (g) Plot of photoresponse of the device with change in intensity obtained at +60 V gate bias. Reproduced with permission [93]. Copyright 2017 ACS.

The device-regulated dark current value ranged fron 1×10^{-9}A to 100×10^{-9}A. Therefore, the fabricated plastic electronic OTP is a promising device for photodetectors.

iii. Wearable Devices

PVDF and PVDF-TrFE are also potential materials for future neuromorphic devices. In an FeFET-based synaptic device, the Postsynaptic Current (PSC), i.e., the source-to-drain current (I_{SD}), is measured from the corresponding presynaptic pulse in the form of a pulse-form gate voltage (V_G), light pulse, or pressure pulse. The ratio of two consecutive PSC represents Pulse Pair Facilitation (PPR), and the difference between I_{SD} before and after a given number of presynaptic pulses represents the synaptic weight (SW) [94]. SW strengthens with increasing frequency of applied stimuli and weakens when the frequency decreases which, can be represented using short-term plasticity (STP) and long-term plasticity (LTP) [95].

Various researchers have envisioned next-generation wearable and intelligent electronics based on ultrathin flexible artificial synaptic devices made of free-standing ferroelectric transistor [94–99]. Lee *et al.* presented a study on an artificial intrinsic-synaptic tactile sensory organ

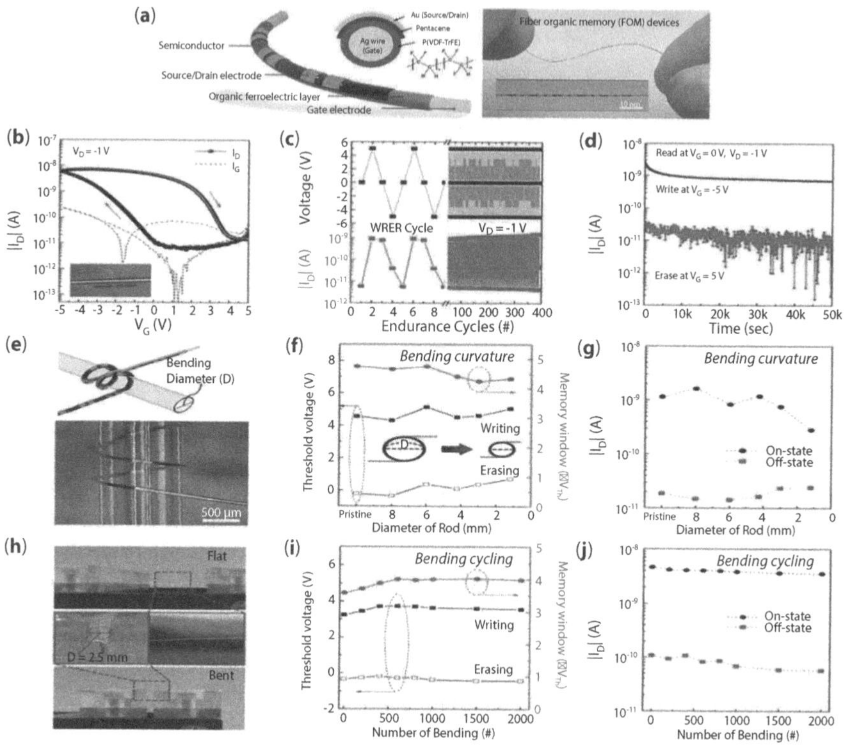

Figure 2.11 (a) Graphical image of fibriform organic transistor memory FOM (left) and optical image of flexible array device of FOM (right). Fabricated FOM device (b) $I_D - V_G$ hysteretic transfer curves, (c) write/erase endurance performance, and (d) data retention test. (e) Schematic illustration of the optical image of the FOM twisted on a glass tube. (f) Threshold voltage shifts and (g) on/off current states of FOM devices at $V_D = -1$ V. (h) Optical image of FOM in bending cycling test. (i) Threshold voltage shifts and (j) on/off current states of FOM for bend testing at a diameter of 2.5 mm. Reproduced with permission [99]. Copyright 2019 ACS.

(AiS-TSO) gated by a triboelectric-capacitive coupling effect that mimics the synapse-like connections of Merkel cell-neurite complexes (MCNCs) [96]. Jang *et al.* demonstrated that the prepared device could be moved to any kind of substrate with a large and flexible surface morphology, such as plastic film, jelly, textile, candy, toothbrush, and brain mold [97]. *Kang et al.* first demonstrated the fabrication of fibriform organic transistor memory (FOM) using the ferroelectric polymer P(VDF-TrFE) [99]. The organic transistor has a channel length (L) and width (W) of 15 and 157 μm respectively, formed using P(VDF-TrFE) dielectric layer ~260 nm, pentacene semiconducting layer ~50 nm and Au electrodes ~50 nm. The capillary tube coating system facilitates the development of nanosized organic film over any kind of wire, ensuring consistent device functionality at low voltages. The fabricated FOM device showed good endurance performance and data retention characteristics over continuous cyclic write/erase operations, as shown in Figure 2.11(c). The retention features of the fabricated FOM device are presented in Figure 2.11(d). The lowest operating power of the device for memory functionality is 3×10^4 pW. Bending testing was performed by measuring the threshold voltage shifts and on/off current states of the FOM devices, as shown in Figures 2.11(f,g). The fabricated FOMs exhibited good stability and robustness against mechanical stresses. The memory functionality of the device was measured over 2,000 bending cycles, as shown in Figures 2.11(i,j). Fabricated FOMs display good memory retention times along with exceptional mechanical endurance. The low-voltage-operated stretchable fabric prepared using FOMs exhibited excellent on/off switching stability and retention characteristics under severe mechanical deformation conditions. Given these features, fabricated FOMs show potential for application in the development of low-power smart wearable devices, particularly as data storage media.

An organic FET device with a ferroelectric polymer as a dielectric was incorporated to prepare an electronic skin, as demonstrated by *Lee et al.* in Figure 2.12 [100]. The fabricated device could perceive and learn a wide range of tactile information. The design of the fabricated electronic skin is distinct, with an array of dome-shaped elastomeric tactile FeFETs having top-gated ferroelectric dielectrics. The tactile recognition responsivity of prepared tactile learning ferroelectric skin (ATFES) has a value of >88 kPa^{-1}, and is capable of emulating various significant synaptic operations [100]. The device showed steady long-term adaptability using 10,000 input electrical spikes and displayed spike-dependent adaptability with variations in the magnitude and number of pressure pulses. The device had a low variability of 3.18%, which supports its ability to learn accurate tactile perception. The designed 4×4 ATFES device can distinguish three dissimilar handwriting styles corresponding to four specific alphabetical letters ("D," "P," "N," and "Z") through 2-diemsional spatial learning and recognition process. The test achieved over 99% accuracy even with 10% noise. Therefore, this study proposes an innovative direction towards devising artificial intelligent e-skins with great accuracy to carry out error-free tactile perception learning.

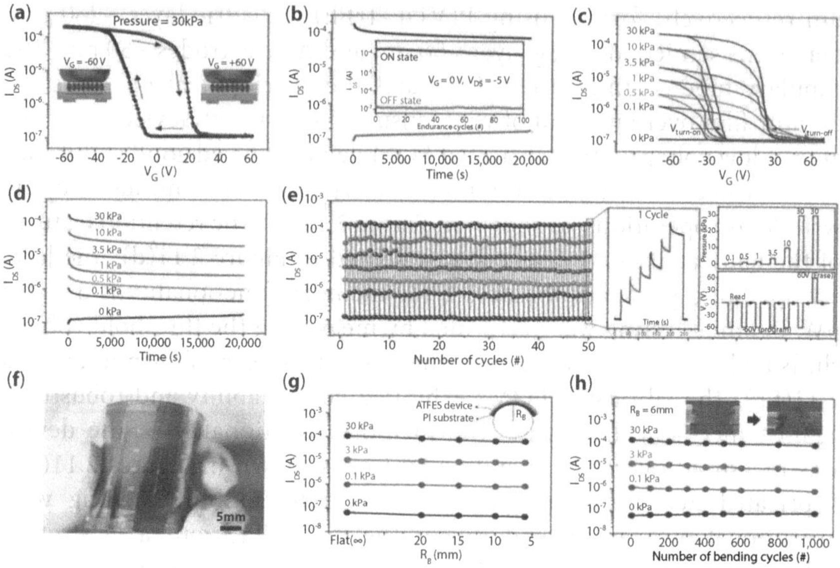

Figure 2.12 (a) I_{DS}–V_G hysteretic transfer curve of ATFES resulting from the polarization states of the P(VDF-TrFE) layer obtained at a pressure of 30 kPa. (b) On- and off-state retention characteristics measured at $V_G = 0$ V and $V_{DS} = -5$ V. Inset showing write/erase endurance of the device. (c) I_{DS}–V_G hysteretic transfer curves and (d) retention characteristics of the device measured at pressures ranging between 0 and 30 kPa. (e) Recurring write/erase switching endurance test measured at seven distinct pressure values. The inset displays the one-cycle real-time response, pressure sequence, and gate-voltage sequence. (f) Optical image of flexible ATFES array on polyimide substrate. Bending test performed (g) with variation in R_B and (h) up to 1,000 cycles at $R_B = 6$ mm (endurance test). Inset showing the bent and unbent states of the device. Reproduced with permission [100]. Copyright 2020 Wiley.

2.5 Summary

Ferroelectric polymer-based electronic devices offer immense opportunities for various applications, including nonvolatile memory (NVM), photodetectors, energy harvesting, and pressure sensors. Their distinctive properties, such as piezoelectric and ferroelectric properties and ease of processing, make these organic materials promising materials for the growth of future electronic devices. Flexible ferroelectric polymer-based FET devices with memory functionalities will be immensely beneficial for various next-generation applications ranging from wearable electronics to information management and communication. Ferroelectric wearable technology designed for tactile interaction could significantly advance as embedded electronic skin (e-skin), facilitating human-interactive interfaces to connect diverse future Internet of Things (IoT) devices. Memory functionality-based optoelectronic devices can greatly advance the features of video and biomedical imaging, security applications, various sensing processes, and motion detection. Advancements in such devices could make significant contributions to the development of modern health-monitoring devices. Therefore, ferroelectric materials have promising applications and should be the focus of future studies.

Acknowledgments

D.R. thanks IIT Guwahati for their support through an IPDF grant. D.R. thanks SERB-INDIA for their support through the NPDF grant-project file no. PDF/2015/000111.The authors declare no competing financial interests. D.R. and M.C. thank UGC-DAE-CSR for support through the project grant file no.CRS/2022-23/01/657.

References

1. Heiji, K., The Piezoelectricity of Poly (vinylidene Fluoride). *Japan. J. Appl. Phys.*, 8, 975, 1969.
2. Gregorio, R. and Ueno, E.M., Effect of crystalline phase, orientation and temperature on the dielectric properties of poly (vinylidene fluoride) (PVDF). *J. Mater. Sci.*, 34, 489–500, 1999.
3. Gregorio, Jr., R. and Cestari, M., Effect of crystallization temperature on the crystalline phase content and morphology of poly(vinylidene fluoride). *J. Polym. Sci. B Polym. Phys.*, 32, 859–870, 1994.
4. Song, D.D., Yang, D.C., Feng, Z.L., Formation of β-phase microcrystals from the melt of PVF2-PMMA blends induced by quenching. *J. Mater. Sci.*, 25, 57–64, 1990.
5. Lund, A. and Hagstrom, B., Melt spinning of poly(vinylidene fluoride) fibers and the influence of spinning parameters on β-phase crystallinity. *J. Appl. Polym. Sci.*, 116, 2685–93, 2010.
6. Martins, P., Lopes, A.C., Lanceros-Mendez, S., Electroactive phases of poly(vinylidene fluoride): determination, processing and applications. *Prog. Polym. Sci.*, 39, 683–706, 2014.
7. Bar-Cohen, Y., *Electroactive Polymer (EAP) Actuators as Artificial Muscles: Reality, Potential, and Challenges 2nd edn*, SPIE Press, Bellingham, WA, 2004.
8. Salimi, A. and Yousefi, A.A., Analysis Method: FTIR studies of β-phase crystal formation in stretched PVDF films. *Polym. Test.*, 22, 699–704, 2003.

9. Septiyani Arifin, D.K. and Ruan, J.J., Study on the curie transition of P(VDF-TrFE) copolymer. *IOP Conf. Ser.: Mater. Sci. Eng.*, 299, 012056, 2018.
10. Lee, G.G., Fujisaki, Y., Ishiwara, H., Tokumitsu, E., Low-Voltage Operation of Ferroelectric Gate Thin Film Transistors Using Indium Gallium Zinc Oxide-Channel and Ferroelectric Polymer Poly-(Vinylidene Fluoride-Trifluoroethylene). *Appl. Phys. Express*, 4, 091103, 2011.
11. Baeg, K.J., Khim, D., Kim, J., Han, H., Jung, S.W., Kim, T.W., Kang, M., Facchetti, A., Hong, S.K., Kim, D.Y., Controlled Charge Transport by Polymer Blend Dielectrics in Top-Gate Organic Field Effect Transistors for Low-Voltage-Operating Complementary Circuits. *ACS Appl. Mater. Interfaces*, 4, 6176–6184, 2012.
12. Hwang, S.K., Bae, I., Kim, R.H., Park, C., Flexible Non-Volatile Ferroelectric Polymer Memory with Gate-Controlled Multilevel Operation. *Adv. Mater.*, 24, 5910–5914, 2012.
13. Kim, R.H., Kang, S.J., Bae, I., Choi, Y.S., Park, Y.J., Park, C., Thin Ferroelectric Poly(Vinylidene Fluoride-Chlorotrifluoro Ethylene) Films for Thermal History Independent Non-Volatile Polymer Memory. *Org. Electron.*, 13, 491–497, 2012.
14. Adhikari, J.M., Gadinski, M.R., Li, Q., Sun, K.G., Reyes Martinez, M.A., Iagodkine, E., Briseno, A.L., Jackson, T.N., Wang, Q., Gomez, E.D., Controlling Chain Conformations of High-k Fluoropolymer Dielectrics to Enhance Charge Mobilities in Rubrene SingleCrystal Field-Effect Transistors. *Adv. Mater.*, 28, 10095–10102, 2016.
15. Wang, B., Huang, W., Chi, L., Al-Hashimi, M., Marks, T.J., Facchetti, A., High- k Gate Dielectrics for Emerging Flexible and Stretchable Electronics. *Chem. Rev.*, 118, 5690–5754, 2018.
16. Auciello, O., Scott, J.F., Ramesh, R., The Physics of Ferroelectric Memories. *Phys. Today*, 51, 7, 22, 1998.
17. Reece, T.J., Ducharme, S., Sorokin, A.V., Poulsen, M., Nonvolatile memory element based on a ferroelectric polymer Langmuir–Blodgett film. *Appl. Phys. Lett.*, 82, 142, 2003.
18. Naber, R.C.G., Tanase, C., Blom, P.W.M., Gelinck, G.H., Marsman, A.W., Touwslager, F.J., Setayesh, S., De Leeuw, D.M., HighPerformance Solution-Processed Polymer Ferroelectric Field-Effect Transistors. *Nat. Mater.*, 4, 243–248, 2005.
19. Park, Y.J., Kang, S.J., Park, C., Kim, K.J., Lee, H.S., Lee, M.S., Chung, U.I., Park, I.J., Irreversible extinction of ferroelectric polarization in P(VDF-TrFE) thin films upon melting and recrystallization. *Appl. Phys. Lett.*, 88, 242908, 2006.
20. Naber, R.C.G., Blom, P.W.M., Marsman, A.W., de Leeuw, D.M., Low voltage switching of a spin cast ferroelectric polymer. *Appl. Phys. Lett.*, 85, 2032, 2004.

21. Noda, K., Ishida, K., Kubono, A., Horiuchi, T., Yamada, H., Matsushige, K., Remnant polarization of evaporated films of vinylidene fluoride oligomers. *J. Appl. Phys.*, 93, 2866, 2003.
22. Xia, F., Xu, H., Fang, F., Razavi, B., Cheng, Z.-Y., Lu, Y., Xu, B., Zhang, Q.M., Thickness dependence of ferroelectric polarization switching in poly (vinylidene fluoride–trifluoroethylene) spin cast films. *J. Appl. Phys.*, 78, 1122, 2001.
23. Xia, F., Xu, H.H., Razavi, B., Cheng, Z.-Y., Zhang, Q.M., Dependence of threshold thickness of crystallization and film morphology on film processing conditions in poly(vinylidene fluoride–trifluoroethylene) copolymer thin. *J. Appl. Phys.*, 92, 3111, 2002.
24. Urayama, K., Tsuji, M., Neher, D., Layer-thinning effects on ferroelectricity and the ferroelectric-to-paraelectric phase transition of vinylidene fluoride–trifluoroethylene copolymer layers. *Macromol.*, 33, 8269–8279, 2000.
25. Roy, D., Chakraborty, M., Pattader, P.S.G.., Islam, A.K.M.M., Bandyopadhyay, D., Role of Annealing with Electric Field Toward Improvement of Ferroelectric and Electroactive Properties of PVDF Copolymer and Terpolymer Thin Films. *Macromol. Rapid Commun.*, 2400496, 2024.
26. Li, Z.-M., Li, S.-Q., Cheng, Z.-Y., Crystalline structure and transition behavior of recrystallized-irradiated P(VDF-TrFE) 65/35 copolymer. *J. Appl. Phys.*, 97, 014102, 2005.
27. Li, Z., Wang, Y., Cheng, Z.-Y., Electromechanical properties of poly (vinylidene-fluoride-chlorotrifluoroethylene) copolymer. *Appl. Phys. Lett.*, 88, 062904, 2006.
28. Park, Y.J., Kang, S.J., Lotz, B., Brinkmann, M., Thierry, A., Kim, K.J., Park, C., Ordered ferroelectric PVDF-TrFE thin films by high throughput epitaxy for nonvolatile polymer memory. *Macromol.*, 41, 8648–8654, 2008.
29. Fujisaki, S., Ishiwara, H., Fujisaki., Y., Low-voltage operation of ferroelectric poly(vinylidene fluoride-trifluoroethylene) copolymer capacitors and metal-ferroelectric-insulator-semiconductor diodes. *Appl. Phys. Lett.*, 90, 162902, 2007.
30. Noh, S.-H., Choi, W., Oh, M.-S., Hwang, D.K., Lee, K., Im, S., ZnO-based nonvolatile memory thin-film transistors with polymer dielectric/ferroelectric double gate insulators. *Appl. Phys. Lett.*, 90, 253504, 2007.
31. Yamauchi, N., A metal-insulator-semiconductor (MIS) device using a ferroelectric polymer thin film in the gate insulator. *Jpn. J. Appl. Phys.*, 25, 590, 1986.
32. Lim, S.H., Rastogi, A.C., Desu, S.B., Electrical properties of metal-ferroelectric-insulator-semiconductor structures based on ferroelectric polyvinylidene fluoride copolymer film gate for nonvolatile random access memory application. *J. Appl. Phys.*, 96, 5673, 2004.
33. Kim, K.L., Lee, W., Hwang, S.K., Joo, S.H., Cho, S.M., Song, G., Cho, S.H., Jeong, B., Hwang, I., Ahn, J.-H., Yu, Y.-J., Shin, T.J., Kwak, S.K., Kang, S.J.,

Park, C., Epitaxial growth of thin ferroelectric polymer films on graphene layer for fully transparent and flexible nonvolatile memory. *Nano Lett.*, 16, 334, 2016.

34. Gerber, A., Kohlstedt, H., Fitsilis, M., Waser, R.T., Reece, J., Ducharme, S., Rije, E., Low-voltage operation of metal-ferroelectric-insulator-semiconductor diodes incorporating a ferroelectric polyvinylidene fluoride copolymer Langmuir-Blodgett film. *J. Appl. Phys.*, 100, 024110, 2006.
35. Unni, K.N.N., Dabos-Seignon, S., Nunzi, J.M., Improved performance of pentacene field-effect transistors using a polyimide gate dielectric layer. *J. Phys. D.*, 38, 1148, 2005.
36. Lee, J.S., Prabu, A.A., Chang, Y.M., Kim, K.J., Ultrathin Ferroelectric, P., (VDF/TrFE) Copolymer Film in Low-Cost Non-Volatile Data Storage Applications. *Macromol. Symp.*, 13, 249–250, 2007.
37. Lee, Y.T., Kwon, H., Kim, J.S., Kim, H.H., Lee, Y.J., Lim, J.A., Song, Y.W., Yi, Y., Choi, W.K., Hwang, D.K., Im, S., Nonvolatile ferroelectric memory circuit using black phosphorus nanosheet-based field-effect transistors with P (VDF-TrFE) polymer. *ACS Nano*, 9, 10394, 2015.
38. Roy, D., Sinha, S., Wang, C.-H., Yang, Y.-W., Mukherjee., M., Interfacial Interaction of Absorbate Copper Phthalocyanine with PVDF Based Ferroelectric Polymer Substrates: A Spectroscopic Study. *Langmuir*, 36, 17, 4607–4618, 2020.
39. Lee, J.H., Lee, K.Y., Gupta, M.K., Kim, T.Y., Lee, D.Y., Oh, J., Ryu, C., Yoo, W.J., Kang, C.Y., Yoon, S.J., Yoo, J.B., Kim, S.W., Highly stretchable piezoelectric-pyroelectric hybrid nanogenerator. *Adv. Mater.*, 26, 765, 2014.
40. Roy, D., Mishra, T., Parashar, C.K., Murmu, K., Chakraborty, M., Effect of Poling on b-Phase Structure of Electrospun PVDF-TrFE Nanofiber Film. *J. Mater. Eng. Perform.*, 33, 55395545, 2024.
41. Bune, A.V., Fridkin, V.M., Ducharme, S., Blinov, L.M., Palto, S.P., Sorokin, A.V., Yudin, S., Zlatkin, A., Two-dimensional ferroelectric films. *Nature*, 391, 6670, 874, 1998.
42. Huitema, H.E.A., Gelinck, G.H., van der Putten, J.B.P.H., Kuijk, K.E., Hart, C.M., Cantatore, E., Herwig, T., van Breemen, A.J.J.M., de Leeuw, D.M., Cantatore, E., Herwig, T., van Breemen, A.J.J.M., de Leeuw, D.M., Plastic transistors in active-matrix displays. *Nature*, 414, 599, 2001.
43. Rogers, J.A., Bao, Z., Baldwin, K., Dodabalapur, A., Crone, B., Raju, V.R., Kuck, V., Katz, H., Amundson, K., Ewing, J., Drzaic, P., Paper-like electronic displays: Large-area rubber-stamped plastic sheets of electronics and microencapsulated electrophoretic inks, *Proc. Natl. Acad. Sci.*, U.S.A., 98, 9, 4835–4840, 2001.
44. Asadi, K., De Leeuw, D.M., De Boer, B., Blom, P.W., Organic non-volatile memories from ferroelectric phase-separated blends. *Nat. Mater.*, 7, 7, 547, 2008.
45. Scott, J.F., *Ferroelectric Memories*, pp. 179–184, Springer-Verlag Berlin Heidelberg, New York, 2000.

46. Yin, Z., Tian, B., Zhu, Q., Duan, C., Characterization and Application of PVDF and Its Copolymer Films Prepared by Spin-Coating and Langmuir–Blodgett Method. *Polymers*, 11, 2033, 2019.
47. Naber, R., Blom, P., Gelinck, G., Marsman, A., De Leeuw, D., An Organic Field-Effect Transistor with Programmable Polarity. *Adv. Mater.*, 17, 2692–2695, 2005.
48. Naber, R., De Boer, B., Blom, P., De Leeuw, D., Low-Voltage Polymer Field-Effect Transistors for Nonvolatile Memories. *Appl. Phys. Lett.*, 87, 203509, 2005.
49. Jung, S.-W., Baeg, K.-J., Yoon, S.-M., You, I.-K., Lee, J.-K., Kim, Y.-S., Noh, Y.-Y., Low-voltage-operated top-gate polymer thin-film transistors with high capacitance poly (vinylidene fluoride-trifluoroethylene)/poly (methyl methacrylate) dielectrics. *J. Appl. Phys.*, 108, 102810, 2010.
50. Kang, S., Bae, I., Park, Y., Park, T., Sung, J., Yoon, S., Kim, K., Choi, K., Park, C., Nonvolatile Ferroelectric Poly(vinylidene fluoride-cotrifluoroethylene) Memory Based on a Single-Crystalline Triisopropylsilylethynyl Pentacene Field-Effect Transistor. *Adv. Funct. Mater.*, 19, 1609–1616, 2009.
51. Kang, S., Park, Y., Bae, I., Kim, K., Kim, H., Bauer, S., Thomas, E., Park, C., Printable Ferroelectric PVDF/PMMA Blend Thin films with Ultralow Roughness for Low Voltage Nonvolatile Polymer Memory. *Adv. Funct. Mater.*, 19, 2812–2818, 2009.
52. Lee, K., Lee, G., Lee, K., Oh, M., Im, S., Yoon, S., High-Mobility Nonvolatile Memory Thin Film Transistors with a Ferroelectric Polymer Interfacing ZnO and Pentacene Channels. *Adv. Mater.*, 21, 4287–4291, 2009.
53. Guo, Y., Yu, G., Liu, Y., Functional Organic Field-Effect Transistors. *Adv. Mater.*, 22, 4427–4447, 2010.
54. Naber, R.C., Asadi, K., Blom, P.W., de Leeuw, D.M., de Boer, B., Organic nonvolatile memory devices based on ferroelectricity. *Adv. Mater.*, 22, 9, 933–945, 2010.
55. Caraveo-Frescas, J., Khan, M., Alshareef, H., Polymer Ferroelectric Field-Effect Memory Device with SnO Channel Layer Exhibits Record Hole Mobility. *Sci. Rep.*, 4, 5243, 2015.
56. Petermann, J. and Gohil, R.M., A new method for the preparation of high modulus thermoplastic films. *J. Mat. Sci.*, 14, 2260, 1979.
57. Yang, D.C. and Thomas, E.L., An electron microscopy and X-ray diffraction study of the microstructures of melt-drawn polyethylene films. *J. Mater. Sci.*, 3, 929, 1984.
58. Yang, D.C. and Thomas, E.L., Effect of PMMA on the morphology and $\alpha \rightarrow \beta$ phase transition of oriented PVF2/PMMA blends. *J. Mater. Sci. Lett.*, 6, 593, 1987.
59. Kaufmann, W., Petermann, J., Reynolds, N., Thomas, E.L., Hsu, S.L., Morphological and infra-red studies on highly oriented poly(vinylidene fluoride)/poly(methyl methacrylate) blends. *Polymer*, 30, 2147, 1989.

60. Matsushige, K., Nagata, K., Imada, S., Takemura, T., The II-I crystal transformation of poly(vinylidene fluoride) under tensile and compressional stresses. *Polymer*, 21, 1391, 1980.
61. Wang, T.T., Piezoelectricity in β-phase poly (vinylidene fluoride) having a "single-crystal" orientation. *J. Appl. Phys.*, 50, 6091, 1979.
62. Kang, S.J., Park, Y.J., Hwang, J., Jeong, H.J., Lee, J.S., Kim, K.J., Kim, H.-C., Huh, J., Park, C., Localized pressure-induced ferroelectric pattern arrays of semicrystalline poly(vinylidene fluoride) by microimprinting. *Adv. Mater.*, 19, 581, 2007.
63. Park, Y.J., Kang, S.J., Lotz, B., Brinkmann, M., Thierry, A., Kim, K.J., Park, C., Molecular and Crystalline Microstructure of Ferroelectric Poly(vinylidene fluoride-co-trifluoroethylene) Ultrathin Films on Bare and Self-Assembled Monolayer-Modified Au Substrates. *Macromol.*, 41, 8648–8654, 2008.
64. Xia, W. and Zhang, Z., PVDF-based dielectric polymers and their applications in electronic materials. *IET Nanodielectr.*, 1, 17–31, 2018.
65. Mulaosmanovic, H., Ocker, J., Muller, S., Noack, M., Muller, J., Polakowski, P., Mikolajick, T., Slesazeck, S., Novel ferroelectric FET based synapse for neuromorphic systems. *Symposium on VLSI Technology*, Kyoto, Japan, pp. T176–T177, 2017.
66. Mulaosmanovic, H., Chicca, E., Bertele, M., Mikolajick, T., Slesazeck, S., Mimicking biological neurons with a nanoscale ferroelectric transistor. *Nanoscale*, 10, 21755, 2018.
67. Singh, P., Jha, R.K., Singh, R.K., Singh, B.R., Electrical properties of $Pb[Zr_{0.35}Ti_{0.65}]O_3$ on PEALD Al_2O_3 for NVM applications. *Microelectron. Int.*, 35, 189, 2018.
68. Gong, N. and Ma, T.-P., Why is FE–HfO_2 more suitable than PZT or SBT for scaled nonvolatile 1-T memory cell? A retention perspective. *IEEE Electron Device Lett.*, 37, 1123, 2016.
69. Banerjee, W., Challenges and Applications of Emerging Nonvolatile Memory Devices. *Electronics*, 9, 1029, 2020.
70. Chen, A., A review of emerging non-volatile memory (NVM) technologies and applications. *Solid State Electron. Lett.*, 125, 25–38, 2016.
71. Yamauchi, N., Kato, K., Wada, T., Observation of Ferroelectricity in Very Thin Vinylidene Fluoride Trifluoroethylene Copolymer[P(VDF·TrFE)] Films by High Frequency C-V Measurements of Al-SiO_2-P(VDF·TrFE)-SiO_2-Si Capacitors. *Jpn. J. Appl. Phys.*, 23, L671, 1984.
72. Kang, S.J., Bae, I., Shin, Y.J., Park, Y.J., Huh, J., Park, S.-M., Kim, H.-C., Park, C., Nonvolatile Polymer Memory with Nanoconfinement of Ferroelectric Crystals. *Nano Lett.*, 11, 138–144, 2011.
73. Roy, D., Pattader, P.S.G., Bandyopadhyay, D., Chakraborty, M., Wang, C.-H., Yang, Y.-W., Mukherjee, M., Dipolar Alignment in a Ferroelectric Dielectric Layer of FeFETs to Boost Charge Mobility and Nonvolatile Memory. *ACS Appl. Electron. Mater.*, 2, 10, 3187–3198, 2020.

74. Sun, H., Wang, Q., Li, Y., Lin, Y.F., Wang, Y., Yin, Y., Xu, Y., Liu, C., Tsukagoshi, K., Pan, L., Wang, X., Boost up carrier mobility for ferroelectric organic transistor memory via buffering interfacial polarization fluctuation. *Sci. Rep.*, 4, 7227, 2014.
75. Liu, C., Xu, Y., Noh, Y.Y., Contact engineering in organic field effect transistor. *Mater. Today*, 18, 2, 79, 2015.
76. Herlogsson, L., Crispin, X., Tierney, S., Berggren, M., Polyelectrolyte-gated organic complementary circuits operating at low power and voltage. *Adv. Mater.*, 23, 40, 4684, 2011.
77. Sun, H., Yin, Y., Wang, Q., Jun, Q., Wang, Y., Tsukagoshi, K., Wang, X., Hu, Z., Pan, L., Zheng, Y., Shi, Y., Reducing contact resistance in ferroelectric organic transistors by buffering the semiconductor/dielectric interface. *Appl. Phys. Lett.*, 117, 23, 1233, 2015.
78. Tobjörk, D. and Österbacka, R., Paper Electronics. *Adv. Mater.*, 23, 1935, 2011.
79. Shin, C., Lee, G.G., Han, D.H., Han, S.P., Tokumitsu, E., Ohmi, S., II, Kim, D.J., Ishiwara, H., Park, M., Kim, S.H., Lee, W.G., Hwang, Y.J., Park, B.E., Experimental demonstration of a ferroelectric FET using paper substrate. *IEICE Electron. Express*, 11, 20140447, 2014.
80. Khan, M.A., Bhansali, U.S., Alshareef, H.N., High-performance non-volatile organic ferroelectric memory on banknotes. *Adv. Mater.*, 24, 2165, 2012.
81. Kim, R.H., Kim, H.J., Bae, I., Hwang, S.K., Velusamy, D.B., Cho, S.M., Takaishi, K., Muto, T., Hashizume, D., Uchiyama, M., Andre, P., Mathevet, F., Heinrich, B., Aoyama, T., Kim, D.-E., Lee, H., Ribierre, J.-C., Park, C., Non-volatile organic memory with sub-millimetre bending radius. *Nat. Commun.*, 5, 3583, 2014.
82. Yoon, S.M., Yang, S., Byun, C.W., Jung, S.W., Ryu, M.K., Park, S.H.K., Kim, B., Oh, H., Hwang, C.S., Yu, B.G., Nonvolatile memory thin-film transistors using an organic ferroelectric gate insulator and an oxide semiconducting channel. *Semicond. Sci. Technol.*, 26, 3, 034007, 2011.
83. Yoon, S.-M., Yang, S., Byun, C., Park, S.-H.K., Cho, D.-H., Jung, S.-W., Kwon, O.-S., Hwang, C.-S., Fully Transparent Non-volatile Memory Thin-Film Transistors Using an Organic Ferroelectric and Oxide Semiconductor Below 200 °C. *Adv. Funct. Mater.*, 20, 6, 2010.
84. Petti, L., Münzenrieder, N., Salvatore, G.A., Zysset, C., Kinkeldei, T., Büthe, L., Tröster, G., Influence of mechanical bending on flexible InGaZnO-based ferroelectric memory TFTs. *IEEE Trans. Elect. Devices*, 61, 4, 2014.
85. Cao, F., Tian, W., Meng, L., Wang, M., Li, L., Ultrahigh-Performance Flexible and Self-Powered Photodetectors with Ferroelectric P(VDF-TrFE)/Perovskite Bulk Heterojunction. *Adv. Funct. Mater.*, 29, 1808415, 2019.
86. Park, C., Lee, K., Koo, M., Park, C., Soft Ferroelectrics Enabling High-Performance Intelligent Photo Electronics. *Adv. Mater.*, 33, 47, 2004999, 2020.

87. Huang, H., Wang, X.D., Wang, P., Ferroelectric polymer tuned two dimensional layered $MoTe_2$ photodetector. *RSC Adv.*, 6, 87416, 2016.
88. Chen, Y., Wang, X.D., Wang, P., Huang, H., Wu, G.J., Tian, B.B., Hong, Z.C., Wang, Y.T., Sun, S., Shen, H., Wang, J.L., Hu, W.D., Sun, J.L., Meng, X.J., Chu, J.H., Optoelectronic Properties of Few-Layer MoS_2 FET Gated by Ferroelectric Relaxor Polymer. *ACS Appl. Mater. Interfaces*, 8, 32083, 2016.
89. Wang, J. and Hu, W., Recent progress on integrating two-dimensional materials with ferroelectrics for memory devices and photodetectors. *Chin. Phys. B*, 26, 037106, 2017.
90. Wu, G.J., Wang, X.D., Wang, P., Huang, H., Chen, Y., Sun, S., Shen, H., Lin, T., Wang, J.L., Tian, B.B., Sun, J.L., Meng, X.J., Chu, J.H., Visible to short wavelength infrared In_2Se_3-nanoflake photodetector gated by a ferroelectric polymer. *Nanotechnology*, 27, 364002, 2016.
91. Zheng, D., Wang, J., Hu, W., Liao, L., Fang, H., Guo, N., Wang, P., Gong, F., Wang, X., Fan, Z., Wu, X., Meng, X., Chen, X., Lu, W., When Nanowires Meet Ultrahigh Ferroelectric Field–High Performance Full-Depleted Nanowire Photodetectors. *Nano Lett.*, 16, 2548, 2016.
92. Wang, X.D., Wang, P., Wang, J.L., Hu, W.D., Zhou, X.H., Guo, N., Huang, H., Sun, S., Shen, H., Lin, T., Ultrasensitive and Broadband MoS_2 Photodetector Driven by Ferroelectrics. *Adv. Mater.*, 27, 6575–6581, 2015.
93. Zhao, Q., Wang, H., Jiang, L., Zhen, Y., Dong, H., Hu, W., Solution-Processed Flexible Organic Ferroelectric Phototransistor. *ACS Appl. Mater. Interfaces*, 9, 43880–43885, 2017.
94. Thakuria, N., Saha, A.K., Thirumala, S.K., Schulman, D., Das, S., Gupta, S.K., Polarization-induced strain-coupled TMD FETs (PS FETs) for nonvolatile memory applications. *Device Research Conference - Conference of Digest DRC*, IEEE, 2020.
95. Rachmuth, G., Shouval, H.Z., Bear, M.F., Poon, C.S., A biophysically-based neuromorphic model of spike rate- and timing-dependent plasticity. *Proc. Natl. Acad. Sci.*, 108, E1266, 2011.
96. Lee, Y.R., Trung, T.Q., Hwang, B.-U., Lee, N.-E., A flexible artificial intrinsic-synaptic tactile sensory organ. *Nat. Comm.*, 11, 2753, 2020.
97. Jang, S., Jang, S., Lee, E.-H., Kang, M., Wang, G., Kim, T.-W., Ultrathin Conformable Organic Artificial Synapse for Wearable Intelligent Device Applications. *ACS Appl. Mater. Interfaces*, 11, 1071–1080, 2019.
98. Yang, K., Yuan, S., Huan, Y., Wang, J., Tu, L., Xu, J., Zou, Z., Zhan, Y., Zheng, L., Seoane, F., Tunable flexible artificial synapses: a new path toward a wearable electronic system. *npj Flex. Electron.*, 2, 20, 2018.

99. Kang, M., Lee, S.A., Jang, S., Hwang, S., Lee, S.K., Bae, S., Hong, J.M., Lee, S.H., Jeong, K.U., Lim, J.A., Kim, T.W., Low-Voltage Organic Transistor Memory Fiber with a Nanograined Organic Ferroelectric Film. *ACS Appl. Mater. Interfaces*, 11, 22575–22582, 2019.
100. Lee, K., Jang, S., Kim, K.L., Koo, M., Park, C., Lee, S., Lee, J., Wang, G., Park, C., Artificially Intelligent Tactile Ferroelectric Skin. *Adv. Sci.*, 7, 2001662, 2020.

3

Ferroelectric Applications in Novel Devices

Keshav Kumar[1]* and Umesh Chandra Bind[2]

[1]International Centre for Materials Science, and School of Advanced Materials, Jawaharlal Nehru Centre for Advanced Scientific Research, Jakkur, Bengaluru, India
[2]Department of Applied Science and Humanities, Institute of Engineering and Technology, Dr. A. P. J. A. Kalam Technical University, Lucknow, Uttar Pradesh, India

Abstract

Advancements in next-generation technologies demand multifunctional devices with ultrafast speed, low cost, and small sizes. Therefore, immense efforts are being made to understand the complex and cooperative behavior of the current, charge, electric dipole moments, and/or spin degrees of freedom of materials. For example, ferroelectrics (a non-centrosymmetric material with switchable spontaneous polarization) near or below the size of 3.2 nm are appealing for exploring their potential application in next-generation devices. Ferroelectric materials are generally used as tunable capacitors and data storage devices in various electronic systems, such as mobiles, computers, space centers, and satellites. They can also be utilized as crucial electronic components, such as phase shifters, electrocaloric effect-based coolers, memory, and energy harvesters in satellites and/or space flight-based advance devices. Extensive studies are required to improve the functionality of these materials further. Here, we briefly present the basic concept of ferroelectricity, its necessity, facile and advanced synthesis techniques, and potential applications of ferroelectric materials in novel devices.

Keywords: Ferroelectricity, piezoelectricity, next-generation devices, energy harvesters

**Corresponding author*: keshavkumar@jncasr.ac.in

Balwinder Raj, Shiromani Balmukund Rahi and Nandakishor Yadav (eds.) FeFET Devices, Trends, Technology and Applications, (57–76) © 2025 Scrivener Publishing LLC

3.1 Introduction

The phenomenon of ferroelectricity was first observed by Valasek in 1921 while investigating the Rochelle salt ($KNaC_4H_4O_6$–$4H_2O$) [1]. To date, more than 1,000 ferroelectric materials have been discovered, and further quests for new members are ongoing. The P–E hysteresis observed in the Rochelle salt is shown schematically in Figure 3.1(a). The value "Ps" and "Ec" represents the spontaneous polarization (Ps) and coercive electric field which is simply the electric field required to reverse the direction of polarization. This hysteresis resembles the behavior of the M–H loop of a ferromagnetic material. In addition to hysteresis, other characteristic properties of ferroelectrics, such as switchable spontaneous polarization (Ps) and Curie temperature (T_c), are similar to those of ferromagnetic materials. Due to this similarity, the term ferro in ferroelectrics is due to ferromagnetism. After the discovery of ferroelectricity in $BaTiO_3$ during the Second World War, the scientific community has realized the technological potential of this phenomenon. Currently, ferroelectric-based electronic components (e.g., memory and capacitor) can be found in almost every electronic device.

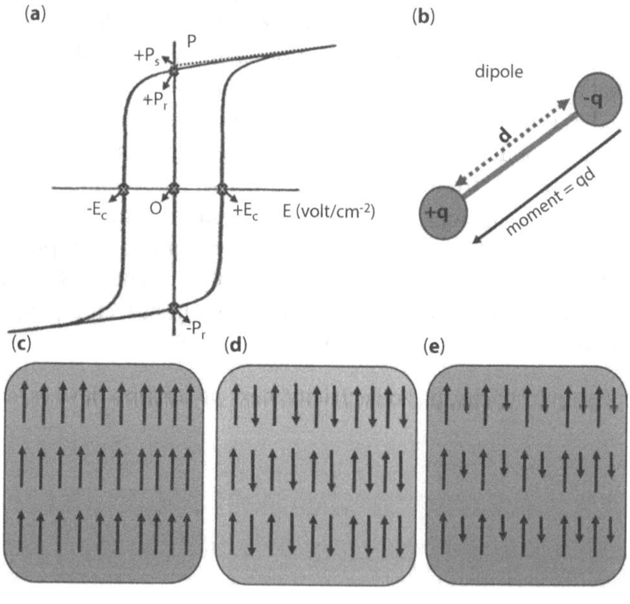

Figure 3.1 Schematic of the (a) hysteresis loop, (b) electric dipole. The dipole arrangements in long-range ordered (c) ferroelectric, (d) antiferroelectric, and (e) ferrielectric materials.

Recently, the scientific community celebrated the 100th anniversary of the discovery of ferroelectricity, and a summary of the various developments and impacts of this discovery on science and technology can be found in refs. [2, 3]. In this chapter, we briefly summarize the basic concepts and origin/mechanism of ferroelectricity. This is correlated with the crystal symmetry. We also briefly discuss a few potential applications of ferroelectrics and their facile and advanced synthesis techniques that are being used to fabricate ferroelectric materials.

3.2 General Concepts of Ferroelectrics

A material is said to be ferroelectric if it exhibits spontaneous polarization, the direction of which can be reversed by applying an external electric field. The fundamental quantity responsible for the ferroelectricity is the electric dipoles, which are simply the two equal and opposite charges of magnitude 'q' separated by a distance 'd' (see Figure 3.1(b)). Mathematically, an electric dipole moment which is a vector quantity can be defined as the product of 'q' and 'd,' i.e., qd. The amount of dipole available in a unit volume is referred to as polarization (P). The dipole moments in a material can organize themselves in a variety of ways to exhibit a long-range ordered state below a critical temperature known as the Curie temperature (T_c). Depending on the type of dipole ordering, materials can be broadly categorized as ferroelectric (dipoles of (un)equal magnitude aligned parallel to each other), antiferroelectric (dipoles of equal magnitude aligned antiparallel to each other), and ferrielectric (dipoles of unequal magnitude align antiparallel to each other) (see Figures 3.1(c–e)). Rochelle salts ($KNaC_4H_4O_6$–$4H_2O$) and $BaTiO_3$ are classic examples of ferroelectric materials [1, 4]. $PbZrO_3$ and $Sr_{1-x}Ca_xTiO_3$ are known for their antiferroelectric and ferrielectric properties, respectively [5, 6]. The ordering of the dipoles in these phases is shown schematically in Figures 3.1(c–e). Thermal fluctuations, quantum fluctuations, and various types of interactions between dipoles at low temperatures play a key role in attaining the long-range ordered state, and sometimes a certain situation leads to dipole glass, incipient ferroelectrics, and complex and exotic phases at low temperatures. $SrTiO_3$ and $BaFe_{12}O_{19}$ are examples of such materials [7, 8].

These phases have fundamental technological importance. The high-temperature phase for all these orderings is termed a paraelectric phase, where electric dipole moments are no longer present in the system, unlike paramagnets, where spin moments survive and orient themselves randomly.

However, a few ferroelectrics are exceptional; for example, KH_2PO_4 retains electric dipoles in its paraelectric phase. In the literature, much attention has been focused on exploring the mechanisms responsible for the electric dipole moments. For example, refer to (i) a dipole can be created within an atom (i.e., the centers of positive and negative charges can be separated) via a strong electric field, (ii) a molecule can have a dipole moment due to different electronegativities of its constituent elements, (iii) charge ordering in an ionic crystal, (iv) rotation and tilting of the polyhedra, and (v) local distortions can lead to dipole moment [9]. In some materials, such as $TbMnO_3$, polarization emerges as a result of special spiral magnetic spin ordering [10]. Such materials are referred to as improper ferroelectrics that provide an additional degree of freedom (i.e., magnetization), which can be used to manipulate ferroelectric properties; therefore, they could be quite useful for advanced spintronic devices.

The crystal symmetry also has some implications for the emergence of ferroelectricity. The symmetry of all the crystalline materials can be explained using a space group of 230. If one excludes the translational symmetry elements, then the number becomes a 32-point group. Among them, only the 10-point groups have a polar axis that can exhibit ferroelectric properties. A ferroelectric material can be neither centrosymmetric nor glass [11]. Interestingly, all ferroelectric materials exhibit piezoelectric and pyroelectric properties; however, the converse is not true. For example, ZnO is pyroelectric and does not exhibit ferroelectric properties [11].

3.3 Ferroelectric Materials Processing for Device Applications

The performance of ferroelectric materials is closely related to their processing method. Therefore, substantial research has been devoted to developing and improving the quality of polycrystalline, single-crystalline, composite, and thin-film ferroelectrics [12–18]. In this section, a few important and commonly used synthetic methods are discussed.

3.3.1 Bulk Synthesis

A few of the facile and commercialized methods for the processing of polycrystalline ferroelectrics are the solid-state reaction, coprecipitation, and sol–gel methods.

3.3.1.1 Solid State Reaction Method

This method requires constituents in the form of oxides or carbonates as starting materials. For example, $BaCO_3$ and TiO_2 can be taken as an initial precursor to synthesize $BaTiO_3$ [19–26]. The various steps involved in the reaction are shown in the schematic below.

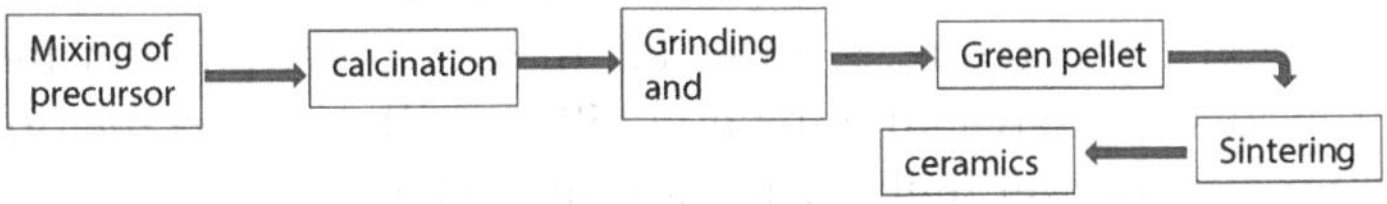

Because this method requires high-temperature calcination and sintering steps, there is no control over the distribution of particle size, agglomeration, morphology, and microstructure of the sample. The synthesis of Pb- and Bi-based ferroelectrics, which are frequently used in ferroelectric devices, using this method led to the loss of Bi and Pb components due to high-temperature sintering, which eventually affects the stoichiometry [27–35]. In such a scenario, the useful properties of ferroelectrics can deteriorate. This method is very facile and cheap and is therefore being used in academia for research purposes and in industries, as well as for bulk production. The solid-state reaction method can produce polycrystalline samples with particle sizes of approximately 1–100 μm [36–43].

3.3.1.2 Coprecipitation Method

This is a chemical method. It utilizes insoluble compounds, such as hydroxide, chloride, or oxalate, to precipitate from a solution of metal salts. The precipitate thus obtained is calcined to get the desired polycrystalline sample [44–51]. For example, Ba and Ti chlorides in oxalic acid have been used to obtain high-purity $BaTiO_3$ samples. The advantages of this method over the solid-state reaction synthesis route include lower calcination temperatures, higher purity, and smaller particle size (<0.5 μm). Ferroelectrics with smaller-particle-size powders could be useful for the miniaturization of ferroelectric devices. However, this method has certain shortcomings, such as: (i) it involves repeated washing of the precipitate to remove contamination that can originate from the precursors themselves, and (ii) it can produce a small quantity of samples in one cycle. Therefore, a huge effort and time are required to produce bulk samples at a large scale [52–58].

3.3.1.3 Sol–Gel Method

This is a wet chemical synthesis method. It utilizes metal alkoxides that allow it to react with water in a nonaqueous solution to produce metal hydrous oxides or metal anhydrous oxides. The (an)hydrous oxides obtained were calcined to obtain the required polycrystalline ferroelectrics. Because metal mixing takes place at the atomic level in the solution the calcination temperature is lower (as in the coprecipitation method) than in the solid-state reaction method [59–66]. The sol–gel method provides better control over high purity, stoichiometry, morphology, and particle size (5–50 nm). This method can also be used to coat films. Because the precursor materials used in these methods are very expensive and very sensitive to environmental moisture, light, and heat, one must be careful during synthesis [67–74].

3.3.2 Thin Films

The current demand for miniaturized and integrated circuit-based ferroelectric devices is increasing daily to meet the requirements of next-generation technologies. This has led to the development of a wide variety of thin-film growth techniques to fabricate high-quality heterostructured films. The growth techniques developed so far can be broadly divided into two categories: (i) chemical vapor deposition (CVD) and (ii) physical vapor deposition (PVD). Among them, advanced and commonly used techniques such as chemical vapor deposition, sputtering, pulse laser deposition (PLD), and molecular beam epitaxy (MBE) are briefly discussed here [75–79].

3.3.2.1 Chemical Vapor Deposition (CVD)

This is a powerful technique for coating and producing high-quality thin films. In this method, a thin film is produced by the chemical reactions of gas-phase precursors on a heated substrate in a closed vacuum chamber [80–86]. For example, the chemical reaction of titanium isopropoxide $Ti(O^iPr)_4$ and bis[tri(i-propyl)cyclopentadienyl] barium $Ba(^iPr_3Cp)_2$ under optimum flow conditions led to the formation of pure polycrystalline $BaTiO_3$ thin films on MgO and/or $SrTiO_3$ substrates at 400 °C [12] A detailed discussion of the film-growth mechanism and instrumentation can be found in refs. [12, 13].

3.3.2.2 Sputtering Method

This is a widely used physical vapor deposition technique. It uses the plasma of a novel gas to sputter the constituents of the target material to be deposited on the substrate. Because different elements of a complex ferroelectric sputter at different rates, control over the rate of deposition is key to obtaining a film in a stoichiometric ratio [87–92]. In some ferroelectrics, such as $Pb(Zr,Ti)O_3$ with volatile components, the energy transferred to the target by the incident ions also plays a key role. This method has been successfully used to grow high-quality thin films of various complex ferroelectrics such as $Pb(Zr,Ti)O_3$ [14, 15], $(Ba,Sr)TiO_3$ [16], and $BiFeO_3$ [17].

3.3.2.3 Pulsed Laser Deposition (PLD) Technique

Another most widely used physical vapor deposition technique utilizes a pulsed laser instead of plasma to sputter the target constituents. The deposition of $Pb(Zr,Ti)O_3$, $(Ba,Sr)TiO_3$, and $BaTiO_3/SrTiO_3$ superlattices, etc., ferroelectric films on various substrates (such as MgO, $SrTiO_3$, Pt-coated Si, and GaAs) has been demonstrated using the PLD method [18–20]. One of the main concerns of this technique is that it cannot be used to fabricate thin films on a large scale [93–98].

3.3.2.4 Molecular Beam Epitaxy (MBE)

MBE is a sophisticated physical vapor deposition-based film-growth technique. It utilizes an ultra-high vacuum to grow films with atomic-layer precision through the flux control of thermally evaporated beams of target constituents [99–103]. This technique provides *in situ* control of growth at the atomic level. MBE allows one to (i) deposit ultrathin films, (ii) grow films of controlled thickness, (iii) grow films with various interfaces, and (iv) control the composition, morphology, and surface roughness. The growth of ultrathin films of $Pb(Zr,Ti)O_3$ and $BaTiO_3$-based ferroelectrics has been successfully demonstrated using MBE techniques [21, 22].

3.4 Advanced Application of Ferroelectric Materials

Since the discovery of ferroelectricity, numerous technological manifestations (such as high-permittivity supercapacitors, memory and display devices, electro-optic modulators, night imaging, microelectronics, and microelectromechanical systems) of ferroelectric materials have been demonstrated. Interested readers can find more application-related details in refs. [23, 24]. In this section, some of the potential applications of ferroelectrics in future technologies are presented [104–107].

3.4.1 Memory Devices

As discussed in *Section 3.3*, ferroelectric materials exhibit a reversible spontaneous polarization. The upward polarization can be designated as '1' state while the downward can be '0' state or visa-versa. These two states of the polarization i.e., '0' and '1' can be utilized to save information. Depending upon the readout protocols, ferroelectric-based non-volatile memory devices can be categorized as (i) ferroelectric random-access memory (FeRAM), (ii) ferroelectric field-effect transistors (FeFETs), and (iii) ferroelectric tunnel junctions (FTJs) [25]. In the literature, various models (such as 1T1C and 2T2C) have been proposed to improve the functionality of these devices [26, 27]. Ferroelectric-based memory devices are attractive as they offer read/write operations at low voltages compared to dynamic random-access memory (DRAM), which is also volatile. From a materials point of view, $Pb(Zr,Ti)O_3$ thin films exhibit superior properties (such as (i) up to 10^{14} read/write cycles, (ii) same state and opposite state retention for up to 10^3 h, and (iii) stable operation in the range of −40 to 125 °C) for data storage [28, 29]. However, other ferroelectrics, such as $(Ba,Sr)TiO_3$, $Bi_4Ti_3O_{12}$, $Pb_5Ge_3O_{11}$, and $(Pb,La)(Zr,Ti)O_3$, exhibit inferior properties ($<10^{12}$ read/write cycles, low retention times, and limited temperature stabilities) than $Pb(Zr,Ti)O_3$. $Pb(Zr,Ti)O_3$-based devices contain Pb, which is volatile, toxic, and environmentally unfriendly. Therefore, there is an ongoing search for alternative materials with similar characteristics. HfO_2 has demonstrated its potential as a more appropriate choice, but much effort is still required for its commercialization [27]. FeRAM, FeFET, and FJT devices utilize the ferroelectric properties of the materials to store high-density data. Using *ab initio* modeling, it has been shown that thin-film materials retain their ferroelectric properties down to a diameter of 3.2 nm in nanodiscs and rods owing to the combined effects of depolarization and size [30]. The size effect, along with the interplay between the anisotropic energy and depolarization energies, gives rise to various exotic phases, such as closure domains, vortices, and polar skyrmions. These exotic phases have been observed in ferroelectric materials like $Pb(Zr,Ti)O_3$, $PbTiO_3$, and $(PbTiO_3)_m/(SrTiO_3)_n$ [31–33]. The different dipole arrangements for a few exotic phases are shown schematically in Figure 3.2. The vortices and skyrmion phases formed by the electric dipoles are chiral

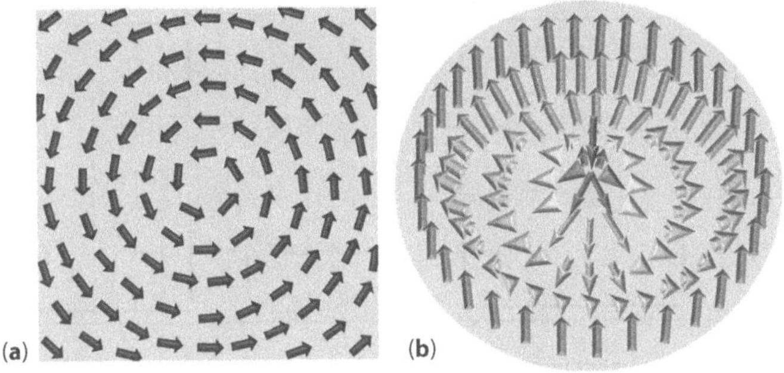

Figure 3.2 (a) Vortex and (b) skyrmion.

and can be tuned using an electric field. Similarly, the chirality of skyrmions formed by magnetic spins can be changed by using magnetic fields [108–113]. Interestingly, magnetoelectric materials offer an additional degree of chirality modulation the, i.e., the required state of skyrmions can be achieved using magnetic and/or electric fields. This opens the possibility for a new type of memory device in which data can be stored using individual chirality and/or vortices [34].

3.4.2 Energy Harvester

Another key application of ferroelectric materials is in energy harvesting. Almost all portable electronic devices and wireless sensors are equipped with chemical batteries that require replacement and/or electrical energy for charging after a certain period. Because all ferroelectrics exhibit piezoelectricity, they can also be utilized to generate electricity in the nano- to milliwatts range [114–121]. The phenomenon of charge development on the material's surface in response to tensile or compressive stress is known as the direct piezoelectric effect. Furthermore, ferroelectric materials develop strain when subjected to an electric field; this behavior is known as the inverse piezoelectric effect. Both the direct and inverse behaviors are schematically shown in Figures 3.3(a,b). The best piezoelectric materials are PVDF, $Pb(Zr,Ti)O_3$, $PbTiO_3$, $BaTiO_3$, $LiNbO_3$, and $LiTaO_3$. They can be synthesized using the techniques discussed in *Section 3.3*. A schematic of a device that can be used to generate electricity using piezoelectric materials is shown in Figure 3.4. To generate electricity, piezoelectric materials need to be placed at the natural source of vibration (for example, vibration generated by engines of heavy vehicles or airplanes) and the generated electricity can be stored in a capacitor [122–129]. The electricity stored can be

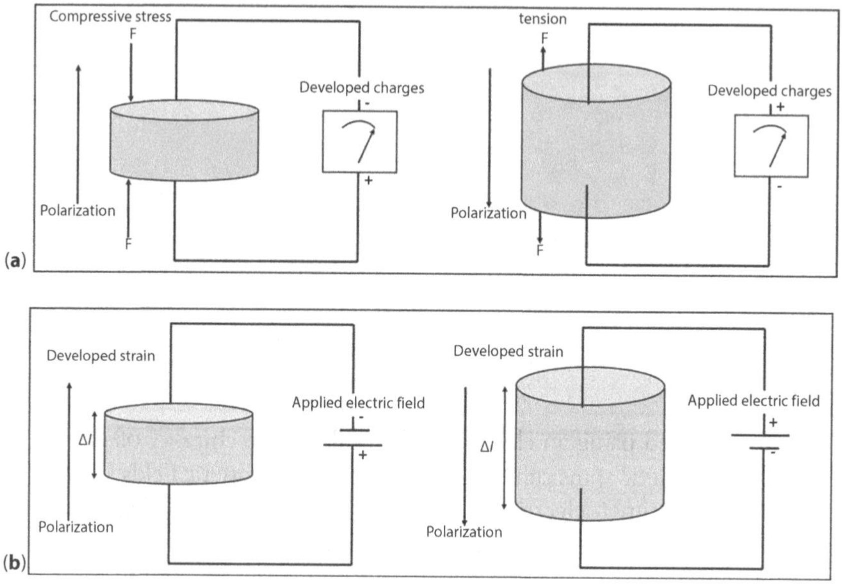

Figure 3.3 A schematic diagram of (a) direct and (b) indirect piezoelectric effects.

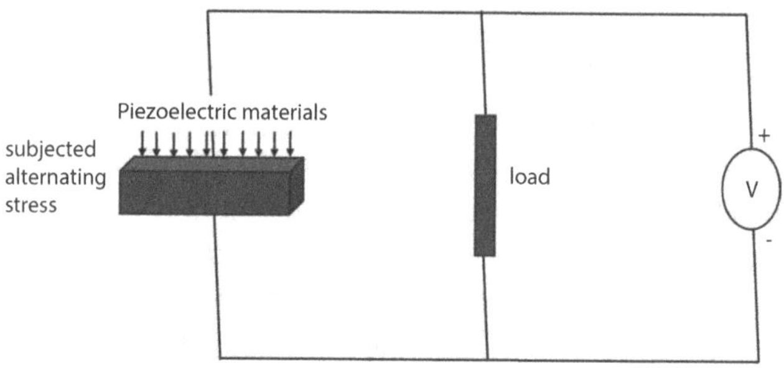

Figure 3.4 Schematic of piezoelectric device for energy harvesting.

utilized to power small electronic devices, such as sensors and in-ear hearing aids. The efficiency of this device can be increased by improving the piezoelectric coefficient, coupling coefficient, and mechanical quality factor, and by decreasing the dielectric loss of the piezoelectric materials [35]. However, some studies suggest that energy-harvesting efficiency can also be improved by rectifying the piezoelectric effect [36]. Currently, this concept of generating electricity to power small electric components is being used in various areas, such as: (i) industrial process control and monitoring, (ii) aerospace and defense, (iii) automotive and transportation, (iv). health and wellness, and (v) consumer electronics.

3.4.3 Space Flight/Satellite Electronics and Other Applications

Electronic components such as normal memory devices that outperform the Earth's surface for space flight/satellites can be damaged in space owing to the presence of high-energy radiation. This has led to the search for high-tolerance and radiation-hardened materials. Ferroelectric materials have shown their potential to survive in space. Ferroelectric materials have inherent radiation-hardened properties and can be utilized in memory devices and phase shifters for satellite communication [37–39]. Furthermore, using the electrocaloric concept, ferroelectric materials can be used to cool mainframes and microelectric motors [11].

3.5 Summary and Outlook

We briefly discuss the fundamental concepts of ferroelectricity, facile synthesis techniques, and a few advanced applications of ferroelectric devices (such as memory devices based on skyrmions/vortices and energy harvesters) for the next-generation technologies. Ultrathin film ferroelectrics with sizes less than 3.2 nm exhibit exotic phases (such as closer flux, vortex, skyrmions, and other exotic phases). In addition to the film size, the combined effects of depolarization and heterojunction interface also contribute to stabilizing these phases. These exotic phases can be extremely useful for shaping next-generation devices. However, the realization of the potential applications of these emerging phases in ferroelectrics for future technologies is challenging and a subject of intensive research.

References

1. Valasek, J., Piezo-electric and allied phenomena in Rochelle salt. *Phys. Rev.*, **17**, 4, 475–481, 1921.
2. Whatmore, R.W., You, Y.M., Xiong, R.G., Eom, C.B., 100 years of ferroelectricity—A celebration. *APL Mater.*, **9**, 7, 070401, 2021.
3. Bhalla, A.S. and Saxena, A., Ferroelectricity: 100 years on. *Phys. World*, **33**, 11, 38–41, 2020.
4. Hippel, A., Breckenridge, R.G., Chesley, F.G., Tisza, L., High dielectric constant ceramics. *Ind. Eng. Chem.*, **38**, 11, 1097–1109, 1946.

5. Sawaguchi, E., Shirane, G., Takagi, Y., Phase transition in Lead Zirconate. *J. Phys. Soc. Jpn.*, **6**, 5, 333–339, 1951.
6. Mishra, S.K. and Pandey, D., Low Temperature X-ray Diffraction Study of the Phase Transitions in $Sr_{(1-x)}Ca_xTiO_3$ (x=0.02,0.04): Evidence for Ferrielectric Ordering. *Appl. Phys. Lett.*, **95**, 232910, 2009.
7. Muller, K.A. and Burkard, H., $SrTiO_3$: An Intrinsic Quantum Paraelectric Below 4 K. *Phys. Rev. B*, **19**, 3593–3602, 1979.
8. Kumar, K. and Pandey, D., Quantum phase transitions in $Ba_{(1-x)}Ca_xFe_{12}O_{19}$ $(0 \leq x \leq 0.10)$. *Phys. Rev. B*, **96**, 2, 024102, 2017.
9. Khomskii, D., Classifying Multiferroics: Mechanisms and Effects. *Physics*, **2**, 20, 2009.
10. Kimura, T., Spiral Magnets as Magnetoelectrics. *Annu. Rev. Mater. Res.*, **37**, 387–413, 2007.
11. Scott, J.F., Applications of modern ferroelectrics. *Science*, **315**, 5814, 954–959, 2007.
12. Reinke, M., Kuzminykh, Y., Eltes, F., Abel, S., LaGrange, T., Neels, A., Fompeyrine, J., Hoffmann, P., Low-Temperature Epitaxial Barium Titanate Thin Film Growth in High Vacuum CVD. *Adv. Mater. Interfaces*, **4**, 18, 1700116, 2017.
13. Sun, L., Yuan, G., Gao, L., Yang, J., Chhowalla, M., Gharahcheshmeh, M.H., Gleason, K.K., Choi, Y.S., Hong, B.H., Liu, Z., Chemical vapour deposition. *Nat. Rev. Methods Primers*, **1**, 1, 1–20, 2021.
14. Krupanidhi, S.B., Maffei, N., Sayer, M., El-Assal, K., rf planar magnetron sputtering and characterization of ferroelectric $Pb(Zr,Ti)O_3$ films. *J. Appl. Phys.*, **54**, 11, 6601–6609, 1983.
15. Takayama, R. and Tomita, Y., Preparation of epitaxial $Pb(Zr_xTi_{1-x})O_3$ thin films and their crystallographic, pyroelectric, and ferroelectric properties. *J. Appl. Phys.*, **65**, 4, 1666–1670, 1989.
16. Im, J., Auciello, O., Baumann, P.K., Streiffer, S.K., Kaufman, D.Y., Krauss, A.R., Composition-control of magnetron-sputter-deposited $(Ba_xSr_{1-x})Ti_{1+y}O_{3+z}$ thin films for voltage tunable devices. *Appl. Phys. Lett.*, **76**, 625–627, 2000.
17. Das, R.R., Kim, D.M., Baek, S.H., Eom, C.B., Zavaliche, F., Yang, S.Y., Ramesh, R., Chen, Y.B., Pan, X.Q., Ke, X., Rzchowski, M.S., Streiffer, S.K., Synthesis and ferroelectric properties of epitaxial BiFeO3 thin films grown by sputtering. *Appl. Phys. Lett.*, **88**, 24, 242904, 2006.
18. Horwitz, J.S., Grabowski, K.S., Chrisey, D.B., Leuchtner, R.E., *In situ* deposition of epitaxial PbZrxTi(1-x)O3 thin films by pulsed laser deposition. *Appl. Phys. Lett.*, **59**, 13, 1565–1567, 1991.
19. Minami, H., Itaka, K., Ahmet, P., Komiyama, D., Chikyow, T., Lippmaa, M., Koinuma, H., Rapid Synthesis and Scanning Probe Analysis of $Ba_xSr_{1-x}TiO_3$ Composition Spread Films on a Temperature Gradient Si(100) Substrate. *Jpn. J. Appl. Phys.*, **41**, L149–L151, 2002.

20. Tabata, H., Tanaka, H., Kawai, T., Formation of artificial $BaTiO_3/SrTiO_3$ superlattices using pulsed laser deposition and their dielectric properties. *Appl. Phys. Lett.*, **65**, 15, 1970–1972, 1994.
21. Choi, K.J., Biegalski, M., Li, Y.L., Sharan, A., Schubert, J., Uecker, R., Reiche, P., Chen, Y.B., Pan, X.Q., Gopalan, V., Che, L.Q., Schlom, D.C., Eom, C.B., Enhancement of ferroelectricity in strained BaTiO3 thin films. *Science*, **306**, 5698, 1005–1009, 2004.
22. Anderson, P.S., Guerin, S., Hayden, B.E., Khan, M.A., Bell, A.J., Han, Y., Pasha, A., Whittle, K.R., Reaney, I.M., Synthesis of the ferroelectric solid solution, $Pb(Zr_{1-x}Ti_x)O_3$ on a single substrate using a modified molecular beam epitaxy technique. *Mater. Res. Soc. Symp. Proc.*, **1034**, 188–193, 2008.
23. Lines, M.E. and Glass, A.M., *Principle and Application of Ferroelectrics and Related Materials*, Clarendon Press, Claredon, Oxford, 1977.
24. Qiao, H., Wang, C., Choi, W.S., Park, M.H., Kim, Y., Ultra-thin ferroelectrics. *Mater. Sci. Eng. R Rep.*, **145**, May, 100622, 2021.
25. Mikolajick, T., Schroeder, U., Slesazeck, S., The Case for Ferroelectrics in Future Memory Devices. *2021 5th IEEE Electron Devices Technology and Manufacturing Conference*, vol. *2021*, EDTM, pp. 1–3, 2021.
26. Ishiwara, H., Current status and prospects of ferroelectric memories. *Technical Digest - International Electron Devices Meeting*, pp. 725–728, 2001.
27. Mueller, S., Slesazeck, S., Mikolajick, T., Müller, J., Polakowski, P., Flachowsky, S., Next-generation ferroelectric memories based on FE-HfO_2. *2015 Joint IEEE International Symposium on the Applications of Ferroelectric, International Symposium on Integrated Functionalities and Piezoelectric Force Microscopy Workshop, ISAF/ISIF/PFM 2015*, pp. 233–236, 2015.
28. Rodriguez, J., Remack, K., Gertas, J., Wang, L., Zhou, C., Boku, K., Rodriguez-Latorre, J., Udayakumar, K.R., Summerfelt, S., Moise, T., Reliability of Ferroelectric Random Access Memory Embedded within 130nm CMOS. *2010 IEEE International Reliability Physics Symposium*, pp. 1–9, 2010.
29. Wang, Y., Chen, W., Wang, B., Zheng, Y., Ultrathin ferroelectric films: Growth, characterization, physics and applications. *Materials*, **6**, 9, 6377–6485, 2014.
30. Scott, J.F., Ferroelectrics: Novel geometric ordering of ferroelectricity. *Nat. Mater.*, **4**, 1, 13–14, 2005.
31. Yadav, A.K., Nelson, C.T., Hsu, S.L., Hong, Z., Clarkson, J.D., Schlepuëtz, C.M., Damodaran, A.R., Shafer, P., Arenholz, E., Dedon, L.R., Chen, D., Vishwanath, A., Minor, A.M., Chen, L.Q., Scott, J.F., Martin, L.W., Ramesh, R., Observation of polar vortices in oxide superlattices. *Nature*, **530**, 198–201, 2016.
32. Das, S., Tang, Y.L., Hong, Z., Gonçalves, M.A.P., McCarter, M.R., Klewe, C., Nguyen, K.X., Gómez-Ortiz, F., Shafer, P., Arenholz, E., Stoica, V.A., Hsu, S.L., Wang, B., Ophus, C., Liu, J.F., Nelson, C.T., Saremi, S., Prasad, B., Mei, A.B., Schlom, D.G., Íñiguez, J., García-Fernández, P., Muller, D.A., Chen, L.Q., Junquera, J., Martin, L.W., Ramesh, R., Observation of room-temperature polar skyrmions. *Nature*, **568**, 368–372, 2019.

33. Stoica, V.A., Laanait, N., Dai, C., Hong, Z., Yuan, Y., Zhang, Z., Lei, S., McCarter, M.R., Yadav, A., Damodaran, A.R., Das, S., Stone, G.A., Karapetrova, J., Walko, D.A., Zhang, X., Martin, L.W., Ramesh, R., Chen, L.Q., Wen, H., Gopalan, V., Freeland, J.W., Optical creation of a super crystal with three-dimensional nanoscale periodicity. *Nat. Mater.*, **18**, 4, 377–383, 2019.
34. Das, S., Hong, Z., Mccarter, M., Shafer, P., Shao, Y.T., Muller, D.A., Martin, L.W., Ramesh, R., A new era in ferroelectrics. *APL Mater.*, **8**, 12, 120902, 2020.
35. Pradeesh, E.L., Udhayakumar, S., Vasundhara, M.G., Kalavathi, G.K., A review on piezoelectric energy harvesting, *Microsyst. Technol.*, 28, 8, 1797–1830, 2022.
36. Umeda, M., Nakamura, K., Ueha, S., Energy storage characteristics of a piezo-generator using impact-induced vibration. *Jpn. J. Appl. Phys.*, **36**, 3146–3151, 1997.
37. Romanofsky, R.R., Advances in scanning reflectarray antennas based on ferroelectric thin-film phase shifters for deep-space communications. *Proc. IEEE*, **95**, 10, 1968–1975, 2007.
38. Sayyah, R., MacLeod, T.C., Ho, F.D., Radiation-hardened electronics and ferroelectric memory for space flight systems. *Ferroelectrics*, **413**, 1, 170–175, 2011.
39. Chandra, P., Lonzarich, G.G., Rowley, S.E., Scott, J.F., Jiang, P., Huang, X., Prospects and applications near ferroelectric quantum phase transitions: A Key Issues Review. *Rep. Prog. Phys.*, **80**, 112502, 2017.
40. Gopal, M. and Raj, B., Low Power 8T SRAM Cell Design for High Stability Video Applications. *ITSI Trans. Electr. Electron. Eng.*, 1, 5, 91–97, 2013.
41. Raj, B., Mitra, J., Bihani, D.K., Rangharajan, V., Saxena, A.K., Dasgupta, S., Analysis of Noise Margin, Power and Process Variation for 32 nm FinFET Based 6T SRAM Cell. *J. Comput. (JCP)*, Academy Publisher FINLAND, 5, 6, 1–8, 2010.
42. Sharma, D., Mehra, R., Raj, B., Comparative Analysis of Photovoltaic Technologies for High Efficiency Solar Cell Design. *Superlattices Microstruct.*, Elsevier, 153, 106861, May 2021.
43. Kaur, P., Buttar, A.S., Raj, B., A comprehensive Analysis of Nanoscale Transistor based Biosensor: A Review. *Indian J. Pure Appl. Phys.*, 59, 304–318, April 2021.
44. Yadav, D., Raj, B., Raj, B., Design and Simulation of Low Power Microcontroller for IoT Applications. *J. Sens. Lett.*, ASP, 18, 401–409, May 2020.
45. Singh, S. and Raj, B., A 2-D Analytical Surface Potential and Drain current Modeling of Double-Gate Vertical t-shaped Tunnel FET. *J. Comput. Electron.*, Springer, 19, 1154–1163, Apl 2020.
46. Singh, J. and Raj, B., An Accurate and Generic Window function for Nonlinear Memristor Model. *J. Comput. Electron.*, Springer, 18, 2, 640–647, June 2019.

47. Kaur, M., Gupta, N., Kumar, S., Raj, B., Singh, A.K., Comparative RF and Crosstalk Analysis of Carbon Based Nano Interconnects. *IET Circuits Devices Syst.*, 15, 6, 493–503, Feb 2021.
48. Kandasamy, N., Ahmad, F., Ajitha, D., Raj, B., Telagam, N., Quantum Dot Cellular Automata based Scan Flip Flop and Boundary Scan Register. *IETE J. Res.*, 66, 535–548, 2020.
49. Singh, J. and Raj, B., Design and Investigation of 7T2M NVSARM with Enhanced Stability and Temperature Impact on Store/Restore Energy. *IEEE Trans. Very Large Scale Integr. Sys.*, 27, 6, 1322–1328, June 2019.
50. Bhardwaj, A.K., Gupta, S., Raj, B., Singh, A., Impact of Double Gate Geometry on the Performance of Carbon Nanotube Field Effect Transistor Structures for Low Power Digital Design. *Comput. Theor. Nanosci.*, ASP, 16, 1813–1820, 2019.
51. Jain, N. and Raj, B., Thermal Stability Analysis and Performance Exploration of Asymmetrical Dual-k underlap Spacer (ADKUS) SOI FinFET for Security and Privacy Applications. *Indian J. Pure Appl. Phys. (IJPAP)*, 57, 352–360, 2019.
52. Jain, N. and Raj, B., Dual-k Spacer Region Variation at the Drain Side of Asymmetric SOI FinFET Structure: Performance Analysis towards the Analog/RF Design Applications. *J. Nanoelectron. Optoelectron.*, American Scientific Publishers, 14, 349–359, March 2019.
53. Singh, J., Sharma, S., Raj, B., Khosla, M., Analysis of barrier layer thickness on performance of In1-xGaxAs based Gate Stack Cylindrical Gate Nanowire MOSFET. *JNO*, ASP, 13, 1473–1477, Oct 2018.
54. Jain, N. and Raj, B., Analysis and Performance Exploration of High-k SOI FinFETs Over the Conventional Low-k SOI FinFET toward Analog/RF Design. *J. Semicond. (JoS)*, IOP Science, 39, 12, 124002-1-7, Dec 2018.
55. Goyal, C., Ubhi, J.S., Raj, B., A reliable leakage reduction technique for approximate full adder with reduced ground bounce noise. *J. Math. Probl. Eng.*, Hindawi, 2018, 16, 15 Oct 2018. Article ID 3501041.
56. Anuradha, Singh, J., Raj, B., Khosla, M., Design and Performance Analysis of Nano-scale Memristor-based Nonvolatile SRAM. *J. Sens. Lett.*, American Scientific Publishers, 16, 798–805, Oct 2018.
57. Wadhwa, G. and Raj, B., Parametric Variation Analysis of Charge-Plasma-based Dielectric Modulated JLTFET for Biosensor Application. *IEEE Sens. J.*, 18, 15, AUGUST 1, 6070 - 6077, 2018 (SCI).
58. Singh, J. and Raj, B., Comparative Analysis of Memristor Models for Memories Design. *JoS*, IoP, 39, 7, 074006-1-12, July 2018.
59. Yadav, D., Chouhan, S.S., Vishvakarma, S.K., Raj, B., Application Specific Microcontroller Design for IoT based WSN. *Sens. Lett.*, ASP, 16, 374–385, May 2018.
60. Singh, G., Sarin, R.K., Raj, B., Fault-Tolerant Design and Analysis of Quantum-Dot Cellular Automata Based Circuits. *IEEE/IET Circuits Devices Syst.*, 12, 638–64, 2018.

61. Singh, J. and Raj, B., Modeling of Mean Barrier Height Levying Various Image Forces of Metal Insulator Metal Structure to Enhance the Performance of Conductive Filament Based Memristor Model. *IEEE Nanotechnol.*, 17, 2, 268–267, March 2018.
62. Jain, A., Sharma, S., Raj, B., Analysis of Triple Metal Surrounding Gate (TM-SG) III-V Nanowire MOSFET for Photosensing Application. *Opto-Electron. J.*, Elsevier, 26, 2, 141–148, May 2018.
63. Jain, A., Sharma, S., Raj, B., Design and Analysis of High Sensitivity Photosensor Using Cylindrical Surrounding Gate MOSFET For Low Power Sensor Applications. *Eng. Sci. Technol. Int. J.*, Elsevier's, 19, 4, 1864–1870, December 2016.
64. Kumar, G.S., Singh, A., Raj, B., Design and Analysis of Gate All Around CNTFET based SRAM cell Design. *J. Comput. Electron.*, Springer, 17, 1, 138–145, March 2018.
65. SinghSohi, G.B.S. and Balwinder, R., Material Properties Analysis of Graphene Base Transistor (GBT) for VLSI Analog Circuits. *Indian J. Pure Appl. Phys. (IJPAP)*, 55, 896–902, Dec-2017.
66. Kumar, S. and Raj, B., Estimation of Stability and Performance metric for Inward Access Transistor based 6T SRAM Cell Design using n-type/p-type DMDG-GDOV TFET. *IEEE VLSI Circuits Syst. Lett.*, 3, 2, 25–39, June 2017.
67. Sharma, S., Kumar, A., Pattanaik, M., Raj, B., Forward Body Biased Multimode Multi-Threshold CMOS Technique for Ground Bounce Noise Reduction in Static CMOS Adders. *Int. J. Inf. Electron. Eng.*, 3, 3, 567–572, 2013.
68. Singh, H., Kumar, P., Raj, B., Performance Analysis of Majority Gate SET Based 1-bit Full Adder. *Int. J. Comput. Commun. Eng. (IJCCE)*, IACSIT Press Singapore, 2, 4, 2933–2940, 2013. ISSN: 2010-3743.
69. Bhardwaj, A.K., Gupta, S., Raj, B., Investigation of Parameters for Schottky Barrier (SB) Height for Schottky Barrier Based Carbon Nanotube Field Effect Transistor Device. *J. Nanoelectron. Optoelectron.*, ASP, 15, 783–791, July 2020.
70. Bansal, P. and Raj, B., MEMRISTOR: A VERSATILE NONLINEAR MODEL FOR DOPANT DRIFT AND BOUNDARY ISSUES. *JCTN*, American Scientific Publishers, 14, 5, 2319–2325, May 2017.
71. Jain, N. and Raj, B., An Analog and Digital Design Perspective Comprehensive Approach on Fin-FET (Fin-Field Effect transistor) Technology - A Review. *Rev. Adv. Sci. Eng. (RASE)*, ASP, 5, 1–14, 2016.
72. Raj, B., Saxena, A.K., Dasgupta, S., Analytical Modeling for the Estimation of Leakage Current and Subthreshold Swing Factor of Nanoscale Double Gate FinFET Device. *Microelectron.*, International UK, 26, 53–63, 2009.
73. SoniyaSingh, S., Wadhwa, G., Raj, B., An Analytical Modeling for Dual Source Vertical Tunnel Field Effect Transistor. *Int. J. Recent Technol. Eng. (IJRTE)*, 8, 2, 603–608, July 2019.

74. Singh, S. and Raj, B., Design and analysis of hetrojunction Vertical T-shaped Tunnel Field Effect Transistor. *J. Electron. Mater.*, Springer, 48, 10, 6253–6260, October 2019.
75. Goyal, C., Ubhi, J.S., Raj, B., A Low Leakage CNTFET based Inexact Full Adder for Low Power Image Processing Applications. *Int. J. Circuit Theory Appl.*, Wiley, 47, 9, 1446–1458, September 2019.
76. Raj, B., Saxena, A.K., Dasgupta, S., A Compact Drain Current and Threshold Voltage Quantum Mechanical Analytical Modeling for FinFETs. *J. Nanoelectron. Optoelectron. (JNO)*, USA, 3, 2, 163–170, 2008.
77. Wadhwa, G. and Raj, B., An Analytical Modeling of Charge Plasma based Tunnel Field Effect Transistor with Impacts of Gate underlap Region. *Superlattices Microstruct.*, Elsevier, 142, 106512, June 2020.
78. Singh, S. and Raj, B., Modeling and Simulation analysis of SiGe hetrojunction Double GateVertical t-shaped Tunnel FET. *Superlattices Microstruct.*, Elsevier, 142, 106496, June 2020.
79. Jain, N. and Raj, B., Device and Circuit Co-Design Perspective Comprehensive Approach on FinFET Technology - A Review. *J. Electron Devices*, 23, 1, 1890–1901, 2016.
80. Kumar, S. and Raj, B., Analysis of I_{ON} and Ambipolar Current for Dual-Material Gate-drain Overlapped DG-TFET. *J. Nanoelectron. Optoelectron.*, American Scientific Publishers USA, 11, 323–333, June 2016.
81. Anjum, N., Bali, T., Raj, B., Design And Simulation Of Handwritten Multiscript Character Recognition. *Int. J. Adv. Res. Comput. Commun. Eng.*, 2, 7, 2544–2549, July 2013.
82. Singh, K. and Raj, B., Performance and analysis of temperature dependent Multi-walled Carbon Nanotubes as Global Interconnects at different technology nodes. *J. Comput. Electron.*, Springer, 14, 2, 469–476, 2015.
83. Kumar, S. and Raj, B., Compact channel potential analytical Modeling of DG-TFET based on Evanescent-mode Approach. *J. Comput. Electron.*, Springer, 14, 2, 820–827, July 2015.
84. Singh, K. and Raj, B., Temperature Dependent Modeling and Performance Evaluation of Multi-Walled CNT and Single-Walled CNT as Global Interconnects. *J. Electron. Mater.*, Springer, 44, 12, 4825–4835, Dec 2015.
85. Sharma, V.K., Pattanaik, M., Raj, B., INDEP approach for leakage reduction in nanoscale CMOS circuits. *Int. J. Electron.*, Taylor & Francis, 102, 2, 200–215, 2014.
86. Singh, K. and Raj, B., Influence of Temperature on MWCNT bundle, SWCNT bundle and Copper interconnects for Nanoscaled Technology nodes. *J. Mater. Science Mater. Electron.*, Springer, 26, 8, 6134–6142, 2015.
87. Anjum, N., Bali, T., Raj, B., Design and Simulation of Handwritten Gurumukhi and Devanagri Numerical Recognition. *Int. J. Comput. Appl.*, Published by Foundation Computer Science New York USA, 73, 12, 16–21, 2013.

88. Khandelwal, S., Gupta, V., Raj, B., Gupta, R.D., Process Variability Aware Low Leakage Reliable Nano Scale DG-FinFET SRAM Cell Design Technique. *J. Nanoelectron. Optoelectron.*, 10, 6, 810–817, 2015.
89. Sharma, V.K., Pattanaik, M., Raj, B., ONOFIC Approach: Low Power High Speed Nanoscale VLSI Circuits Design. *Int. J. Electron.*, Taylor & Francis, 101, 1, 61–73, 2014.
90. Khandelwal, S., Raj, B., Gupta, R.D., FinFET based 6T SRAM Cell Design: Analysis of Performance Metric, Process Variation and Temperature Effect. *J. Comput. Theor. Nanosci.*, ASP USA, 12, 2500–2506, 2015.
91. Singh, S., Yadav, S., Rahul, J., Srivastava, A., Raj, B., Impact of HfO_2 in Graded Channel Dual Insulator Double Gate MOSFET. *J. Comput. Theor. Nanoscience*, American Scientific Publishers, 12, 6, 950–953, April 2015.
92. Sharma, V.K., Pattanaik, M., Raj, B., PVT variations aware low leakage INDEP approach for nanoscale CMOS Circuits. *Microelectron. Reliab.*, Elsevier, 54, 90–99, 2014.
93. Raj, B., Saxena, A.K., Dasgupta, S., Quantum Mechanical Analytical Modeling of Nanoscale DG FinFET: Evaluation of Potential, Threshold Voltage and Source/Drain Resistance. *Elsevier's J. Mater. Sci. Semicond. Process.*, Elsevier, 16, 4, 1131– 1137, 2013.
94. Gopal, M., Prasad, S.S.D., Raj, B., 8T SRAM Cell Design for Dynamic and Leakage Power Reduction. *Int. J. Comput. Appl.,* Published by Foundation Computer Science New York USA, 71, 9, 43–48, June 2013.
95. Pattanaik, M., Raj, B., Sharma, S., Kumar, A., Diode Based Trimode Multi-Threshold CMOS Technique for Ground Bounce Noise Reduction in Static CMOS Adders. *Adv. Mater. Res.,* Trans Tech Publications Switzerland, 548, 885–889, 2012.
96. RajSaxena, B.A.K. and Dasgupta, S., Nanoscale FinFET Based SRAM Cell Design: Analysis of Performance metric, Process variation, Underlapped FinFET and Temperature effect. *IEEE Circuits Syst. Mag.*, 11, 2, 38–50, 2011.
97. Sharma, V.K., Pattanaik, M., Raj, B., Leakage Current ONOFIC Approach for Deep Submicron VLSI Circuit Design. *Int. J. Elec., Comp., Elec. Commun. Eng.*, World Academy of Sciences, Engineering and Technology, 7, 4, 239–244, 2013.
98. Kaur, P., Gill, S.S., Raj, B., Comparative Analysis of OFETs Materials and Devices for Sensor Applications. *J. Silicon*, Springer, 14, 4463–4471, 2022.
99. Sharma, S.K., Kumar, P., Raj, B., Raj, B., $In_{1-x}Ga_xAs$ Double Metal Gate-Stacking Cylindrical Nanowire MOSFET for highly sensitive Photo detector. *J. Silicon*, Springer, 14, 3535–3541, 2022.
100. Raj, B., Saxena, A.K., Dasgupta, S., Analytical Modeling of Quasi Planar Nanoscale Double Gate FinFET with Source/Drain Resistance and field dependent carrier mobility: A quantum mechanical study. *J. Comput. (JCP),* Academy Publisher FINLAND, 4, 9, 1–8, 2009.
101. Bhushan, S., Khandelwal, S., Raj, B., Analyzing Different Mode FinFET based Memory Cell at Different Power Supply for Leakage Reduction.

Seventh International Conference on Bio-Inspired Computing: Theories and Application, (BIC-TA 2012) Advances in Intelligent Systems and Computing, vol. 202, pp. 89–100, 20122013.
102. Singh, J. and Raj, B., Temperature Dependent Analytical Modeling and Simulations of Nanoscale Memristor. *J. Eng. Sci. Technol. Int. J.*, Elsevier's, 21, 862–868, Oct 2018.
103. Singh, S., Bala, S., Raj, B., Raj, B., Improved Sensitivity of Dielectric Modulated Junctionless Transistor for Nanoscale Biosensor Design. *Sens. Lett.*, ASP, 18, 328–333, Apl 2020.
104. Kumar, V., Vishvakarma, S.K., Raj, B., Design and Performance Analysis of ASIC for IoT Applications. *Sens. Lett.*, ASP, 18, 31–38, Jan 2020.
105. Jaiswal, A., Sarin, R.K., Raj, B., Sukhija, S., A Novel Circular Slotted Microstrip-fed Patch Antenna with Three Triangle Shape Defected Ground Structure for Multiband Applications. *Adv. Electromagn.*, 7, AEM3, 56–63, Aug 2018.
106. Wadhwa, G. and Raj, B., Label Free Detection of Biomolecules using Charge-Plasma-Based Gate Underlap Dielectric Modulated Junctionless TFET. *J. Electron. Mater. (JEMS)*, Springer, 47, 8, 4683–4693, August 2018.
107. SinghSarin, G.R.K. and Raj, B., Design and Performance Analysis of a New Efficient Coplanar Quantum-Dot Cellular Automata Adder. *Indian J. Pure Appl. Phys. (IJPAP)*, 55, 97–103, February 2017.
108. Kaur, P., Pandey, V., Raj, B., Comparative Study of Efficient Design, Control and Monitoring of Solar Power using IoT. *Sens. Lett.*, ASP, 18, 419–426, May 2020.
109. Bhardwaj, A.K., Gupta, S., Raj, B., Development & Analysis of compact model for Double Gate Schottky Barrier CNTFET. *J. Nanoelectron. Optoelectron.*, ASP, 15, 1199–1208, Aug 2020.
110. Wadhwa, G., Kamboj, P., Singh, J., Raj, B., Design and Investigation of Junctionless DGTFET for Biological Molecule Recognition. *Trans. Electr. Electron. Mater.*, Springer, 22, 282–289, 2021.
111. Verma, S.K., Singh, S., Wadhwa, G., Raj, B., Detection of Biomolecules using Charge-Plasma based Gate Underlap Dielectric Modulated Dopingless TFET. *Trans. Electr. Electron. Mater. (TEEM)*, Springer, 21, 528–535, June 2020.
112. Jain, N. and Raj, B., Impact of underlap spacer region variation on Electrostatic and analog/RF performance of symmetrical high-k SOI FinFET at 20 nm channel length. *J. Semicond. (JoS)*, IOP Science, 38, 12, 122002, Dec 2017.
113. Singh, S. and Raj, B., Analytical Modeling and Simulation analysis of T-shaped III-V heterojunction Vertical T-FET. *Superlattices Microstruct.*, Elsevier, 147, 106717, Nov 2020.
114. SinghSarin, G.R.K. and Raj, B., Design and Analysis of Area Efficient QCA Based Reversible Logic Gates. *J. Microprocess. Microsys.*, Elsevier, 52, 59–68, May 2017.

115. Singh, S. and Raj, B., Parametric variation analysis on hetero-junction Vertical t-shape TFET for Supressing ambipolar conduction. *Indian J. Pure Appl. Phys.*, 58, 478–485, June 2020.
116. Soniya, Singh, S., Wadhwa, G., Raj, B., Design and Analysis of Dual Source Vertical Tunnel Field Effect Transistor for high performance. *Trans. Electr. Electron. Mater.*, Springer, 21, 74–82, Oct 2019.
117. Kaur, M., Gupta, N., Kumar, S., Raj, B., Singh, A.K., RF Performance Analysis of Intercalated Graphene Nanoribbon Based Global Level Interconnects. *J. Comput. Electron.*, Springer, 19, 1002–1013, June 2020.
118. Wadhwa, G. and Raj, B., Design and Performance Analysis of Junctionless TFET Biosensor for high sensitivity. *IEEE Nanotechnol.*, 18, 567–574, 2019.
119. Singh, J. and Raj, B., Enhanced Nonlinear Memristor Model Encapsulating Stochastic Dopant Drift. *JNO*, ASP, 14, 958–963, 2019.
120. Bhardwaj, A., Kumar, P., Raj, B., Anand, S., Design and Performance Optimization of doping-less Vertical Nanowire TFET using Gate Stack Technique. *J. Electron. Mater. (JEMS)*, Springer, 41, 7, 4005–4013, 2022.
121. Singh, J. and Raj, B., Tunnel Current Model of Asymmetric MIM Structure Levying Various Image Forces to Analyze the Characteristics of Filamentary Memristor. *Appl. Phys. A*, Springer, 125, 3, 203.1 to 203.11, Feb 2019.
122. Goyal, C., Ubhi, J.S., Raj, B., Low Leakage Zero Ground Noise Nanoscale Full Adder using Source Biasing Technique. *J. Nanoelectron. Optoelectron.*, American Scientist Publishers, 14, 360–370, March 2019.
123. SinghSarin, G.R.K. and Raj, B., A Novel Robust Exclusive-OR Function Implementation in QCA Nanotechnology with Energy Dissipation Analysis. *J. Comput. Electron.*, Springer, 15, 2, 455–465, June 2016.
124. Wadhwa, G., Kamboj, P., Raj, B., Design optimisation of junctionless TFET biosensor for high sensitivity. *Adv. Nat. Sci.: Nanosci. Nanotechnol.*, 10, 7, 045001, 2019.
125. Bansal, P. and Raj, B., MEMRISTOR MODELING AND ANALYSIS FOR LINEAR DOPANT DRIFT KINETICS. *J. Nanoeng. Nanomanuf.*, American Scientific Publishers, 6, 1–7, 2016.
126. Ashima, Vaithiyanathan, D., Raj, B., Performance Analysis of Charge Plasma induced Graded Channel Si nanotube. *J. Eng. Res. (JER)*, EMSME Special Issue, 9, 146–154, Aug 2021.
127. Tomar, A.S., Magraiya, V.K., Raj, B., Scaling of Access and Data Transistor for High Performance DRAM Cell Design. *Quantum Matter*, 2, 412–416, Oct. 2013.
128. Jain, N. and Raj, B., Parasitic Capacitance and Resistance Model Development and Optimization of Raised Source/Drain SOI FinFET Structure for Analog Circuit Applications. *J. Nanoelectron. Optoelectron.*, ASP USA, 13, 531–539, Apl 2018.
129. SinghVishvakarma, S.S.K. and Raj, B., Analytical Modeling of Split-Gate Junction-Less Transistor for a Biosensor Application. *Sens. Bio-Sensing*, Elsevier, 18, 31–36, 2018.

4

Optimization of Hetero Buried Oxide Ferro TFET and Its Analysis

Sirisha Meriga[1,2]* and Brinda Bhowmick[1]

[1]National Institute of Technology, Silchar, Assam, India
[2]Chalapathi Institute of Engineering and Technology, Lam, Guntur, Andhra Pradesh, India

Abstract
This chapter introduces a hetero-buried oxide (HBOX)–ferroelectric tunnel FET. Numerous electrical parameters were tuned with respect to the silicon and BOX layer thickness adjustment, including ON/OFF-state currents, and I_{ON}/I_{OFF} ratio. A negative-capacitance-based ferroelectric gate stack is introduced to obtain a steep Subthreshold Slope (SS). To ensure compatibility with a typical FET production approach, a ferroelectric (Fe) layer composed of silicon-doped hafnium oxide was employed at the bottom of the gate. The thickness of the Fe layer was carefully optimized to improve the Analog and DC parameters. A doped pocket (n+) was imparted within the source in conjunction with the gate overlapping and drain underlapping. In addition to the design features, a pocket that improves the electric field around the tunnel junction, makes it more likely for carriers to tunnel towards the channel. The hetero-buried oxide was integrated into the Silicon on Insulator material. A high k-BOX remains at the source and a low k-BOX is exploited at the drain sides to minimize the ambipolar properties of the TFET and increase the ON-state current. The proposed structure's characteristics were optimized to accommodate the changes in its physical dimensions. This arrangement provides a 38.5 mV/dec subthreshold slope and I_{ON}/I_{OFF} ratio of approximately 10^9. The Fe HBOX TFET is known to operate at an impressive speed depending on the capacitance recovered from it. It is suitable for analog/RF applications because it exhibits superior transconductance and a reduced amount of output conductance. A steep subthreshold slope makes it ideal for low-power demand applications. The proposed ferroelectric TFET device was evaluated using Sentaurus TCAD 2D software.

**Corresponding author*: sirisha21_rs@ece.nits.ac.in; ORCID: orcid.org/0000-0001-8904-1674

Balwinder Raj, Shiromani Balmukund Rahi and Nandakishor Yadav (eds.) FeFET Devices, Trends, Technology and Applications, (77–102) © 2025 Scrivener Publishing LLC

Keywords: Hetero buried oxide, negative capacitance, ferroelectric layer, analog and DC parameters, subthreshold slope

4.1 Introduction

Modern nanoelectronic components must function at sub-1.5 V supply voltages with minimal heat immoderation. The shrinking of MOSFETs is caused by unfavorable short-channel consequences (SCEs), notably drain-induced barrier narrowing, punch-through, hot-carrier development, and velocity saturation, which steadily affect the semiconductor devices' capability [1–3]. In the future, tunnel FETs, which employ the band-to-band tunnelling (B2B) approach, will deliver sub-threshold inclines (SS) smaller than 60 mV/dec that will supersede MOSFETs in nanoelectronic contraptions [4–6]. BTBT refrains from the SCEs in tunnelling FETs and enhances the transistor's SS compared to thermionic emission-operated MOSFETs. Much research has been conducted in the context of TFET development [7–11].

For the continuous advancement of microelectronics, the introduction of conventional TFETs and their numerous structural and material enhancements have made TFETs an adequate MOSFET replacement. Ambipolar conduction and a low on-state current are essential characteristics when designing devices. For silicon TFETs, the electric field inside the device could be generated by an on-state current that may be promoted by adding an n+ pocket at the device's source terminal. Such devices can function at meager power supplies, have a steeper subthreshold slope, and have a higher level of reliability in comparison to standard p-i-n TFETs while maintaining an excellent improved ON-state current [12, 13].

The TFET's inherent ambipolar conduction nature limits their feasibility in DC circuit applications [14, 15]. Several transistor topologies have been investigated, including gate-drain underlap, low-k spacers, wide-bandgap semiconductors at the output side, and later on lateral heterostructures [16, 17]. However, reduced ambipolar current, on-state current and fabrication intricacy cause higher series resistance in the drain, which results in an elevated cost of manufacture [18, 19]. Sahay *et al.* and Sirisha *et al.* found that by placing a hetero-box underneath the silicon substrate to lower ambipolarity, the breadth of the tunnelling barrier on the side of the drain channel edge could be adjusted [20, 21].

Extensive research has been conducted on ferroelectric FETs and Tunnel FETs with multiple geometries. The ferroelectric material's negative quantum capacitance, which depends on the density of states, provides the potential for improved overall capacitance. There is a different temperature behavior in Fe materials [22–27]. Landau's hypothesis states that the proportional dielectric permittivity of Fe compounds near the vicinity of the Curie temperature (Tc) is optimum [3]. The ferroelectric's negative capacitance will allow for an escalating voltage, lowering the subthreshold swing to less than 60 mV/decade, and permitting low-voltage/low-power applications. Negative capacitance in ferroelectric materials has been proposed to reduce power dissipation in electronics beyond fundamental limits [28–35].

The device will continue to operate perfectly at all temperatures with B2B tunnelling. Incorporating ferroelectric materials into transistor architectures is reasonably straightforward and follows CMOS technology for chips, which will ultimately result in improved efficiency when developing nanoelectronic systems. Ferroelectric gate stacks must be present in ferroelectric TFETs to boost their output current. Silicon-doped hafnium oxide (Si: HfO2) exhibits better ferroelectric abilities [22, 23] and represents a good match for the CMOS manufacturing method. The Si:HfO2-based ferroelectric gate pile-up is considered to have with a thickness of 6 nm throughout the study [36–42].

The current research developed a firmly doped n+ pocket over the source end of a HBOX coating beneath the silicon surface. To overcome customary TFETs' flaws, gate-to-drain underlap and gate-source overlap have also been formulated to underrate the ambipolar current.

A summary of this chapter is as follows: The designed HBOX-Fe TFET and the simulation methodology are described in *Section 4.2. Section 4.3* illustrates the Results and discussion. In this, *Sections 4.3.1 and 4.3.2* demonstrate the essence of polarization and DC analysis of the TFET, respectively [43–48]. The analog/RF effectiveness analysis of the specified TFET is further discussed in *Section 4.3.3.* Finally, conclusions are presented in *Section 4.4.*

4.2 Mechanism of the Device and Method of Simulation

Figure 4.1 depicts the HBOX pocket-doped ferroelectric gate TFET architecture. To suppress the ambipolar current and boost the on-state current, a low-k oxide layer with a thiclness of 30 nm and length of 50 nm was layered at the drain end of the transistor, and a high-k oxide layer of a similar thickness was laid down at the source surface [21]. A 5 nm thick strongly

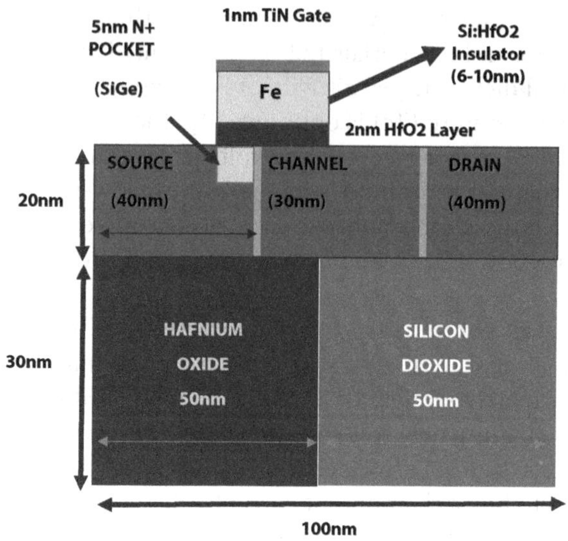

Figure 4.1 HBOX-Fe TFET architecture.

doped n+ pocket doping concentration is assumed to be $5 \times 10^{19}/cm^3$ at the source, modulating the band gap between the source and channel [49–53]. This also generated a stronger electric field. The 20 nm thick source and drain concentrations were considered as $1 \times 10^{20}/cm^3$ and $5 \times 10^{18}/cm^3$, respectively. The thickness of the ferroelectric layer was considered as 6 nm. The 20 nm-thick boron doped channel was $1 \times 10^{16}/cm^3$.

The length of the a) source, b) channel, and c) drain was 40 nm, 30 nm, and 30 nm, respectively. Low-k oxide (SiO2) has been suggested as the drain side, and high-k oxide (HfO2) is considered the source end of a hetero-buried oxide in this arrangement. In addition to a gate-to-source overlap length of 5 nm for reducing the ambipolar current, a gate-to-drain underlap distance of 15 nm was considered. The ideal hafnium oxide layer is two nm thick and has a 4.33 eV work function TiN gate stack with a 6 nm thick Fe oxide layer at the gate stack. The HBOX-Fe TFET's overall length is 100 nm [54–59].

Simulations of the suggested HBOX TFET were accomplished using 2D-Sentaurus TCAD [24]. To implement minority recombination, Shockly–Read–Hall doping, relying upon the recombination model, was employed. Due to the highly doped materials used in the source, drain, and pocket, a doping-dependent mobility scheme was adopted. It is necessary to consider the influence of polarization of the ferroelectric substance, and the inter-band tunnelling operation, nonlocal B2B tunnelling, and the

Table 4.1 The following factors and their implications are investigated for simulation studies.

Parameter	Value	Parameter	Value
Coercive Field Ec (MV/cm) [23]	1.1	Remnant Polarization Pr (μc/cm²) [23]	11.2
Saturation Polarization Ps (μc/cm²) [23]	11.5	Relative permittivity of Si:HfO2	32.5
α (cm/F) [23]	-4.32×10^{10}	β(cm⁵/FC²) [23]	3.87×10^{19}

Landau–Khalatnikov calculation are explored. Furthermore, the bandgap narrowing model was used. Table 4.1 summarizes the specifications used to analyze the functioning of the ferroelectric TFET attributes.

4.3 Results and Discussions

In this section, the HBOX-Fe TFET's polarization behavior as well as the DC and analog/RF interpretation parameters are explored.

4.3.1 The Fundamental Nature of Polarization

The TFET's polarization is shown in Figure 4.2. The Fe material present over the input gate layer exhibited polarization owing to the negative capacitance impact, as shown in the plot of the related electric field. The configuration of the dipoles in ferromaterials governs the extent of polarization (P) [60–63]. As stated in Equation (4.1), the relationship is with the electric field. In a ferroelectric material, the Landau–Khalatnikov equation is provided by [23] Equation (4.1).

$$E=2\alpha P+a\beta P^{3}+6\gamma P^{5}+\rho\frac{\partial P}{\partial t} \tag{4.1}$$

$$\text{The rate of change}\frac{\partial P}{\partial t}=0 \tag{4.2}$$

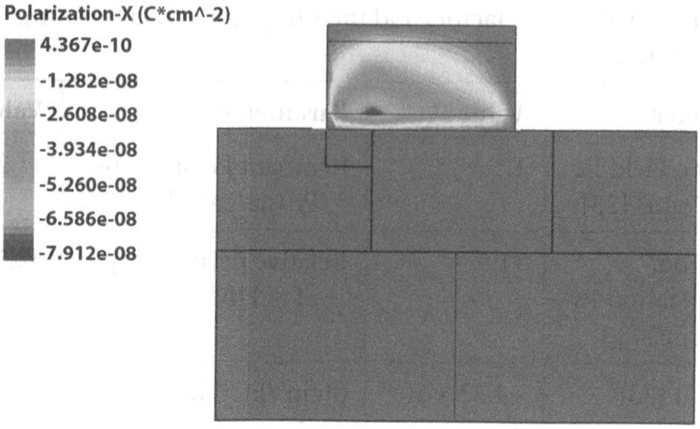

Figure 4.2 Polarization of the proposed HBOX Fe-TFET.

$$\text{therefore } \rho\frac{\partial P}{\partial t}=0 \tag{4.3}$$

$$\text{and } Vfe = tfe\,(2\alpha Q + 4\beta Q^3) \tag{4.4}$$

$$\alpha=-\frac{3\sqrt{3}}{4}\frac{Ec}{Pr} \tag{4.5}$$

$$\beta=\frac{3\sqrt{3}}{8}\frac{Ec}{P_r^3} \tag{4.6}$$

The ferroelectric capacitance decreased in the negative-capacitance area as the gate voltage increased. The negative capacitance slant is depicted as an S-curve, as shown in Figure 4.3, which reveals that as the input gate electrical voltage develops, the ferroelectric layer's capacitance decreases [64–67]. Equation (4.4) expresses the amount of voltage travelling through a ferroelectric capacitor [23].

In order to observe the variations in the ON-state current, the ferroelectric thickness, or "T_{FE}," is adjusted between 6 nm and 10 nm. The surface charge density is represented by Q, and the anisotropy constants α and β are given by Equations (4.5) and (4.6), respectively. The remnant and saturation polarizations are denoted by the symbols Pr and Ps, respectively,

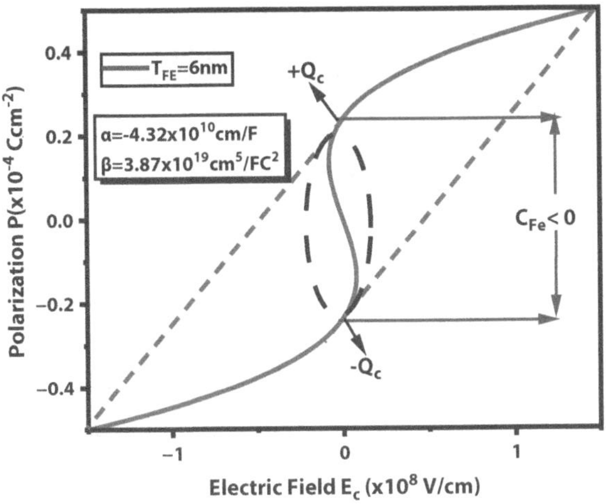

Figure 4.3 S-curve.

where Ec is the coercive electric field. Table 4.1 provides these values for the parameters.

At the gate stack, the total capacitance in the negative-capacitance transistors is given by the conventional oxide capacitance Cox becoming series with ferroelectric oxide capacitance Cfe.

$$C_{Total} = \frac{C_{fe}C_{ox}}{C_{ox}+C_{fe}} \tag{4.7}$$

Ferroelectric materials positioned at the gate oxide layer of the device were used to provide (∂Vgs)/(∂φs) <1. This makes it possible to optimize the steep slope and TFET architecture [22, 23].

4.3.2 DC Analysis

The suggested ferroelectric TFET parameters are thoroughly examined in the following section using a range of exploitations, including variations in ferroelectric thickness.

Figure 4.4 depicts the transfer characteristics of HBOX-Fe TFETs with and without the Fe layer. Based on the characteristics, the SS is 44.5 mV/dec in the absence of ferroelectric substance, and the threshold potential Vth, I_{ON}, and I_{OFF} currents were determined as 0.72 V, 10^{-13} A/μm, and 10^{-4} A/μm, accordingly. The threshold voltage and off-state current fell from 0.72 V to

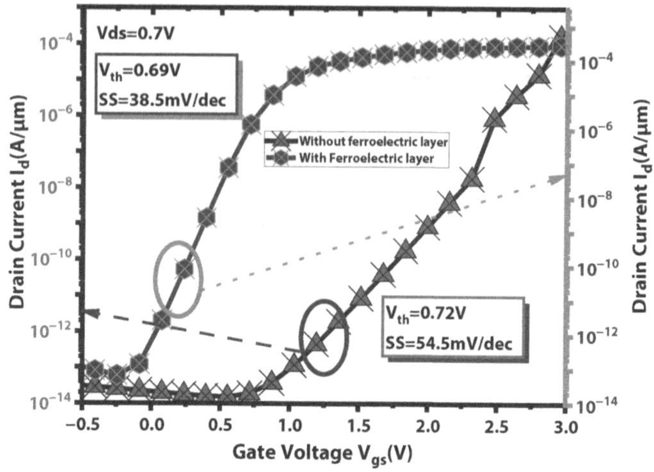

Figure 4.4 Id-Vgs with and without ferroelectric layer.

0.69 V and 10^{-14} A/μm respectively, when the Fe substance was positioned at the gate stack [68–72]. Enhancement of the subthreshold slope (SS) to 38.5 mV/dec increases the on-state current to 5.5×10^{-4} A/μm. This is due to a boost in the potential at the surface "φs" across the gate stack. The voltage at the gate terminal is increased for ultra-low-power operations via the negative capacitance phenomenon [23].

The transfer characteristic alterations in the diverse drain-to-source voltages are shown in Figure 4.5. Figure 4.6 ahead shows the output

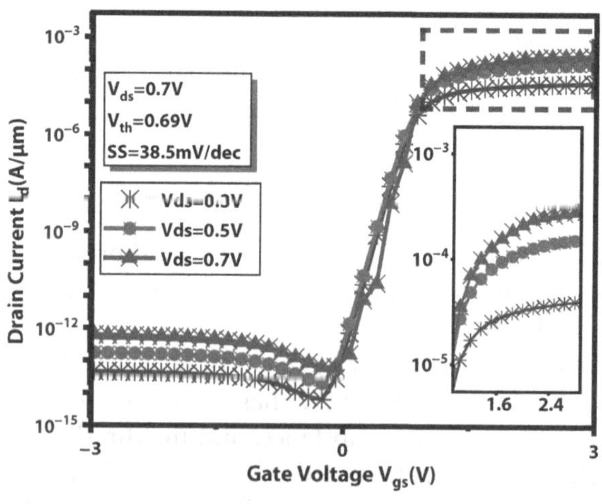

Figure 4.5 Id-Vgs characteristics.

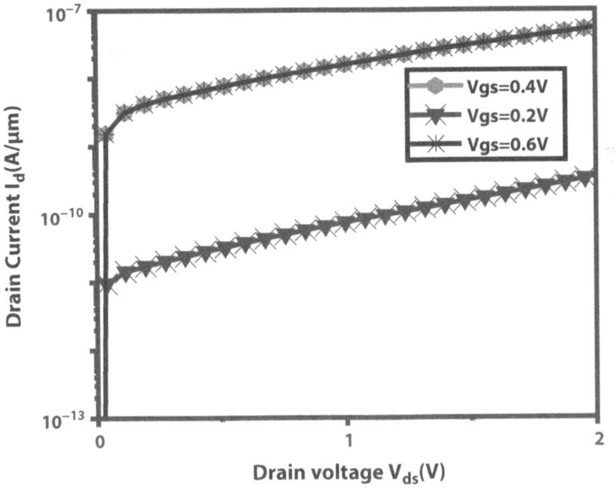

Figure 4.6 Output characteristics.

properties of the HBOX Fe-TFET. When the input gate voltage is increased more thoroughly, from 0.4 V to 0.6 V, the output current is predicted to remain constant. Ferroelectric materials such as silicon-doped hafnium oxide were introduced via a TiN metal gate pile in the TFET's construction. Pr is pictured when no electric field is applied, as Fe has considerable dipoles [73–78].

The Fe width is essential to the device's functioning. The thickness of the Fe "T_{FE}" is changed from 6 nm to 10 nm to represent this, and the corresponding procedure is shown in Figure 4.7. It is clear from the features that as the electric field varies, the OFF-state and ON-state currents are altered in relation to the T_{FE} [22]. This results from the modification in the ferroelectric layer's dipole configuration [23]. While there is less of a shift in the ON state, there is an enormous variation in the OFF current [79–83].

A thin layer of Fe is essential for the accurate operation of the device; the Fe thickness is tampered with from 6 nm to 10 nm, and the associated action is analyzed. Figure 4.7 portrays these characteristics; the electric field changes the OFF and ON-state currents [23] as the dipole setup in the Fe material is altered. The OFF-state shift was significant, whereas the ON-state change was minimal. This could be attributed to the dominant band-to-band tunnelling. Permissible OFF- and ON-state currents were observed at a ferroelectric layer of 6 nm.

Therefore, throughout this study, the thickness of the Fe oxide is presumed to be 6 nm. The OFF- and ON-state currents and their ratios improve as t_{fe}

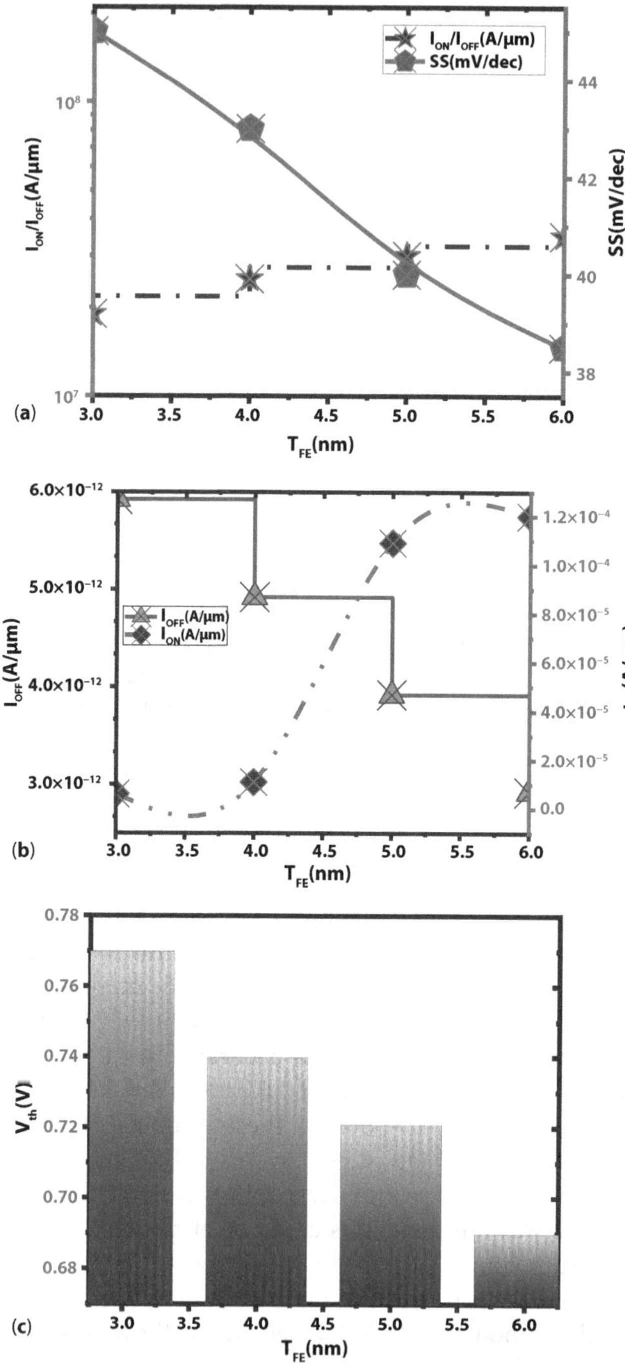

Figure 4.7 Parameters w.r.to ferro thickness (a) I_{ON}/I_{OFF} and SS (b) I_{ON} and I_{OFF} and (c) Vth.

increases. Figures 4.7(a, b) show that the device's performance improved in addition to the SS. Figure 4.7(c) also depicts the Vth refinements that follow the change in t_{fe} and the Off-state current is reduced. However, I_{ON} remains stable at 10^{-4} A/μm and a favorable Vth and SS [84–90].

4.3.3 Analog/RF Performance

The electrical characteristics of HBOX Fe TFET have been previously reported before. The ferroelectric HBOX TFET's Analog/RF performance is discussed in the following subsection. The ferroelectric thickness was selected as 6 nm throughout the study due to its exceptional linear behavior [91–96].

The device's operating speed is controlled by its capacitors. The Miller capacitance influences speed. The gate capacitance is shown in Figure 4.8. The combined source and drain capacitances, Cgs and Cgd, are represented by Cgg. The source-channel tunnelling area in traditional TFETs diminishes Cgs, whereas Cgd increases. When it comes to overall capacitance, Cgd is more dominant than Cgs [25–27].

$$C_{gg}=C_{gs}+C_{gd} \tag{4.8}$$

Compared to traditional TFETs, electrostatic bonding can generate lower capacitances, which follow the channel's path due to the negative capacitance exposed by the Si:HfO2 layer coating at the gate's exterior. From now on, the device performance in Fe-TFETs will be exceptional.

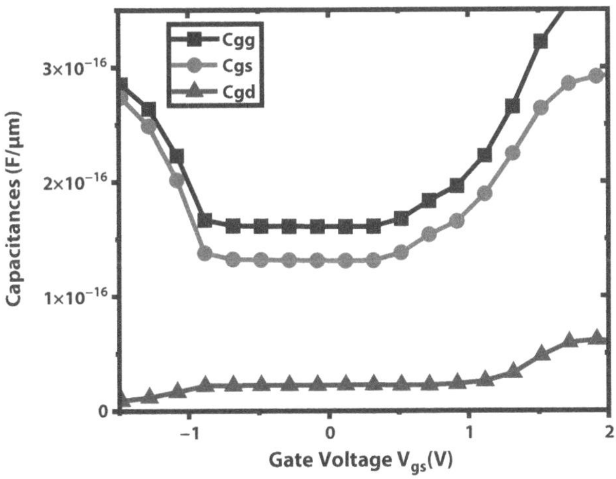

Figure 4.8 Various capacitances.

It is apparent that the transconductance gm in Figure 4.9. This metric defines device efficiency. The TFET exhibited an outstanding gm value. The transconductance is correlated with the intrinsic gain of the device. As a result, the TFET exhibits the most remarkable (highest) amplification [97–102].

This prominence is revealed by introducing HBOX with an n+ pocket at the source and a ferroelectric layer on the side of the gate. The enhanced electric field was due to the negative capacitance of Si:HfO2 result and an n+ pocket that was privileged inside the source terminal. As mentioned above, the above parameters have now been added to enable the HBOX-Fe TFETs' optimal performance for both DC and analog/RF applicability.

The output conductance (gd) is another indicator of merit. Figure 4.10 shows that the output conductance is highest in the linear zone and lowest in the saturation region. The recommended Fe HBOX TFET is perfect for Analog/RF operations regarding its high transconductance and output conductance [102–108].

A comparison of the projected Fe HBOX TFET with the distinct TFETs in the literature is presented in Table 4.2. From the performance parameters, it is evident that the characteristics of the proposed device are superior.

Equation (4.9) describes the device's signal enhancement capacity across an envisaged frequency range, ft, which is the cut-off frequency. The transit time exposed by the proposed device is portrayed in Figure 4.11 and it has been noticed that the device's operating speed is high.

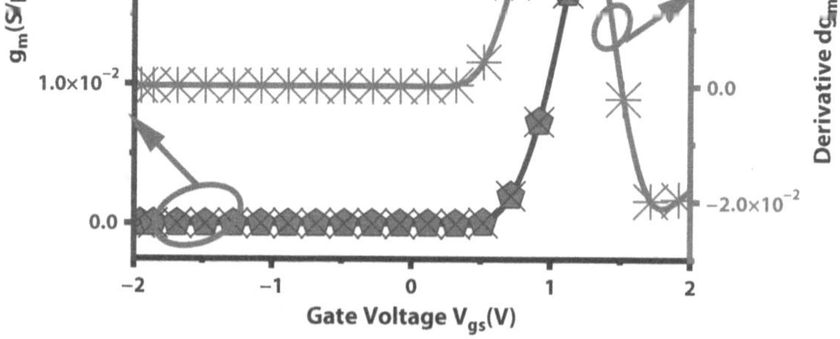

Figure 4.9 Transconductance.

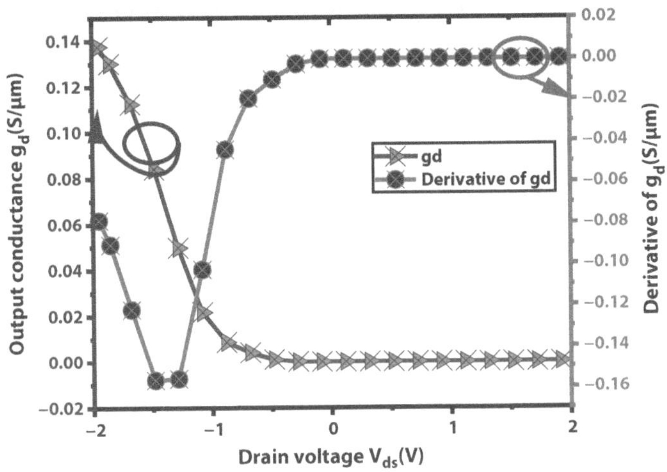

Figure 4.10 Output transconductance.

Table 4.2 Analyzing the retrieved properties of the abovementioned ferroelectric gate—HBOX—and comparing them to those of the other TFETs.

Parameters ➡	I_{OFF} (A/μm)	I_{ON} (A/μm)	SS (mV/dec)	I_{ON}/I_{OFF}
[28]	10^{-17}	10^{-6}	36	10^{11}
[29]	10^{-13}	10^{-6}	–	10^{7}
[30]	2.9×10^{-15}	1.4×10^{-6}	45	0.5×10^{9}
Proposed TFET	7.3×10^{-14}	5.5×10^{-4}	38.5	0.75×10^{10}

$$f_t = \frac{g_m}{2\pi C_{gg}} \tag{4.9}$$

Figure 4.12 depicts the diagram of the cut-off frequency. Owing to their lower capacitance levels and greater transconductance amounts, electronics can operate without attenuation in the 10^{14} Hz frequency range [109–115]. Similarly, another important element for analog/RF applications is the

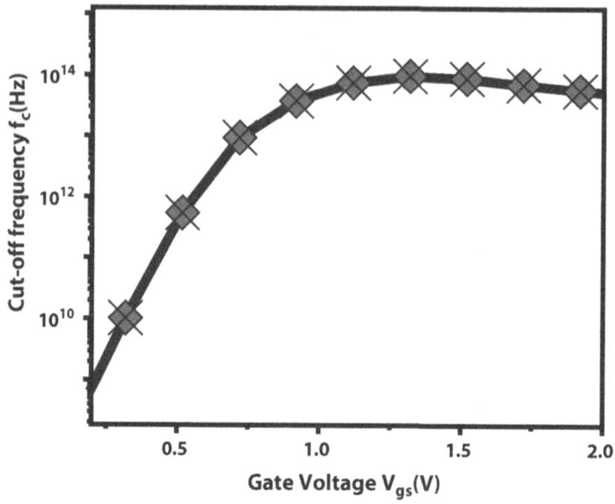

Figure 4.11 Cut-off frequency.

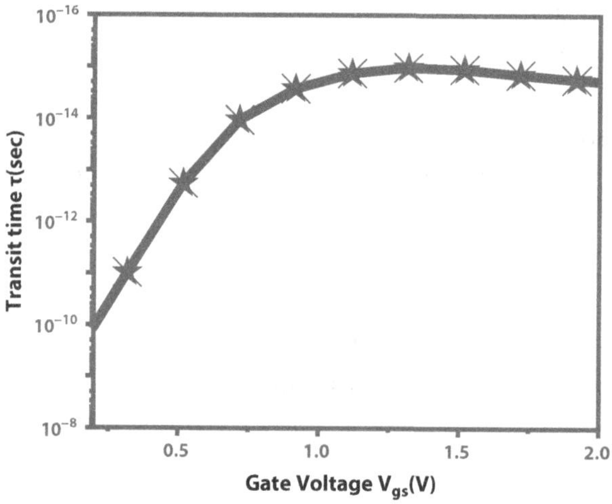

Figure 4.12 Transit time.

transit time τ, which is inversely related to ft and is given by Equation (4.10).

$$\tau = \frac{1}{20 \times \pi \times f_t} \tag{4.10}$$

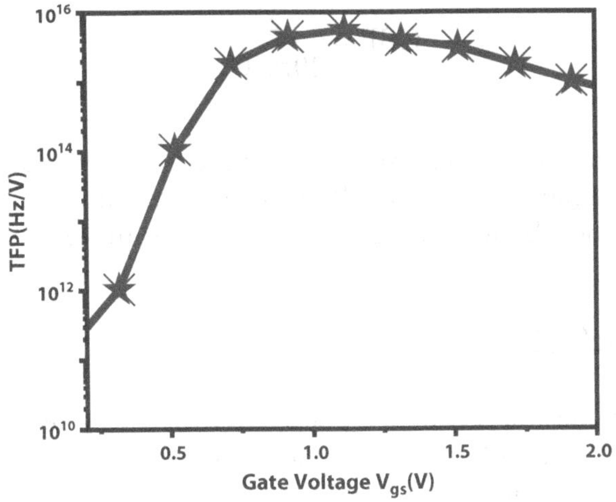

Figure 4.13 TFP.

The Transistor Frequency Product (TFP) is an essential element in evaluating the Analog/RF performance of Tunnel FETs in the high frequency region of operation. It is presented as [30–34]. The TFP exhibited by the proposed device is portrayed in Figure 4.13 and it has been observed that the device operates at high frquencies.

Figure 4.14 portrays the Gain Bandwidth Product (GBP) that represents the relationship between bandwidth and gain. This component appears as Equation (4.11) [25].

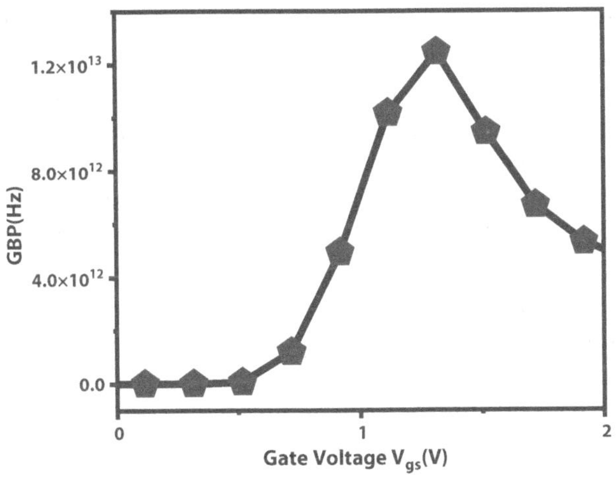

Figure 4.14 GBP.

$$GBP = \frac{g_m}{20\pi XC_{gd}} \tag{4.11}$$

From the above characteristics such as the Cut-off frequency, Transit time, TFP and gain, the stated Fe HBOX TFET can be used not only for DC applications, but also for Analog/RF applications [116–120].

4.4 Conclusion

This chapter illustrates a doped-pocket ferroelectric gate HBOX-TFET structure that is ideal for both DC and analog/RF applications. The projected TFET presented an SS of 38.5 mV/dec, ON-state current ratio of 0.74×10^{10}, OFF-state current of 7.39×10^{-14} A/μm, and ON-state current of 5.5×10^{-4} A/μm. The operational speed is excellent because of the reduced parasitic capacitances (2.4×10^{-15} F/μm). The device's conductance helps assess the performance of analog and radio frequency systems. This device can now be used for admiringly low-power DC and analog/RF applications.

4.4.1 Summary

This chapter introduces a Hetero-Buried Oxide (HBOX)–Ferroelectric Tunnel FET. Various electrical characteristics were tuned in relation to silicon and BOX layer thickness adjustments, including ON/OFF-state currents, and the I_{ON}/I_{OFF} ratio. A negative-capacitance-based ferroelectric gate stack is utilized to achieve a steep Subthreshold Slope (SS). To maintain compatibility with the standard FET manufacturing process, a ferroelectric (Fe) layer made of silicon-doped hafnium oxide was used at the bottom of the gate. The thickness of the Fe layer was carefully adjusted for optimum Analog and DC performances. The transfer properties of the HBOX-Fe TFETs with and without an Fe layer were explored. The subthreshold slopes are 44.5 mV/dec in the absence of ferroelectric material, and the threshold voltage Vth, I_{ON}, and I_{OFF} currents were obtained as 0.72 V, 10^{-13}A/μm, and 10^{-4}A/μm, correspondingly. Compared to traditional TFETs, electrostatic bonding can generate lesser capacitances, which follow the channel's path due to the negative capacitance exposed by the Si:HfO2 layer coating at the gate's exterior. From now on, the device's performance in Fe-TFETs will be exceptional.

When Fe material was added to the gate stack, the threshold voltage and off-state current lowered from 0.72 V to 0.69 V and 10^{-14}A/μm, respectively. By boosting the subthreshold slope (SS) to 38.5 mV/dec boosts the on-state current to 5.5×10^{-4} A/μm. This was due to the elevated surface potential (φs) throughout the gate stack. For ultralow-power operations, the voltage at the gate terminal is boosted via negative capacitance phenomena. The comparison of the projected Fe HBOX TFET with the distinct TFETs in the literature are compared and are projected in Table 4.2. From the performance parameters, it is evident that the characteristics of the proposed device are superior.

References

1. Mohankumar, N., Syamal, B., Sarkar, C.K., Influence of channel and gate engineering on the analog and RF performance of DG MOSFETs. *IEEE Electron Device Lett.*, 57, 4, 820–826, 2010.
2. Kilchytska, V., *et al.*, Influence of device engineering on the analog and RF performances of SOI MOSFETs. *IEEE Trans. Electron Devices*, 50, 3, 577–588, 2003.
3. Colinge, J.-P. (Ed.), *FinFETs and other multi-gate transistors*, vol. 73, p. 82, New York Springer, 2008.
4. Bangsaruntip, S., *et al.*, Universality of short-channel effects in undoped-body silicon nanowire MOSFETs. *IEEE Electron Device Lett.*, 31, 9, 903–905, 2010.
5. Koswatta, S.O., Lundstrom, M.S., Nikonov, D.E., Performance comparison between pin tunneling transistors and conventional MOSFETs. *IEEE Trans. Electron Devices*, 56, 3, 456–465, 2009.
6. Reddick, W.M. and Amaratunga, G.A.J., Silicon surface tunnel transistor. *Appl. Phys. Lett.*, 67, 4, 494–496, 1995.
7. Choi, W.Y., *et al.*, Tunneling field-effect transistors (TFETs) with subthreshold swing (SS) less than 60 mV/dec. *IEEE Electron Device Lett.*, 28, 8, 743–745, 2007.
8. Ionescu, A.M. and Riel, H., Tunnel field-effect transistors as energy-efficient electronic switches. *Nature*, 479, 7373, 329–337, 2011.
9. Boucart, K. and Ionescu, A.M., Double-gate tunnel FET with high-$\kappa $ gate dielectric. *IEEE Trans. Electron Devices*, 54, 7, 1725–1733, 2007.
10. Agarwal, S., Klimeck, G., Luisier, M., Leakage-reduction design concepts for low-power vertical tunneling field-effect transistors. *IEEE Electron Device Lett.*, 31, 6, 621–623, 2010.

11. Ahmad, S., Alam, N., Hasan, M., Robust TFET SRAM cell for ultra-low power IoT applications. *AEU Int. J. Electron. Commun.*, 89, 70–76, 2018.
12. Chattopadhyay, A. and Mallik, A., Impact of a spacer dielectric and a gate overlap/underlap on the device performance of a tunnel field-effect transistor. *IEEE Trans. Electron Devices*, 58, 3, 677–683, 2011.
13. Chen, S., Liu, H., Wang, S., Li, W., Wang, X., Zhao, L., Analog/RF Performance of T-Shape Gate Dual-Source Tunnel Field-Effect Transistor. *Nanoscale Res. Lett.*, 13, 1, 321, 2018.
14. Abdi, D.B. and Kumar, M.J., In-built N+ pocket p-n-p-n tunnel field-effect transistor. *IEEE Electron Device Lett.*, 35, 12, 1170–1172, 2014.
15. Krishnamohan, T., Kim, D., Raghunathan, S., Saraswat, K., Doublegate strained-ge heterostructure tunneling FET (TFET) with record high drive currents and <60 mV/dec subthreshold slope. *IEDM Tech. Dig.*, San Francisco, CA, USA, pp. 1–3, 2008.
16. Hraziia, Vladimirescu, A., Amara, A., Anghe, C., An analysis on the ambipolar current in Si double-gate tunnel FETs. *Solid State Electron.*, 70, 67–72, 2012.
17. Wan, J., Le Royer, C., Zaslavsky, A., Cristoloveanu, S., Tunneling FETs on SOI" Suppression of ambipolar leakage, low-frequency noise behavior, and modeling. *Solid State Electron.*, 65–66, 226–233, 2011.
18. Ohno, T., *et al.*, Experimental 0.25-/spl mu/m-gate fully depleted CMOS/SIMOX process using a new two-step LOCOS isolation technique. *IEEE Trans. Electron Devices*, 42, 8, 1481–1486, 1995.
19. Verreck, D., *et al.*, Quantum mechanical performance predictions of pnin versus pocketed line tunnel field-effect transistors. *IEEE Trans. Electron Devices*, 60, 7, 2128–2134, 2013.
20. Sahay, S. and Kumar, M.J., Controlling the Drain Side Tunnelling Width to Reduce Ambipolar Current in Tunnel FETs Using Hetero-dielectric box. *IEEE Trans. Electron Devices*, 62, 11, 82, 2015.
21. Meriga, S. and Bhowmick, B., Optimization of Hetero Buried Oxide Doped-Pocket Gate-Engineered Tunnel FET Structures, in: *Advanced Field-Effect Transistors*, pp. 111–126, CRC Press, 2023.
22. Meriga, S. and Bhowmick, B., Robustness of Raised Buried Oxide Ferro Electric Tunnel FET in presence of Temperature and Traps and its Analog/RF Performance. *Appl. Phys. A*, 130, 4, 223, 2024.
23. Meriga, S. and Bhowmick, B., Deep Insight into Raised Buried Oxide SOI-Fe TFET and It's Analog/RF and Linearity Performance Parameters. *Trans. Electr. Electron. Mater.*, 24, 6, 589–602, 2023.
24. TCAD Sentaurus Device User's Manual, Synopsys, Mountain View, CA, USA, 2010.
25. Meriga, S. and Bhowmick, B., Investigation of a dual gate pocket-doped drain engineered tunnel FET and its reliability issues. *Appl. Phys. A*, 129, 2, 104, 2023.

26. Meriga, S. and Bhowmick, B., Compact drain current model of a double-gate raised buried oxide TFET for integrated circuit application. *J. Comput. Electron.*, 22, 5, 1443–1452, 2023.
27. Meriga, S. and Bhowmick, B., Schrodinger Wave equation-based Drain Current Model for Nano Scale device. *IEEE 3rd Global Conference for Advancement in Technology (GCAT)*, pp. 1–4, 2022, Oct. 7.
28. Sahay, S. and Kumar, M.J., Controlling the drain side tunneling width to reduce ambipolar current in tunnel FETs using heterodielectric BOX. *IEEE Trans. Electron Devices*, 62, 11, 3882–3886, 2015.
29. Beniwal, S. and Saini, G., L-shaped tunnelling field effect transistor with hetero-gate dielectric and hetero dielectric box. *2019 3rd International Conference on Trends in Electronics and Informatics (ICOEI)*, IEEE, 2019.
30. Pindoo, I.A. and Sinha, S.K., Hetero-gate dielectric with hetero dielectric BOX for suppressing ambipolar current in tunnel FETs. *2020 International Conference on Intelligent Engineering and Management (ICIEM)*, 2020.
31. Singh, S. and Raj, B., A 2-D Analytical Surface Potential and Drain current Modeling of Double-Gate Vertical t-shaped Tunnel FET. *J. Comput. Electron.*, Springer, 19, 1154–1163, Apl. 2020.
32. Singh, J. and Raj, B., An Accurate and Generic Window function for Non-linear Memristor Model. *J. Comput. Electron.*, Springer, 18, 2, 640–647, June 2019.
33. Kaur, M., Gupta, N., Kumar, S., Raj, B., Singh, A.K., Comparative RF and Crosstalk Analysis of Carbon Based Nano Interconnects. *IET Circuits Devices Syst.*, 15, 6, 493–503, Feb. 2021.
34. Kandasamy, N., Ahmad, F., Ajitha, D., Raj, B., Telagam, N., Quantum Dot Cellular Automata based Scan Flip Flop and Boundary Scan Register. *IETE J. Res.*, 66, 535–548, 2020.
35. Singh, J. and Raj, B., Design and Investigation of 7T2M NVSARM with Enhanced Stability and Temperature Impact on Store/Restore Energy. *IEEE Trans. Very Large Scale Integr. Syst.*, 27, 6, 1322 – 1328, June 2019.
36. Bhardwaj, A.K., Gupta, S., Raj, B., Singh, A., Impact of Double Gate Geometry on the Performance of Carbon Nanotube Field Effect Transistor Structures for Low Power Digital Design. *Comput. Theor. Nanosci.*, ASP, 16, 1813–1820, 2019.
37. Jain, N. and Raj, B., Thermal Stability Analysis and Performance Exploration of Asymmetrical Dual-k underlap Spacer (ADKUS) SOI FinFET for Security and Privacy Applications. *Indian J. Pure Appl. Phys. (IJPAP)*, 57, 352–360, May 2019.
38. Jain, N. and Raj, B., Dual-k Spacer Region Variation at the Drain Side of Asymmetric SOI FinFET Structure: Performance Analysis towards the Analog/RF Design Applications. *J. Nanoelectron. Optoelectron.*, American Scientific Publishers, 14, 349–359, March 2019.

39. Singh, J., Sharma, S., Raj, B., Khosla, M., Analysis of barrier layer thickness on performance of In1-xGaxAs based Gate Stack Cylindrical Gate Nanowire MOSFET. *JNO*, ASP, 13, 1473–1477, Oct. 2018.
40. Jain, N. and Raj, B., Analysis and Performance Exploration of High-k SOI FinFETs Over the Conventional Low-k SOI FinFET toward Analog/RF Design. *J. Semicond. (JoS)*, IOP Science, 39, 12, 124002-1–7, Dec. 2018.
41. Goyal, C., Ubhi, J.S., Raj, B., A reliable leakage reduction technique for approximate full adder with reduced ground bounce noise. *J. Math. Probl. Eng.*, Hindawi, 2018, Article ID 3501041, 16, 15 Oct. 2018.
42. Anuradha, Singh, J., Raj, B., Khosla, M., Design and Performance Analysis of Nano-scale Memristor-based Nonvolatile SRAM. *J. Sens. Lett.*, American Scientific Publishers, 16, 798–805, Oct. 2018.
43. Wadhwa, G. and Raj, B., Parametric Variation Analysis of Charge-Plasma-based Dielectric Modulated JLTFET for Biosensor Application. *IEEE Sens. J.*, 18, 15, 6070 - 6077, AUGUST 1, 2018, (SCI).
44. Singh, J. and Raj, B., Comparative Analysis of Memristor Models for Memories Design. *JoS*, IoP, 39, 7, 074006-1–12, July 2018.
45. Yadav, D., Chouhan, S.S., Vishvakarma, S.K., Raj, B., Application Specific Microcontroller Design for IoT based WSN. *Sens. Lett.*, ASP, 16, 374–385, May 2018.
46. Singh, G., Sarin, R.K., Raj, B., Fault-Tolerant Design and Analysis of Quantum-Dot Cellular Automata Based Circuits. *IEEE/IET Circuits Devices Syst.*, 12, 638–644, 2018.
47. Singh, J. and Raj, B., Modeling of Mean Barrier Height Levying Various Image Forces of Metal Insulator Metal Structure to Enhance the Performance of Conductive Filament Based Memristor Model. *IEEE Nanotechnol.*, 17, 2, 268–267, March 2018.
48. Jain, A., Sharma, S., Raj, B., Analysis of Triple Metal Surrounding Gate (TM-SG) III-V Nanowire MOSFET for Photosensing Application. *Opto-Electron. J.*, Elsevier, 26, 2, 141–148, May 2018.
49. Jain, A., Sharma, S., Raj, B., Design and Analysis of High Sensitivity Photosensor Using Cylindrical Surrounding Gate MOSFET For Low Power Sensor Applications. *Eng. Sci. Technol. Int. J.*, Elsevier's, 19, 4, 1864–1870, December 2016.
50. Saiphani Kumar, G., Singh, A., Raj, B., Design and Analysis of Gate All Around CNTFET based SRAM cell Design. *J. Comput. Electron.*, Springer, 17, 1, 138–145, March 2018.
51. Singh, G.P., Sohi, B.S., Raj, B., Material Properties Analysis of Graphene Base Transistor (GBT) for VLSI Analog Circuits. *Indian J. Pure Appl. Phys. (IJPAP)*, 55, 896–902, Dec. 2017.
52. Kumar, S. and Raj, B., Estimation of Stability and Performance metric for Inward Access Transistor based 6T SRAM Cell Design using n-type/p-type DMDG-GDOV TFET. *IEEE VLSI Circuits Syst. Lett.*, 3, 2, 25–39, June 2017.

53. Sharma, S., Kumar, A., Pattanaik, M., Raj, B., Forward Body Biased Multimode Multi-Threshold CMOS Technique for Ground Bounce Noise Reduction in Static CMOS Adders. *Int. J. Inf. Electron. Eng.*, 3, 3, 567–572, 2013.
54. Singh, H., Kumar, P., Raj, B., Performance Analysis of Majority Gate SET Based 1-bit Full Adder. *Int. J. Comput. Commun. Eng. (IJCCE)*, IACSIT Press Singapore, 2, 4, 84, 2013, ISSN: 2010-3743.
55. Bhardwaj, A.K., Gupta, S., Raj, B., Investigation of Parameters for Schottky Barrier (SB) Height for Schottky Barrier Based Carbon Nanotube Field Effect Transistor Device. *J. Nanoelectron. Optoelectron.*, ASP, 15, 783–791, July 2020.
56. Bansal, P. and Raj, B., MEMRISTOR: A VERSATILE NONLINEAR MODEL FOR DOPANT DRIFT AND BOUNDARY ISSUES. *JCTN*, American Scientific Publishers, 14, 5, 2319–2325, May 2017.
57. Jain, N. and Raj, B., An Analog and Digital Design Perspective Comprehensive Approach on Fin-FET (Fin-Field Effect transistor) Technology - A Review. *Rev. Adv. Sci. Eng. (RASE)*, ASP, 5, 1–14, 2016.
58. Raj, B., Saxena, A.K., Dasgupta, S., Analytical Modeling for the Estimation of Leakage Current and Subthreshold Swing Factor of Nanoscale Double Gate FinFET Device. *Microelectron. Int.*, UK, 26, 53–63, 2009.
59. Badgujjar, S., Singh, S., Wadhwa, G., Raj, B., An Analytical Modeling for Dual Source Vertical Tunnel Field Effect Transistor. *Int. J. Recent Technol. Eng. (IJRTE)*, 8, 2, 603–608, July 2019.
60. Singh, S. and Raj, B., Design and analysis of hetrojunction Vertical T-shaped Tunnel Field Effect Transistor. *J. Electron. Mater.*, Springer, 48, 10, 6253–6260, October 2019.
61. Goyal, C., Ubhi, J.S., Raj, B., A Low Leakage CNTFET based Inexact Full Adder for Low Power Image Processing Applications. *Int. J. Circuit Theory Appl.*, Wiley, 47, 9, 1446–1458, September 2019.
62. Raj, B., Saxena, A.K., Dasgupta, S., A Compact Drain Current and Threshold Voltage Quantum Mechanical Analytical Modeling for FinFETs. *J. Nanoelectron. Optoelectron. (JNO)*, USA, 3, 2, 163–170, 2008.
63. Wadhwa, G. and Raj, B., An Analytical Modeling of Charge Plasma based Tunnel Field Effect Transistor with Impacts of Gate underlap Region. *Superlattices Microstruct.*, Elsevier, 142, 106512, June 2020.
64. Singh, S. and Raj, B., Modeling and Simulation analysis of SiGe hetrojunction Double GateVertical t-shaped Tunnel FET. *Superlattices Microstruct.*, Elsevier, 142, 106496, June 2020.
65. Jain, N. and Raj, B., Device and Circuit Co-Design Perspective Comprehensive Approach on FinFET Technology - A Review. *J. Electron Devices*, 23, 1, 1890–1901, 2016.
66. Kumar, S. and Raj, B., Analysis of I_{ON} and Ambipolar Current for Dual-Material Gate-drain Overlapped DG-TFET. *J. Nanoelectron. Optoelectron.*, American Scientific Publishers, USA, 11, 323–333, June 2016.

67. Anjum, N., Bali, T., Raj, B., Design And Simulation Of Handwritten Multiscript Character Recognition. *Int. J. Adv. Res. Comput. Commun. Eng.*, 2, 7, 2544–2549, July 2013.
68. Singh, K. and Raj, B., Performance and analysis of temperature dependent Multi-walled Carbon Nanotubes as Global Interconnects at different technology nodes. *J. Comput. Electron.*, Springer, 14, 2, 469–476, June 2015.
69. Kumar, S. and Raj, B., Compact channel potential analytical Modeling of DG-TFET based on Evanescent-mode Approach. *J. Comput. Electron.*, Springer, 14, 2, 820–827, July 2015.
70. Singh, K. and Raj, B., Temperature Dependent Modeling and Performance Evaluation of Multi-Walled CNT and Single-Walled CNT as Global Interconnects. *J. Electron. Mater.*, Springer, 44, 12, 4825–4835, Dec. 2015.
71. Sharma, V.K., Pattanaik, M., Raj, B., INDEP approach for leakage reduction in nanoscale CMOS circuits. *Int. J. Electron.*, Taylor & Francis, 102, 2, 200–215, 2014.
72. Singh, K. and Raj, B., Influence of Temperature on MWCNT bundle, SWCNT bundle and Copper interconnects for Nanoscaled Technology nodes. *J. Mater. Sci.: Mater. Electron.*, Springer, 26, 8, 6134–6142, 2015.
73. Anjum, N., Bali, T., Raj, B., Design and Simulation of Handwritten Gurumukhi and Devanagri Numerical Recognition. *Int. J. Comput. Appl.*, Foundation of Computer Science, New York, USA, 73, 12, 16–21, 2013.
74. Khandelwal, S., Gupta, V., Raj, B., Gupta, R.D., Process Variability Aware Low Leakage Reliable Nano Scale DG-FinFET SRAM Cell Design Technique. *J. Nanoelectron. Optoelectron.*, 10, 6, 810–817, December 2015.
75. Sharma, V.K., Pattanaik, M., Raj, B., ONOFIC Approach: Low Power High Speed Nanoscale VLSI Circuits Design. *Int. J. Electron.*, Taylor & Francis, 101, 1, 61–73, 2014.
76. Khandelwal, S., Raj, B., Gupta, R.D., FinFET based 6T SRAM Cell Design: Analysis of Performance Metric, Process Variation and Temperature Effect. *J. Comput. Theor. Nanosci.*, ASP, USA, 12, 2500–2506, 2015.
77. Singh, S., Yadav, S., Rahul, J., Srivastava, A., Raj, B., Impact of HfO_2 in Graded Channel Dual Insulator Double Gate MOSFET. *J. Comput. Theor. Nanosci.*, American Scientific Publishers, 12, 6, 950–953, April 2015.
78. Sharma, V.K., Pattanaik, M., Raj, B., PVT variations aware low leakage INDEP approach for nanoscale CMOS Circuits. *Microelectron. Reliab.*, Elsevier, 54, 90–99, 2014.
79. Raj, B., Saxena, A.K., Dasgupta, S., Quantum Mechanical Analytical Modeling of Nanoscale DG FinFET: Evaluation of Potential, Threshold Voltage and Source/Drain Resistance. *Elsevier's J. Mater. Sci. Semicond. Process.*, Elsevier, 16, 4, 1131–1137, 2013.
80. Gopal, M., Prasad, S.S.D., Raj, B., 8T SRAM Cell Design for Dynamic and Leakage Power Reduction. *Int. J. Comput. Appl.*, Foundation of Computer Science, New York, USA, 71, 9, 43–48, June 2013.

81. Pattanaik, M., Raj, B., Sharma, S., Kumar, A., Diode Based Trimode Multi-Threshold CMOS Technique for Ground Bounce Noise Reduction in Static CMOS Adders. *Adv. Mater. Res.*, Trans Tech Publications, Switzerland, 548, 885–889, 2012.
82. Raj, B., Saxena, A.K., Dasgupta, S., Nanoscale FinFET Based SRAM Cell Design: Analysis of Performance metric, Process variation, Underlapped FinFET and Temperature effect. *IEEE Circuits Syst. Mag.*, 11, 2, 38–50, 2011.
83. Sharma, V.K., Pattanaik, M., Raj, B., Leakage Current ONOFIC Approach for Deep Submicron VLSI Circuit Design. *Int. J. Electr. Comput. Electron. Commun. Eng.*, World Academy of Sciences, Engineering and Technology, 7, 4, 239–244, 2013.
84. Kaur, P., Gill, S.S., Raj, B., Comparative Analysis of OFETs Materials and Devices for Sensor Applications. *J. Silicon*, Springer, 14, 4463–4471, 2022.
85. Sharma, S.K., Kumar, P., Raj, B., Raj, B., $In_{1-x}Ga_xAs$ Double Metal Gate-Stacking Cylindrical Nanowire MOSFET for highly sensitive Photo detector. *J. Silicon*, Springer, 14, 3535–3541, 2022.
86. Raj, B., Saxena, A.K., Dasgupta, S., Analytical Modeling of Quasi Planar Nanoscale Double Gate FinFET with Source/Drain Resistance and field dependent carrier mobility: A quantum mechanical study. *J. Comput. (JCP)*, Academy Publisher, FINLAND, 4, 9, 1–8, 2009.
87. Bhushan, S., Khandelwal, S., Raj, B., Analyzing Different Mode FinFET based Memory Cell at Different Power Supply for Leakage Reduction, in: *Seventh International Conference on Bio-Inspired Computing: Theories and Application, (BIC-TA 2012)*, Advances in Intelligent Systems and Computing, vol. 202, pp. 89–100, 2013.
88. Singh, J. and Raj, B., Temperature Dependent Analytical Modeling and Simulations of Nanoscale Memristor. *Eng. Sci. Technol. Int. J.*, Elsevier's, 21, 862–868, Oct. 2018.
89. Singh, S., Bala, S., Raj, B., Raj, B., Improved Sensitivity of Dielectric Modulated Junctionless Transistor for Nanoscale Biosensor Design. *Sens. Lett.*, ASP, 18, 328–333, Apl. 2020.
90. Kumar, V., Vishvakarma, S.K., Raj, B., Design and Performance Analysis of ASIC for IoT Applications. *Sens. Lett.*, ASP, 18, 31–38, Jan. 2020.
91. Jaiswal, A., Sarin, R.K., Raj, B., Sukhija, S., A Novel Circular Slotted Microstrip-fed Patch Antenna with Three Triangle Shape Defected Ground Structure for Multiband Applications. *Adv. Electromagn. (AEM)*, 7, 3, 56–63, Aug. 2018.
92. Wadhwa, G. and Raj, B., Label Free Detection of Biomolecules using Charge-Plasma-Based Gate Underlap Dielectric Modulated Junctionless TFET. *J. Electron. Mater. (JEMS)*, Springer, 47, 8, 4683–4693, August 2018.
93. Singh, G., Sarin, R.K., Raj, B., Design and Performance Analysis of a New Efficient Coplanar Quantum-Dot Cellular Automata Adder. *Indian J. Pure Appl. Phys. (IJPAP)*, 55, 97–103, February 2017.

94. Kaur, P., Pandey, V., Raj, B., Comparative Study of Efficient Design, Control and Monitoring of Solar Power using IoT. *Sens. Lett.*, ASP, 18, 419–426, May 2020.
95. Bhardwaj, A.K., Gupta, S., Raj, B., Development & Analysis of compact model for Double Gate Schottky Barrier CNTFET. *J. Nanoelectron. Optoelectron.*, ASP, 15, 1199–1208, Aug. 2020.
96. Wadhwa, G., Kamboj, P., Singh, J., Raj, B., Design and Investigation of Junctionless DGTFET for Biological Molecule Recognition. *Trans. Electr. Electron. Mater.*, Springer, 22, 282–289, 2021.
97. Verma, S.K., Singh, S., Wadhwa, G., Raj, B., Detection of Biomolecules using Charge-Plasma based Gate Underlap Dielectric Modulated Dopingless TFET. *Trans. Electr. Electron. Mater. (TEEM)*, Springer, 21, 528–535, June 2020.
98. Jain, N. and Raj, B., Impact of underlap spacer region variation on Electrostatic and analog/RF performance of symmetrical high-k SOI FinFET at 20 nm channel length. *J. Semicond. (JoS)*, IOP Science, 38, 12, 122002, Dec. 2017.
99. Singh, S. and Raj, B., Analytical Modeling and Simulation analysis of T-shaped III-V heterojunction Vertical T-FET. *Superlattices Microstruct.*, Elsevier, 147, 106717, Nov 2020.
100. Singh, G., Sarin, R.K., Raj, B., Design and Analysis of Area Efficient QCA Based Reversible Logic Gates. *J. Microprocess. Microsys.*, Elsevier, 52, 59–68, May 2017.
101. Singh, S. and Raj, B., Parametric variation analysis on hetero-junction Vertical t-shape TFET for Supressing ambipolar conduction. *Indian J. Pure Appl. Phys.*, 58, 478–485, June 2020.
102. Badgujjar, S., Singh, S., Wadhwa, G., Raj, B., Design and Analysis of Dual Source Vertical Tunnel Field Effect Transistor for high performance. *Trans. Electr. Electron. Mater.*, Springer, 21, 74–82, Oct. 2019.
103. Kaur, M., Gupta, N., Kumar, S., Raj, B., Singh, A.K., RF Performance Analysis of Intercalated Graphene Nanoribbon Based Global Level Interconnects. *J. Comput. Electron.*, Springer, 19, 1002–1013, June 2020.
104. Wadhwa, G. and Raj, B., Design and Performance Analysis of Junctionless TFET Biosensor for high sensitivity. *IEEE Nanotechnol.*, 18, 567 - 574, 2019.
105. Singh, J. and Raj, B., Enhanced Nonlinear Memristor Model Encapsulating Stochastic Dopant Drift. *JNO*, ASP, 14, 958–963, 2019.
106. Bhardwaj, A., Kumar, P., Raj, B., Anand, S., Design and Performance Optimization of doping-less Vertical Nanowire TFET using Gate Stack Technique. *J. Electron. Mater. (JEMS)*, Springer, 41, 7, 4005–4013, 2022.
107. Singh, J. and Raj, B., Tunnel Current Model of Asymmetric MIM Structure Levying Various Image Forces to Analyze the Characteristics of Filamentary Memristor. *Appl. Phys. A*, Springer, 125, 3, 203.1 to 203.11, Feb. 2019.

108. Goyal, C., Ubhi, J.S., Raj, B., Low Leakage Zero Ground Noise Nanoscale Full Adder using Source Biasing Technique. *J. Nanoelectron. Optoelectron.*, American Scientific Publishers, 14, 360–370, March 2019.
109. Singh, G., Sarin, R.K., Raj, B., A Novel Robust Exclusive-OR Function Implementation in QCA Nanotechnology with Energy Dissipation Analysis. *J. Comput. Electron.*, Springer, 15, 2, 455–465, June 2016.
110. Wadhwa, G., Kamboj, P., Raj, B., Design optimisation of junctionless TFET biosensor for high sensitivity. *Adv. Nat. Sci.: Nanosci. Nanotechnol.*, 10, 045001, 7, 2019.
111. Bansal, P. and Raj, B., Memristor Modeling and Analysis for Linear Dopant Drift Kinetics. *J. Nanoeng. Nanomanuf.*, American Scientific Publishers, 6, 1–7, 2016.
112. Ashima, Vaithiyanathan, D., Raj, B., Performance Analysis of Charge Plasma induced Graded Channel Si nanotube. *J. Eng. Res. (JER)*, EMSME Special Issue, 9, 146–154, Aug 2021.
113. Tomar, A.S., Magraiya, V.K., Raj, B., Scaling of Access and Data Transistor for High Performance DRAM Cell Design. *Quantum Matter*, 2, 412–416, Oct. 2013.
114. Jain, N. and Raj, B., Parasitic Capacitance and Resistance Model Development and Optimization of Raised Source/Drain SOI FinFET Structure for Analog Circuit Applications. *J. Nanoelectron. Optoelectronins*, ASP, USA, 13, 531–539, Apl. 2018.
115. Singh, S., Vishvakarma, S.K., Raj, B., Analytical Modeling of Split-Gate Junction-Less Transistor for a Biosensor Application. *Sens. Bio-Sens.*, Elsevier, 18, 31–36, April 2018.
116. Gopal, M. and Raj, B., Low Power 8T SRAM Cell Design for High Stability Video Applications. *ITSI Trans. Electr. Electron. Eng.*, 1, 5, 91–97, 2013.
117. Raj, B., Mitra, J., Bihani, D.K., Rangharajan, V., Saxena, A.K., Dasgupta, S., Analysis of Noise Margin, Power and Process Variation for 32 nm FinFET Based 6T SRAM Cell. *J. Comput. (JCP)*, Academy Publisher, FINLAND, 5, 6, 1–8, 2010.
118. Sharma, D., Mehra, R., Raj, B., Comparative Analysis of Photovoltaic Technologies for High Efficiency Solar Cell Design. *Superlattices Microstruct.*, Elsevier, 153, 106861, May 2021.
119. Kaur, P., Buttar, A.S., Raj, B., A comprehensive Analysis of Nanoscale Transistor based Biosensor: A Review. *Indian J. Pure Appl. Phys.*, 59, 304–318, April 2021.
120. Yadav, D., Raj, B., Raj, B., Design and Simulation of Low Power Microcontroller for IoT Applications. *J. Sens. Lett.*, ASP, 18, 401–409, May 2020.

5

Ferroelectric Material-Based Field Effect Transistor and Its Applications

Avinash Kumar* **and Balwinder Raj**

Nanoelectronic Research Lab, Dept. of ECE, NIT Jalandhar, Punjab, India

Abstract

A thorough analysis of ferroelectric materials in cutting-edge nanoelectronic devices is presented herein. Ferroelectric-material-based FETs describe how mitigating the problem arises in conventional CMOS. A ferroelectric polarization field is injected into a field-effect transistor (FET) to control semiconductor carriers, resulting in a ferroelectric field-effect transistor (FeFET). This article compares several ferroelectric materials' characteristics, such as HfO_2, PZT, and other 2-D materials, and shows how ferroelectricity varies with temperature. After outlining the distinct electrical and physical characteristics of ferroelectric materials, they concentrated on different structures of field-effect transistors with ferroelectric materials based on 2D and 3D structures. FeFET's different architectures, such as MFS and MFIS, and applications based on these structures are also given. The various FeFET structures were compared with each other in terms of performance and limitations in the real world. Other structures of field-effect transistors have been proposed based on ferroelectric materials for various applications, such as memory and sensors. To pique the attention of scientists interested in the continued development of this rapidly developing field of study, this study thoroughly analyzes the current and upcoming challenges in ferroelectric-based nanoelectronics. FeFETs are appealing for cutting-edge electrical and optoelectronic applications, such as developing memory, artificial neural networks, high-performance photodetectors, and intelligent sensors due to their ferroelectric and semiconductor flexible interactions. In this review, we also discuss the potential applications of FeFETs in the future for better device operation and which ferroelectric materials have the best FeFET structure and can retain ferroelectricity up to 1 nm technology.

**Corresponding author*: avinashk.ec.23@nitj.ac.in

Balwinder Raj, Shiromani Balmukund Rahi and Nandakishor Yadav (eds.) FeFET Devices, Trends, Technology and Applications, (103–130) © 2025 Scrivener Publishing LLC

Keywords: Ferroelectricity, polarization, gate-all-around, sensor, memory, lead zirconium titanate, curie temperature

5.1 Introduction

Improvements in semiconductor technology have rapidly progressed toward higher packing densities, improved functionality, faster circuit speeds, and lower power dissipation since the introduction of integrated circuits (ICs) in the late 1950s. Moore's first law states that the widely used silicon Complementary Metal Oxide Semiconductor (CMOS) technology will double in capability every 18 months (or halve in feature size by a factor of 1.41). The persistent dominance of CMOS in the Very Large Scale Integration (VLSI) sector has emerged as the foundation of contemporary electronics and opened the way for the expanding market for portable devices. However, scaling MOS technology has been a major challenge. Over the last four decades, field-effect transistors have continued to scale, which has increased chip density and led to integrated circuit performance degradation because of the increased leakage power. Although there has been a decrease in potential, the historical trend of contracting MOS transistors has revolutionized the production of incredibly complex devices. The development of sophisticated fabrication techniques has caused CMOS transistor shrinkage into the deep submicron zone. As a result of this technological advancement, transistor dimensions are being reduced to increase the device speed [1]. Enhancing the gate controllability of nanoscale CMOS devices is critical for overcoming short-channel effects (SCEs) [2, 3]. In conventional MOSFET, obtaining a subthreshold swing (SS) below 60 mV/decade is very difficult because of Boltzmann Tyranny. Therefore, to improve the subthreshold swing, researchers have designed multigate transistors such as double-gate FET and Fin-FET to achieve a low subthreshold slope. As transistor length decreases, achieving the desired characteristics with ordinary materials becomes challenging. To achieve better device performance, the researcher introduced the concept of ferroelectrics, which provide internal voltage amplification to achieve a high drive current within the device. Furthermore, the insertion of ferroelectric negative capacitance (NC) at the gate stack has guaranteed that the device's subthreshold swing (SS) can exceed the Boltzmann limit of 60 mV/decade, and has given rise to an additional method for gate controllability [4]. The structure of this chapter is as follows. *Section 5.1* provides an introduction focusing on the history of MOSFETs and the concept of ferroelectricity. *Section 5.2* goes into great details the properties of ferroelectric materials.

Section 5.3 provides details on ferroelectricity in nanoelectronics and compares several ferroelectric materials. *Section 5.4* discusses the different Fe-FET structures and difficulties faced. *Section 5.5* explains the various applications based on ferroelectric materials and concludes with a look ahead based on recent research.

5.1.1 What is Ferroelectricity?

In most materials, including metals, opposite charges are fascinated by each other and wish to combine. The only movement they perform is when an external force is applied, which moves them to the reverse side. In contrast, spontaneous polarization occurs in ferroelectric materials because the charge separation may be maintained without an external electric field. In addition, the application of an external force has an impact: an electric field in the same direction does not have an effect; an electric field in the opposite direction results in charge-switching sides and maintains their polarization. To eliminate polarization, a coercive field is applied, and the remnant polarization, which indicates the ferroelectric's inherent polarization when no force is applied, is the two variables that characterize the essential feature of any ferroelectric [1]. As a result, ferroelectric materials are employed in many different electronic systems, including radio-frequency identification (RFID) cards, medical ultrasounds, and memory devices. The capacity of ferroelectric materials to preserve the memory state is the primary benefit. The secret to memory applications is a hysteresis window with remnant polarization. The nondestructive nature of the reading operation is the primary benefit of combining ferroelectric materials with FETs [5].

5.1.2 Effects of Temperature on Ferroelectricity

Certain materials exhibit a phenomenon known as ferroelectricity, in which an external electric field can reverse spontaneous electric polarization. Studies on Rochelle salts have led to the discovery of ferroelectric materials [6]. A polarization hysteresis loop characterizes these materials and can exhibit reversible and spontaneous polarization, even in the absence of a field. A ferroelectric material can become Para electric polarized under an electric field above its Curie temperature. This is a result of structural changing at this transition temperature. Conversely, ferroelectric materials exhibit spontaneous polarization below the Curie temperature. The Curie temperature (T_0) is the temperature below which materials commonly exhibit ferroelectricity, and above which they behave like paraelectric materials, as shown in Figure 5.1.

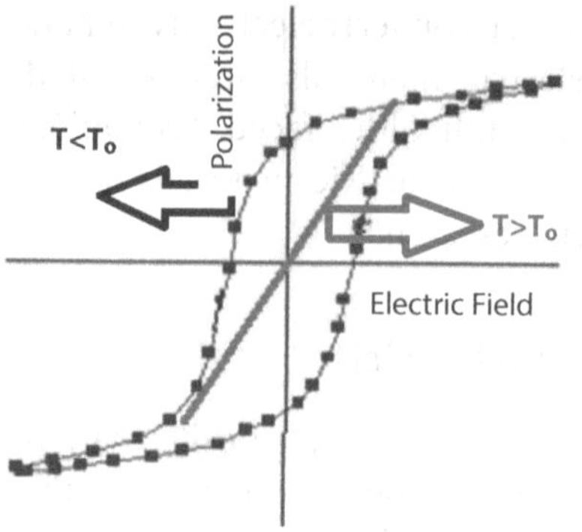

Figure 5.1 Hysteresis of polarization and electric field.

5.2 Ferroelectric Material Properties and Advantages

With the advancement of ferroelectrics with widths of less than 10 nm, nearly unexpected research subjects in nanoelectronics have surfaced recently. Many materials, known as ferroelectrics, have an induced DC electric field, E, that can reverse their spontaneous electrical polarization (P), as shown in Figure 5.2. Ferroelectric materials exhibit spontaneous nonzero polarization even without an applied electric field. Ferroelectric capacitors employed in applications such as ultrasound machines and infrared cameras can be fabricated using ferroelectric materials' nonlinear properties [7].

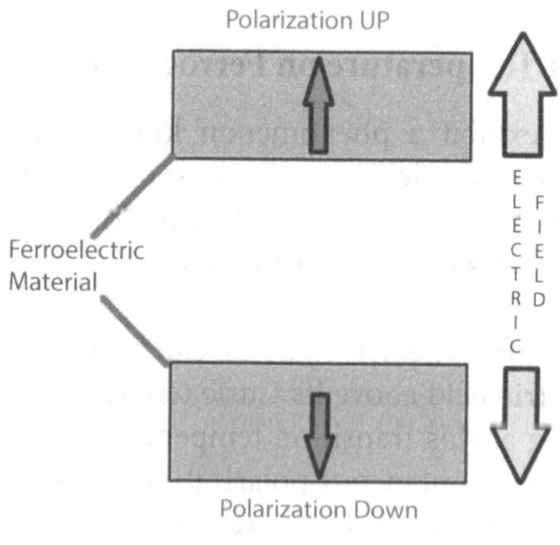

Figure 5.2 The typical response of a ferroelectric material while applying an external electric field in terms of polarization exchange.

Ferroelectrics stand out because of the ability of a strong electric field applied in the opposite direction to reverse their spontaneous polarization, creating a hysteresis loop. Ferroelectric RAM or computer memory can be used with this effect. Ferroelectric capacitors are very helpful in sensor domains such as sonar, fire, gases, and vibration sensors because ferroelectrics are pyroelectric and piezoelectric [8–12].

A perovskite crystal structure is present in the majority of ferroelectric materials. The usual stoichiometry of the perovskite structure is ABO_3, where "A" and "B" symbolize cations and "O" signifies an anion. Figures 5.3 (a, b) show that the central atom possesses two stable low-energy states [13–18]. Polarization occurs when the middle atom travels in the same direction as the electric field when an electric field is applied. Depending on the atom's location, the two states of the center atom are commonly categorized as updown polarizations, as shown in Figures 5.3(a, b). The center atom's two states in a perovskite Barium Titanate Oxide (BTO) ferroelectric material are depicted in the figure below.

Usually, ferroelectric materials exhibit spontaneous polarization, implying that polarization can occur even in the absence of an applied electric field. A permanent dipole moment causes this feature, which is known as ferroelectricity. Domains or specific locations exist in ferroelectric materials where all dipoles point in the same direction. As seen in Figure 5.4(a), the random orientations of the domains cause the net polarization to equal zero without an applied electric field. As shown in Figure 5.4(b), the dipole moments spin to align themselves along an applied external field, generating a larger region with a net dipole moment oriented in a particular direction [19–26].

A coercive field (E_c) is required to return the stored polarization to a minimum value. To change the states, the ferroelectric device's threshold electric field (E_T) must be higher than the coercive field (E_C). There will be

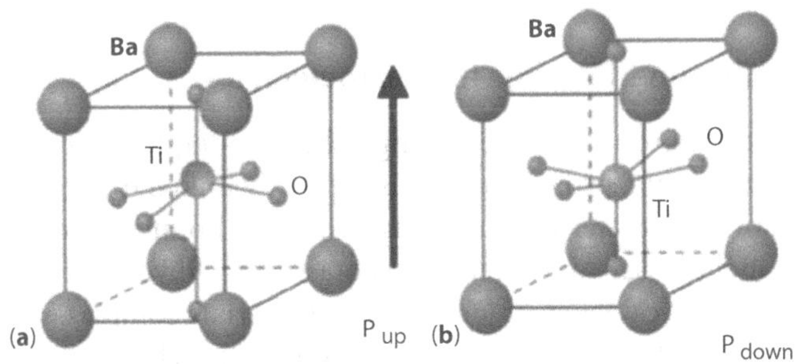

Figure 5.3 Positions of interior atom in BTO (a) P_{up} (b) P_{down} polarization.

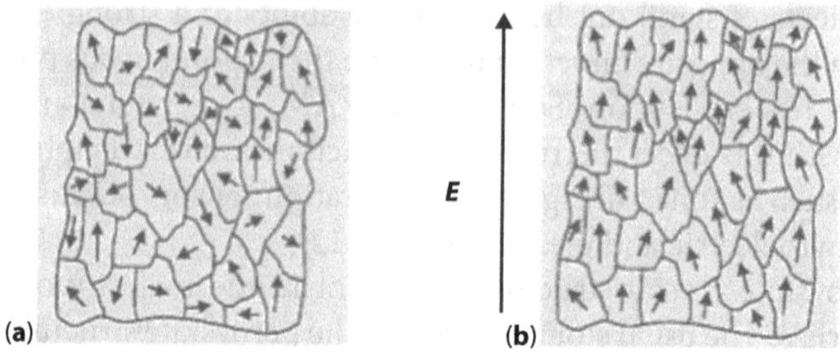

Figure 5.4 (a) Ferroelectric domain with zero polarization (b) with polarization.

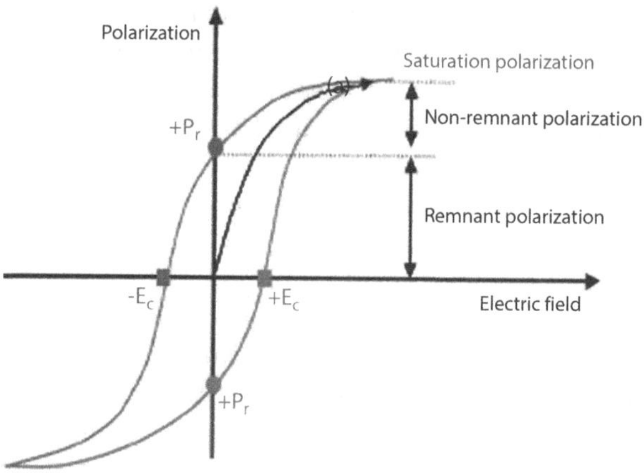

Figure 5.5 Different polarization states.

no switching, and the charge will remain constant when the force is applied in a direction similar to that used previously [2]. The hysteresis curve of the ferroelectric material with different polarization rates is shown in Figure 5.5 with different polarization states.

5.2.1 Common Materials for Ferroelectrics

Three structural groups can be distinguished among ferroelectric materials. In the first group, proton motion is explicitly linked to the ferroelectric characteristics of a material containing hydrogen-bound radicals. Rochelle Examples include salt and KH_2PO_4 (Potassium Dihydrogen Phosphate) [27–32]. The next category of ferroelectric materials comprises ionic crystals with perovskite-like arrangements. Examples include Lead Zirconate

Titanate (PZT), Lead Magnesium Niobate (PMN), $BaTiO_3$, $KTaO_3$, $NaTaO_3$, $KNbO_3$, $PbTiO_3$, and others.

The third group consists of polymers and polymer-based composites such as PVDF [33–37].

5.3 Ferroelectricity in Nanoelectronics

The two recently discovered significant groups of ferroelectrics are the driving forces behind numerous applications in nanoelectronics. The first type is ferroelectrics based on hafnium oxide (HfO_2), which has been the subject of intensive gate dielectric research in recent years because of its compatibility with CMOS technology [38–43]. Conventional ferroelectrics (perovskites and PZT) are incompatible with CMOS, and their ferroelectricity tends to vanish as they are thinned to the nanoscale scale. In further research, hafnium oxide-based ferroelectrics were integrated up to 1 nm in thickness. The subsequent category is characterized by 2D ferroelectric materials [3]. Two well-known examples include semi-metal graphene and semiconductor molybdenum disulfide (MoS_2) [44–49].

5.3.1 HfO_2-Based Ferroelectrics in Nanodevices

The device's dielectric constant (or relative permittivity) must be enhanced to control the leakage current of the device, which mainly consists of several transistors incorporated on a single chip. Doping HfO_2 with elements such as silicon, zirconium, and yttrium facilitates phase transitions and permittivity changes. The tetragonal phase with the highest permittivity (~70) is generated at ambient temperature [50–54].

The ability of high-frequency components to be tuned is essential for several applications, including the Internet of Things (IOTs), radar technologies, and future wireless communications. HfO_2-based ferroelectrics with a high coercive field allow the tunability of high-frequency devices when only 1–3 V from a typical battery is utilized.

Romania's Institute for Research in Micro Technologies created a microwave filter using atomic layer deposition (ALD) techniques. They started with a silicon layer 525 nm thick and a layer of hafnium oxide doped with zirconium (HfZrO) 6 nm thick. The filters are continuously adjustable between 0.1 and 16 GHz when the applied direct current (DC) voltage is swept between 0 and 4 V [55–59].

Another concept that has recently gained popularity is the ferroelectric tunnel junction (FTJ), which comprises two metal electrodes and a nanometer-thick ferroelectric layer. The ferroelectric layer's thickness was adequate for electron tunneling. Researchers have created and manufactured a thick hafnium oxide doped with a zirconium thin film developed directly on nobbled silicon at a technology node of 6 nm, with an upper electrode composed of 100 nm thick gold and 5 nm thick chromium [60–65].

5.3.2 2D Ferroelectric Nanomaterials

Owing to the rapid advancement of nanotechnology, there is growing interest in 2D ferroelectrics, which may be used to create micro- and nanointelligent components with excellent performance and low energy consumption. Numerous 2D van der Waals materials, such as metal triphosphates, MX_2 layer perovskites, transition metal dichalcogenides (TMDs), and indium selenide (In_2Se_3), have been proposed as innately ferroelectric materials. However, only a limited number of 2D ferroelectric materials has been experimentally confirmed. According to a recent paper from the Abdullah College of Science and Innovation, low-temperature ferroelectricity can be demonstrated by stacked In_2Se_3 Nanochips. In 2019, a graphene/In_2Se_3 heterojunction was created, and a $MoTe_2$ 2D nanostructure with ferroelectric characteristics was created after extended periods of successful use as a room-temperature ferroelectric diode [8, 9]. Research conducted by MIT and collaborators has also shown that ferroelectricity and moiré supercross-section frames can be achieved by sandwiching two graphene bilayers between molecularly thin boron nitride layers [66–70]. Neuromorphic computation relies on circuits with artificial neurons and neurotransmitters to simulate various cerebral functions, such as vision and learning. Because 2D materials are delicate, devices typically exhibit limited endurance (a few cycles), dull performance (on/off ratio of less than ten), and unsound activity. Some commonly used 2-D ferroelectric materials are listed in Table 5.1.

The existence of ferroelectric materials is well known. Initially, traditional ferroelectric materials such as perovskite $P(Zr_1{-}xTix)O_3$ (PZT),

Table 5.1 Most known 2D ferroelectric materials.

Material	Behavior	Thickness	Coercive field
SnS [8]	Semiconductor	15 nm	10.7 Kv/cm
α-In_2Se_3 [9]	Semiconductor	10 nm	200 Kv/cm

Table 5.2 Comparison of some ferroelectric materials.

Properties	SBT	PZT [11]	PVDF-TRFE [12] organic	Ferroelectric HfO_2 [12]
E_C (MV/cm)	0.05	0.1	0–1	0–2
Polarization P_R (μC/cm²)	12–25	10–60	~10	30–60
Permittivity	200	1,000	2	~25
Gate voltage swing (v)	-	±20	±60	±3

$SrBi_2Ta_2O_9$ (SBT), and $Bi_4Ti_3O_{12}$ (BTO) garnered significant interest because of their advantages, which include greater dielectric constants, larger E_c and P_R, and favorable thermal stabilities. P(VDF-TrFE) and other organic ferroelectric materials have existed for over 50 years [10]. Their application in ferroelectric memory devices has long been suggested. Table 5.2 provides the ferroelectric performance of several traditional ferroelectrics, such as hybrid systems and inorganic and organic ferroelectrics [71–75]. Ferroelectric ceramics are the most popular and extensively used inorganic ferroelectric materials.

5.4 Structures of Ferroelectric FET

Ferroelectric field-effect transistors are categorized into various forms that are used to design multiple devices to acquire the desired results; some of them are listed below (Figure 5.6) [11].

5.4.1 Metal Ferroelectric Semiconductor Field-Effect Transistor

A notable distinction between MOSFETs and MFSFETs (or related device technologies) is the replacement of a typical oxide layer with a ferroelectric-material layer. Because MFSFETs are frequently employed as memory components, the ferroelectric material's performance properties significantly affect MFSFET's quality. Wu created a mathematical model in tandem with developing the ferroelectric field-effect transistor [13]. As the gate insulator in an MFSFET, a ferroelectric material is employed in place of or in addition to more widely used materials such as silicon dioxide, as

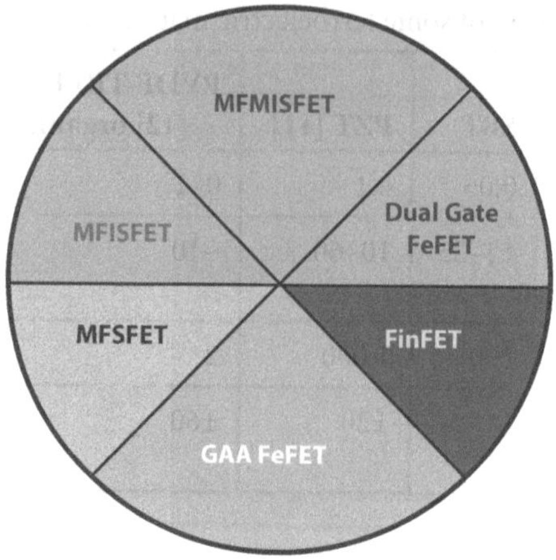

Figure 5.6 Structures of FeFETs.

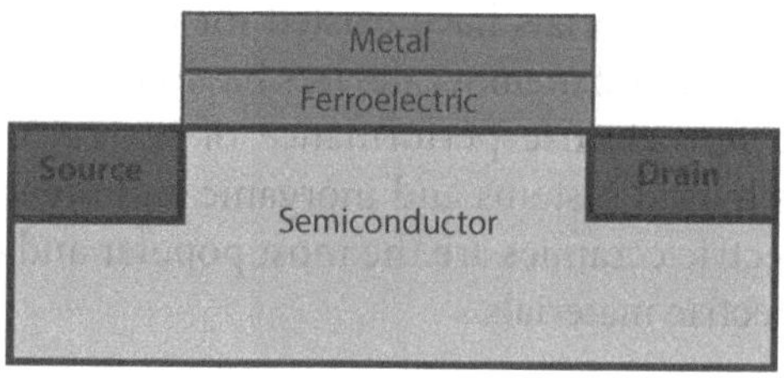

Figure 5.7 (i) Structure of MFSFET.

shown in Figure 5.7(i). We suggest that the ferroelectric material at the MOSFET gate introduces an extra charge equivalent to the polarization of the ferroelectric at the semiconductor-ferroelectric contact to explain its function, given that the MOSFET threshold voltage (V_T) depends on the total charge at the semiconductor-insulator interface [14]. Although the MFSFET manufacturing process is straightforward, it has certain drawbacks such as a significant interface between the materials in its ferroelectric and semiconductor channels.

5.4.2 Metal Ferroelectric Insulator Semiconductor Field-Effect Transistor

Owing to the direct interaction of ferroelectric materials with semiconductors, interface reactivity and inter-diffusion cause serious retention issues, meaning that information is lost over time if ferroelectric thin films are

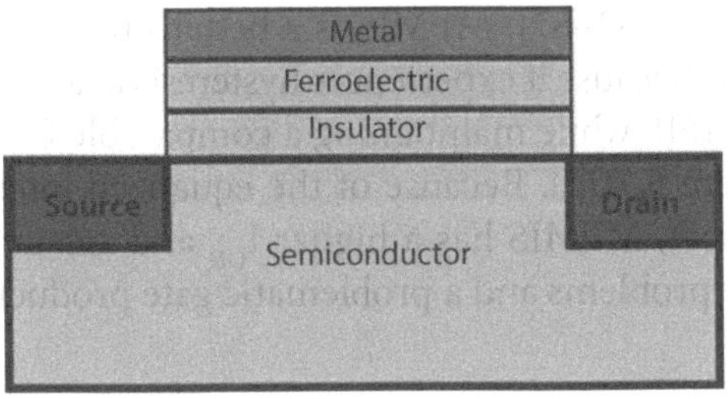

Figure 5.7 (ii) Structure of MFISFET.

directly deposited on silicon [15, 16]. Consequently, an intermediary layer was inserted as a reaction and diffusion barrier between the ferroelectric layers and silicon, such as SiO_2, Al_2O_3, and ZrO_2. This results in the formation of an MFIS structure, as shown in Figure 5.7(ii). Compared to the MFMIS negative capacitance field-effect transistor (NCFET), which has a reduced ferroelectric width, the MFIS NCFET is more likely to experience hysteresis and behave hysterically [76–81]. The unequal internal gate voltage amplification leads to a voltage drop across the channel, which is a shortcoming of MFIS, despite its comparatively simple gate stack structure and ability to overcome the significant interface states between ferroelectric and semiconductor channel materials.

5.4.3 Metal Ferroelectric Metal Insulator Semiconductor Field-Effect Transistor

Except for ferroelectrics with extremely low remnant polarization (P_R) in the high operating voltage domain, MFMIS NCFET offer a larger I_{ON}

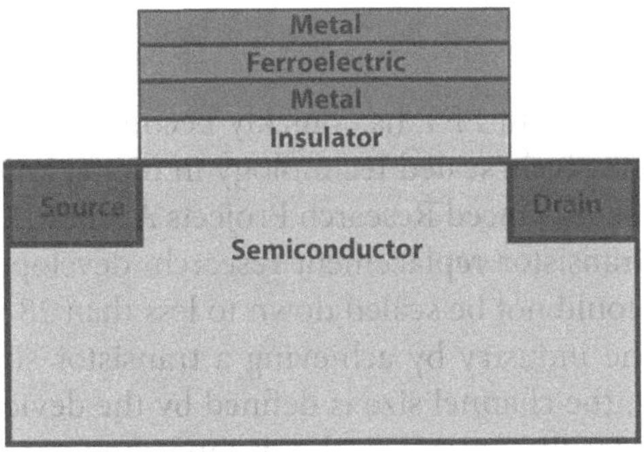

Figure 5.7 (iii) Structure of MFMIS.

than MFIS NCFET [82–87]. MFMIS is a better option for hysteresis-free logic applications because it experiences hysteresis at a greater ferroelectric thickness than MFIS while maintaining a comparable ON current [17], as depicted in Figure 5.7(iii). Because of the equalized internal gate voltage amplification effect, MFMIS has a higher I_{ON} and smaller SS, however, it also has stability problems and a problematic gate production procedure.

5.4.4 Dual Gate FeFET

This section shows the physical layout of the dual-gate FeFET arrangement shown in Figure 5.7 (iv), which provides remarkable control over the channel and improves the current on/off ratio. In the subsequent dual-gate FeFET, the ferroelectric gate functions as a memory element, and the nonferroelectric metal provides a valuable means of adjusting the device's threshold voltage (V_{th}) [88–92].

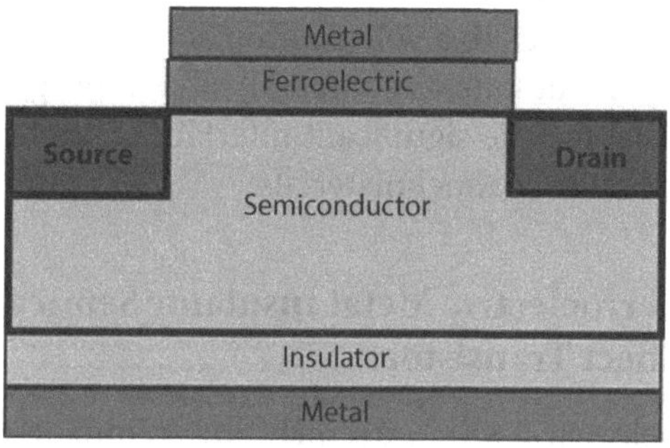

Figure 5.7 (iv) Structure of dual gate FeFET.

5.4.5 Fin FET

Negative capacitance Fin FET has quickly become the most well-known competitor for massively scaled technology in digital and analog applications. The Defense Advanced Research Projects Agency (DARPA), eager to support planar transistor replacement research, developed FinFETs [93–96]. Transistors could not be scaled down to less than 28 nm, but FinFETs revolutionized the industry by achieving a transistor size of just 14 nm [18]. In FinFETs, the channel size is defined by the device width, making it vertical. For a traditional MOS, the distance between the horizontally aligned S–D connectors is known as the channel length.

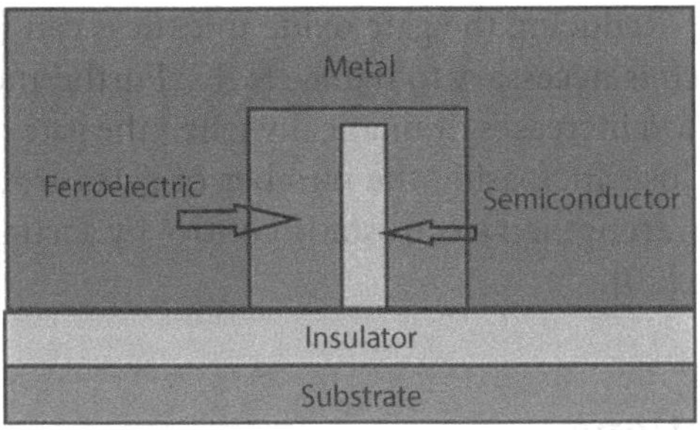

Figure 5.7 (v) Structure of fin FeFET.

$$\textbf{\textit{Width of Channel}} = (2 \times \textbf{\textit{Fin Height}} + \textbf{\textit{Fin Width}})$$

Therefore, improving the fin height and channel width increased the device's driving capabilities. In contrast, planar device designers can change the channel size according to the required power. FinFETs are divided into two groups according to their gate structure: an independent gate (IG) with a four-terminal (4T) FinFET and Shorted gate (SG) with a three-terminal (3T) FinFET. For the SG FinFET, the front and rear gates are physically shorted, whereas for the IG, the gates avoid physical interaction. This provides increased device design options for IG FinFET [19].

5.4.6 Gate All-Around Structure

As the transistor size decreases, it is difficult for the FinFET technology to provide better control for devices below 3 nm. Therefore, if better controllability is required over the channel, a gate-all-around (GAA) structure is proposed, as shown in Figure 5.7(vi). A gate-all-around (GAA) MOSFET provides strong gate controllability over the channel, as shown in Figure 5.7(vi), which has cylindrical source, drain, and channel

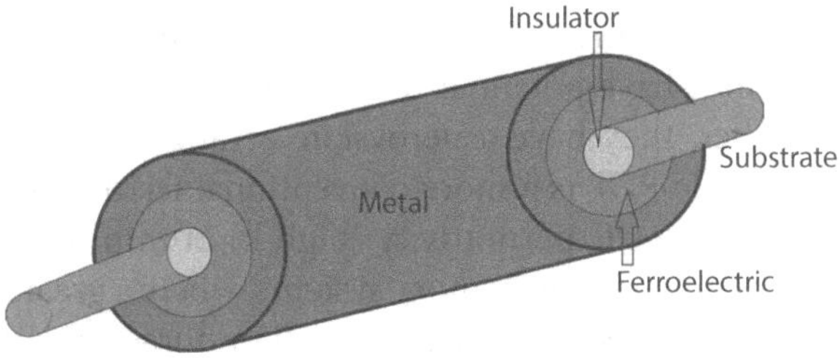

Figure 5.7 (vi) Gate all-around structure.

regions [97–100]. Reducing the gate oxide thickness can increase the gate capacitance, which is necessary to reduce SCEs. Furthermore, the off-state power consumption increases dramatically when the gate oxide is too thin, and reliability is lost. Increasing the number of gates around the channel is another way to strengthen electrostatic control by accumulating distinct gate capacitances [20].

5.5 Applications

5.5.1 Fe-FET-Based Memory Design

The fundamental benefit of ferroelectric materials is their capacity to preserve memory. The secret to memory applications is a hysteresis window with an electric field. The reading process's non-destructive nature is the primary benefit of combining ferroelectric materials with FETs. Different ferroelectric materials have been used to fabricate FeFETs, demonstrating faster read/write speeds and better low-power driving than DRAM, static RAM (SRAM), and flash memory [21]. Ferroelectric memory technology has recently received much attention owing to its potential use in next-generation integrated circuits. Ferroelectric gate memory transistors offer numerous benefits, including nonvolatile data archiving, nondestructive data reading, quick operation, and dense packing [101–106]. After substantial research and development, ferroelectric random access memories (FeRAMs) have used thin films between metal electrodes as their memory medium. Thin films of PZT ($Pb_{1-x}Zr_x$) TiO_3, $SrBi_2Ta_2O_9$ (SBT), and BaTiO3/$GdScO_3$ (BTO) have shown significant potential as ferroelectric memory components. High residual polarization (P_r) and low processing temperature are just two benefits of PZT, whereas $SrBi_2Ta_2O_9$ (SBT) has strong fatigue resistance. Nevertheless, both methods have shortcomings. Their hysteresis characteristics are shown in Figure 5.8 (i), and their relationships are presented in Table 5.3. Once noble metal electrodes are employed, PZT suffers from significant fatigue, and SBT requires a high annealing temperature (750 °C), which most silicon devices cannot withstand [21]. Both have potential uses in FeFETs and FRAM memory [107–112]. Together, they have a perovskite arrangement (SBT is a layered perovskite, while PZT is a more conventional ferroelectric kind). As a result, the high dielectric permittivity (high k) of thin films is approximately 250 [22]. In nonvolatile memory systems, where a greater saturation polarization (P_s) might result in longer retention durations, P_s is a crucial ferroelectric characteristic for memory applications, as depicted in Figure 5.8(ii) [19]. Ferroelectric materials with high P_R or low coercive fields (E_c) are preferable for driving FeFETs. Higher switching voltages are required in a ferroelectric material with a coercive solid field, which results in a larger memory window. The memory window is the breadth of the hysteresis loop. Higher currents are produced in FeFETs by a ferroelectric with a high polarization P_R. However, this ferroelectric cannot be fully polarized using low-voltage operation, necessitating the use of subloops that are less stable. A comparison of the different ferroelectric materials with various parameters is presented in Table 5.3.

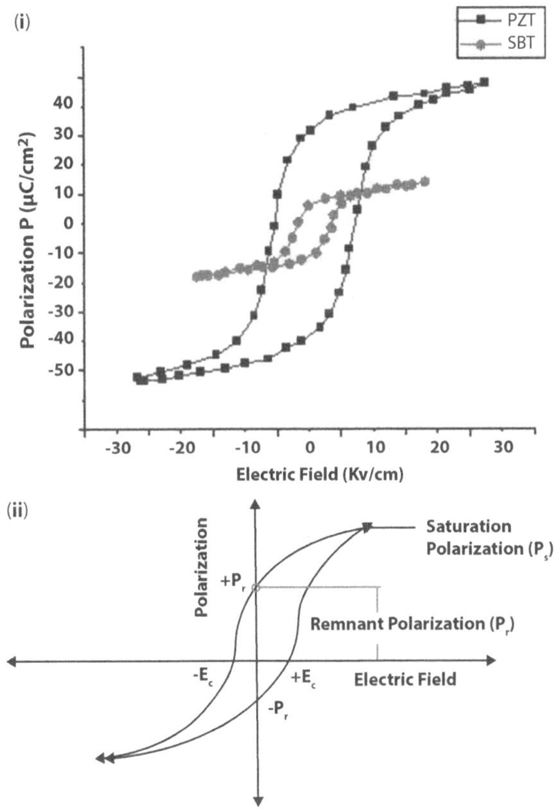

Figure 5.8 (i) Saturated polarization hysteresis loops for SBT and PZT. (ii) Typical ferroelectric hysteresis curve using E_c as the coercive field, P_s as saturated polarization, and P_R as remnant polarization.

Table 5.3 Material parameters for thin film SBT and oriented PZT.

Type	Materials	P_R (μC/cm²)	P_s (μC/cm²)	E_C (kV/cm)
PZT [21]	PZT $(Pb_{1-x}Zr_x)TiO_3$	32	40	70
SBT [21]	$SrBi_2Ta_2O_9$	8	10	30
BTO [11]	$BaTiO_3/GdScO_3$	50	–	80

5.5.2 Fe-FET-Based Sensor Design

The expansion of ferroelectric sensors includes a wide range of devices based on piezoelectric, pyroelectric, and conductive phenomena. Its presentations are in various industries, including communications, healthcare, environmental monitoring, automotive and aerospace applications, and industrial production [22]. Ferroelectric materials are adaptable and have a broad range of potential sensor applications. Current developments are toward miniaturization, higher resolution and precision, applications in harsh environments, novel processing techniques, and better materials. There are several techniques for devising a device that works as a sensor. Some sensor examples that are used to detect biomolecules and hazardous gases are explained here. Sensors are designed using various ferroelectric materials such as HfO_2, CIPS, and doped hafnium oxide with varying structures, as discussed in an earlier section.

The published work [23] is a detailed analysis of the performance of Polarity Controllable Ferroelectric-Ion Sensitive FETs (PC-FE-ISFETs) for pH sensing applications. Compared to the conventional polarity FET, the ferroelectric layer in the gate stack of the PC-FE-ISFET is anticipated to improve sensitivity, in addition to offering sub 60 mV/dec subthreshold swing values and enhanced current driveability. It is equally high when comparing the I_{ON}-to-I_{OFF} current ratio sensitivity (I_{ON}/I_{OFF}) between the proposed PCFE-ISFET and conventional PC-ISFET. The proposed PC-FE-ISFET is shown in Figure 5.9. An undoped semiconductor silicon channel was used to form the structure.

The usefulness of the dielectric-modulated dual-gate ferroelectric field-effect transistor without junction as a biosensor was investigated in another study. Biomolecules are used to assess the device's ability to sense both charged and neutral biomolecules [24]. In this study, the MFIS structure was used to design the biosensor, and the cavity was created below the ferroelectric layer for biomolecule inclusion, as shown in Figure 5.10.

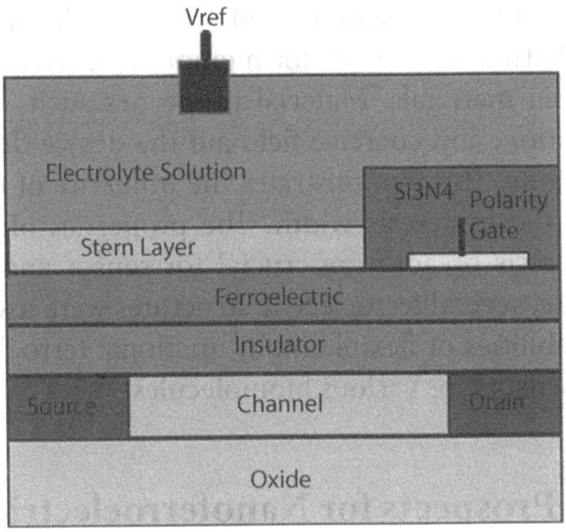

Figure 5.9 PC-FE-ISFETs from a structural perspective.

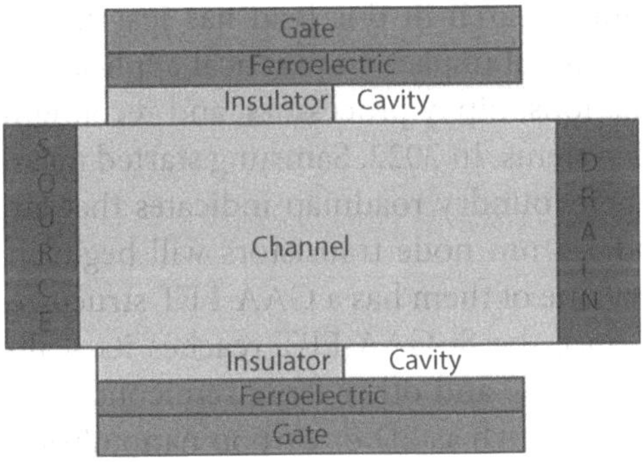

Figure 5.10 Dielectric modulated dual gate ferroelectric junction-less transistor structure.

This study uses a dual gate with a ferroelectric material for superior control over the channel. Once biomolecules enter the cavity, the threshold voltage and drive current, such as the drain current of the device, change. In other devices, the work function is used to change the device parameters to make the device a sensor.

5.6 Conclusion

This chapter compares several ferroelectric materials and FeFET architectures centered on ferroelectric resources, such as PZT, SBT, BTO, and hafnium oxide (HfOx) for memory and sensor applications. Researchers can distinguish between materials for memory and sensor applications by comparing several materials. Material properties such as residue, spontaneous polarization, and coercive field aid the device design process for specific applications. This demonstrates the influence of the material and temperature on the hysteresis width. The properties of negative capacitance and hysteresis breadth are crucial for sensor and memory-based applications. Here, several ferroelectric structures were used to explain the pH-sensing capabilities of flexible multifunctional ferroelectric transistor sensors and biosensors for various biomolecules.

5.7 Future Prospects for Nanoferroelectric Devices

Switches, negative-capacitance transistors, and memristors are a few examples of low-power nanoelectronic devices that increasingly utilize ferroelectrics. Although research in this field has just recently started, it has already had a substantial impact on practical applications, including neuromorphic calculations, nippy processors, and reconfigurable systems for communications systems. In 2022, Samsung started mass-producing 3 nm Si GAA-FETs. Their foundry roadmap indicates that large-scale production of 2 nm and 1.4 nm node transistors will begin in 2025 and 2027, respectively. Every one of them has a GAA-FET structure.

Furthermore, once the Si GAA-FET reaches its scaling limit, a situation affecting the TSMC and other large semiconductor foundries, such as advanced materials such as 2D or carbon nanotubes, may save the day. Additionally, these advancements make Moore's law applicable to sub-1 nm nodes.

Acknowledgments

The authors acknowledge the Nano Electronics Research Lab of the ECE Department at NIT Jalandhar for giving them opportunities to do their work. We thank the Department of ECE at NIT Jalandhar for providing the laboratory space and research settings needed to complete this project. We would especially like to thank the Electronics and Communication Engineering Nano Electronics and VLSI research group for their interest in this work and for their insightful feedback, which helped shape the final version of the chapter.

References

1. Chauhan, V. and Samajdar, D.P., Recent Advances in Negative Capacitance FinFETs for Low Power Applications: A Review. *IEEE Trans. Ultrason. Ferroelectr. Freq. Control.*, 68, 10, 3056–3068, 2021.
2. Kim, S., Lee, K., Lee, J.H., *et al.*, Vertically stacked gate all around structured tunneling-based ternary CMOS. *IEEE Trans. Electron Devices*, 67, 3889–3893, 2020.
3. Herrera, F., Hirano, Y., Miura Mattausch, M., *et al.*, Advanced Short-channel-effect modeling with applicability to device optimization Potentials and scaling. *IEEE Trans. Electron Devices*, 66, 3726–3733, 2019.
4. Dragoman, M., Aldrigo, M., Dragoman, D., Iordanescu, S., Dinescu, A., Modreanu, M., The Rise of Ferroelectricity at Nanoscale: Nanoelectronics is rediscovering the ferroelectricity. *IEEE Nanatechnol. Mag.*, 15, 5, 8–19, 2021.
5. Dawber, M., Rabe, K.M., Scott, J.F., Physics of thin-film ferroelectric oxides. *Rev. Mod. Phys.*, 77, 1083–130, 2005.
6. Qi, L., Ruan, S., Zeng, Y.J., Review on recent developments in 2D ferroelectrics: Theories and applications. *Adv. Mater.*, 33, 13, 2005098, 2021.
7. Valasek, J., Piezoelectric and allied phenomena in Rochelle Salt. *Phys. Rev. J. Arch.*, 17, 475, 1921.
8. Guan, *et al.*, Recent progress in two-dimensional ferroelectric materials. *Adv. Electron. Matter*, 6, 1, 1900818, 2019, doi: 10.1002/aelm.201900818.
9. Gong, L.M. and Zeng, H., Atomically thin α-In2Se3: An emergent two-dimensional room temperature ferroelectric semiconductor. *J. Semicond.*, 40, 061002, 2019.
10. Bune, A.V., Fridkin, V.M., Ducharme, S., Blinov, L.M., Palto, S.P., Sorokin, A.V., Yudin, S.G., Zlatkin, A., Two-dimensional ferroelectric films. *Nature*, 391, 874–877, Feb. 1998, doi: 10.1038/36069.
11. Jiao, H., Wang, X., Wu, S., Chen, Y., Chu, J., Wang, J., Ferroelectric field effect transistors for electronics and optoelectronics. *Appl. Phys. Rev.*, 10, 011310, 2023, https://doi.org/10.1063/5.0090120.
12. Mikolajick, T., Schroeder, U., Slesazeck, S., Hafnium oxide based ferroelectric devices for memories and beyond. *International Symposium on VLSI Technology, Systems and Application (IEEE)*, vol. 1, 2018.
13. Wu, S.Y., A New Ferroelectric Memory Device, Metal-Ferroelectric-Semiconductor Transistor. *IEEE Trans. Electron Devices*, ED-31, 8, 499–504, August, 1974.
14. Schroeder, Park M. H., Mikolajick, T., Hwang, C.S., The fundamentals and applications of ferroelectric HfO_2. *Nature Rev. Mater.* 7, 8, 653–669, 2022.
15. Han, J.-P., Guo, X., Ma, T.P., Memory effects of $SrBi_2Ta_2O_9$ capacitor on silicon with a silicon nitride buffer. *Integr. Ferroelectr.*, 22, 213–221, 1998.

16. Kijima, T. and Matsunaga, H., Preparation of $Bi_4Ti_3O_{12}$ thin films by MOCVD method and electrical properties of metal/ferroelectric/insulator/semiconductor structure. *Jpn. J. Appl. Phys.*, 38, 2281, 1999.
17. Pahwa, G., Dutta, T., Agarwal, A., Chauhan, Y.S., Physical insights on negative capacitance transistor in no hysteresis and hysteresis regime: MFMIS vs. MFS structures. *IEEE Trans. Electron Devices*, 65, 867–873, 2018.
18. Guo, X., Verma, V., Gonzalez-Guerrero, P., Mosanu, S., Stan, M.R., Back to the future : digital circuit design in the FinFET era. *J. Low Power Electron.*, 13, 3, 2017, doi: 10.1166/jolpe.2017.1489.
19. Fulde, M. and Engelst, J.P., Analog circuits using FinFETs : benefits in speed-accuracy-power trade-off and simulation of parasitic effects. *Adv. Radio Sci.*, 5, 285–290, 2007, doi: 10.5194/ars-5-285-2007.
20. Hossain, I.M., Touhid, K., Abu, S., Nur, A.K., Junctionless Gate-all-around Nanowire FET with Asymmetric Spacer for Continued Scaling, 2021.
21. Damjanovic, D., Muralt, P., Setter, N., Ferroelectric Sensors. *IEEE Sens. J.*, 1, 3, 191–206, 2001.
22. Pandey, P. and Kaur, H., Analysis of pH sensing performance of polarity controllable-ferroelectric-ion sensitive FET. *IEEE International Women in Engineering (WIE) Conference on Electrical and Computer Engineering (WIECON-ECE)*, 2022.
23. Yadav, S., Rewari, S., Pandey, R., Sensitivity analysis of dielectric modulated dual gate ferroelectric junctionless transistor-based biosensor. *IEEE International Conference of Electron Devices*, pp. 26–27, Nov 2022.
24. Smaili, I.H., Design and simulation of short channel Si: HfO2 ferroelectric field e field effect transistor (FeFET), pp. 8–15, 2014.
25. Goyal, C., Ubhi, J.S., Raj, B., A reliable leakage reduction technique for approximate full adder with reduced ground bounce noise. *J. Math. Probl. Eng.*, Hindawi, 2018, Article ID 3501041, 16, 15 Oct 2018.
26. Anuradha, Singh, J., Raj, B., Khosla, M., Design and Performance Analysis of Nano-scale Memristor-based Nonvolatile SRAM. *J. Sens. Lett.*, American Scientific Publishers, 16, 798–805, Oct 2018.
27. Wadhwa, G. and Raj, B., Parametric Variation Analysis of Charge-Plasma-based Dielectric Modulated JLTFET for Biosensor Application. *IEEE Sens. J.*, 18, 15, 6070 – 6077, AUGUST 1, 2018, (SCI).
28. Singh, J. and Raj, B., Comparative Analysis of Memristor Models for Memories Design. *JoS*, IoP, 39, 7, 074006-1–12, July 2018.
29. Yadav, D., Chouhan, S.S., Vishvakarma, S.K., Raj, B., Application Specific Microcontroller Design for IoT based WSN. *Sens. Lett.*, ASP, 16, 374–385, May 2018.
30. Singh, G., Sarin, R.K., Raj, B., Fault-Tolerant Design and Analysis of Quantum-Dot Cellular Automata Based Circuits. *IEEE/IET Circuits Devices Syst.*, 12, 638–644, 2018.

31. Singh, J. and Raj, B., Modeling of Mean Barrier Height Levying Various Image Forces of Metal Insulator Metal Structure to Enhance the Performance of Conductive Filament Based Memristor Model. *IEEE Nanotechnol.*, 17, 2, 267, 268, March 2018.
32. Jain, A., Sharma, S., Raj, B., Analysis of Triple Metal Surrounding Gate (TM-SG) III-V Nanowire MOSFET for Photosensing Application. *Opto-Electron. J.*, Elsevier, 26, 2, 141–148, May 2018.
33. Jain, A., Sharma, S., Raj, B., Design and Analysis of High Sensitivity Photosensor Using Cylindrical Surrounding Gate MOSFET For Low Power Sensor Applications. *Eng. Sci. Technol. Int. J.*, Elsevier's, 19, 4, 1864–1870, December 2016.
34. Saiphani Kumar, G., Singh, A., Raj, B., Design and Analysis of Gate All Around CNTFET based SRAM cell Design. *J. Comput. Electron.*, Springer, 17, 1, 138–145, March 2018.
35. Singh, G.P., Sohi, B.S., Raj, B., Material Properties Analysis of Graphene Base Transistor (GBT) for VLSI Analog Circuits. *Indian J. Pure Appl. Phys. (IJPAP)*, 55, 896–902, Dec-2017.
36. Kumar, S. and Raj, B., Estimation of Stability and Performance metric for Inward Access Transistor based 6T SRAM Cell Design using n-type/p-type DMDG-GDOV TFET. *IEEE VLSI Circuits Syst. Lett.*, 3, 2, 25–39, June 2017.
37. Sharma, S., Kumar, A., Pattanaik, M., Raj, B., Forward Body Biased Multimode Multi-Threshold CMOS Technique for Ground Bounce Noise Reduction in Static CMOS Adders. *Int. J. Inf. Electron. Eng.*, 3, 3, 567–572, 2013.
38. Singh, H., Kumar, P., Raj, B., Performance Analysis of Majority Gate SET Based 1-bit Full Adder. *Int. J. Comput. Commun. Eng. (IJCCE)*, IACSIT Press Singapore, 2, 4, 2013, ISSN: 2010-3743.
39. Bhardwaj, A.K., Gupta, S., Raj, B., Investigation of Parameters for Schottky Barrier (SB) Height for Schottky Barrier Based Carbon Nanotube Field Effect Transistor Device. *J. Nanoelectron. Optoelectron.*, ASP, 15, 783–791, July 2020.
40. Bansal, P. and Raj, B., Memristor: A Versatile Nonlinear Model for Dopant Drift and Boundary Issues. *JCTN*, American Scientific Publishers, 14, 5, 2319–2325, May 2017.
41. Jain, N. and Raj, B., An Analog and Digital Design Perspective Comprehensive Approach on Fin-FET (Fin-Field Effect transistor) Technology - A Review. *Rev. Adv. Sci. Eng. (RASE)*, ASP, 5, 1–14, 2016.
42. Raj, B., Saxena, A.K., Dasgupta, S., Analytical Modeling for the Estimation of Leakage Current and Subthreshold Swing Factor of Nanoscale Double Gate FinFET Device. *Microelectron. Int.*, UK, 26, 53–63, 2009.
43. Soniya, Singh, S., Wadhwa, G., Raj, B., An Analytical Modeling for Dual Source Vertical Tunnel Field Effect Transistor. *Int. J. Recent Technol. Eng. (IJRTE)*, 8, 2, 603–608, July 2019.

44. Singh, S. and Raj, B., Design and analysis of hetrojunction Vertical T-shaped Tunnel Field Effect Transistor. *J. Electron. Mater.*, Springer, 48, 10, 6253–6260, October 2019.
45. Goyal, C., Ubhi, J.S., Raj, B., A Low Leakage CNTFET based Inexact Full Adder for Low Power Image Processing Applications. *Int. J. Circuit Theory Appl.*, Wiley, 47, 9, 1446–1458, September 2019.
46. Raj, B., Saxena, A.K., Dasgupta, S., A Compact Drain Current and Threshold Voltage Quantum Mechanical Analytical Modeling for FinFETs. *J. Nanoelectron. Optoelectron. (JNO)*, USA, 3, 2, 163–170, 2008.
47. Wadhwa, G. and Raj, B., An Analytical Modeling of Charge Plasma based Tunnel Field Effect Transistor with Impacts of Gate underlap Region. *Superlattices Microstruct.*, Elsevier, 142, 106512, June 2020.
48. Singh, S. and Raj, B., Modeling and Simulation analysis of SiGe hetrojunction Double GateVertical t-shaped Tunnel FET. *Superlattices Microstruct.*, Elsevier, 142, 106496, June 2020.
49. Jain, N. and Raj, B., Device and Circuit Co-Design Perspective Comprehensive Approach on FinFET Technology - A Review. *J. Electron Devices*, 23, 1, 1890–1901, 2016.
50. Kumar, S. and Raj, B., Analysis of I_{ON} and Ambipolar Current for Dual-Material Gate-drain Overlapped DG-TFET. *J. Nanoelectron. Optoelectron.*, American Scientific Publishers, USA, 11, 323–333, June 2016.
51. Anjum, N., Bali, T., Raj, B., Design And Simulation Of Handwritten Multiscript Character Recognition. *Int. J. Adv. Res. Comput. Commun. Eng.*, 2, 7, 2544–2549, July 2013.
52. Singh, K. and Raj, B., Performance and analysis of temperature dependent Multi-walled Carbon Nanotubes as Global Interconnects at different technology nodes. *J. Comput. Electron.*, Springer, 14, 2, 469–476, June 2015.
53. Kumar, S. and Raj, B., Compact channel potential analytical Modeling of DG-TFET based on Evanescent–mode Approach. *J. Comput. Electron.*, Springer, 14, 2, 820–827, July 2015.
54. Singh, K. and Raj, B., Temperature Dependent Modeling and Performance Evaluation of Multi-Walled CNT and Single-Walled CNT as Global Interconnects. *J. Electron. Mater.*, Springer, 44, 12, 4825–4835, Dec 2015.
55. Sharma, V.K., Pattanaik, M., Raj, B., INDEP approach for leakage reduction in nanoscale CMOS circuits. *Int. J. Electron.*, Taylor & Francis, 102, 2, 200–215, 2014.
56. Singh, K. and Raj, B., Influence of Temperature on MWCNT bundle, SWCNT bundle and Copper interconnects for Nanoscaled Technology nodes. *J. Mater. Sci.: Mater. Electron.*, Springer, 26, 8, 6134–6142, 2015.
57. Anjum, N., Bali, T., Raj, B., Design and Simulation of Handwritten Gurumukhi and Devanagri Numerical Recognition. *Int. J. Comput. Appl.*, Foundation of Computer Science, New York, USA, 73, 12, 16–21, 2013.

58. Khandelwal, S., Gupta, V., Raj, B., Gupta, R.D., Process Variability Aware Low Leakage Reliable Nano Scale DG-FinFET SRAM Cell Design Technique. *J. Nanoelectron. Optoelectron.*, 10, 6, 810–817, December 2015.
59. Sharma, V.K., Pattanaik, M., Raj, B., ONOFIC Approach: Low Power High Speed Nanoscale VLSI Circuits Design. *Int. J. Electron.*, Taylor & Francis, 101, 1, 61–73, 2014.
60. Khandelwal, S., Raj, B., Gupta, R.D., FinFET based 6T SRAM Cell Design: Analysis of Performance Metric, Process Variation and Temperature Effect. *J. Comput. Theor. Nanosci.*, ASP, USA, 12, 2500–2506, 2015.
61. Singh, S., Yadav, S., Rahul, J., Srivastava, A., Raj, B., Impact of HfO_2 in Graded Channel Dual Insulator Double Gate MOSFET. *J. Comput. Theor. Nanosci.*, American Scientific Publishers, 12, 6, 950–953, April 2015.
62. Sharma, V.K., Pattanaik, M., Raj, B., PVT variations aware low leakage INDEP approach for nanoscale CMOS Circuits. *Microelectron. Reliab.*, Elsevier, 54, 90–99, 2014.
63. Raj, B., Saxena, A.K., Dasgupta, S., Quantum Mechanical Analytical Modeling of Nanoscale DG FinFET: Evaluation of Potential, Threshold Voltage and Source/Drain Resistance. *Elsevier's J. Mater. Sci. Semicond. Process.*, Elsevier, 16, 4, 1131–1137, 2013.
64. Gopal, M., Prasad, S.S.D., Raj, B., 8T SRAM Cell Design for Dynamic and Leakage Power Reduction. *Int. J. Comput. Appl.*, Foundation of Computer Science, New York, USA, 71, 9, 43–48, June 2013.
65. Pattanaik, M., Raj, B., Sharma, S., Kumar, A., Diode Based Trimode Multi-Threshold CMOS Technique for Ground Bounce Noise Reduction in Static CMOS Adders. *Adv. Mater. Res.*, Trans Tech Publications, Switzerland, 548, 885–889, 2012.
66. Raj, B., Saxena, A.K., Dasgupta, S., Nanoscale FinFET Based SRAM Cell Design: Analysis of Performance metric, Process variation, Underlapped FinFET and Temperature effect. *IEEE Circuits Syst. Mag.*, 11, 2, 38–50, 2011.
67. Sharma, V.K., Pattanaik, M., Raj, B., Leakage Current ONOFIC Approach for Deep Submicron VLSI Circuit Design. *Int. J. Electr. Comput. Electron. Commun. Eng.*, World Academy of Sciences, Engineering and Technology, 7, 4, 239–244, 2013.
68. Kaur, P., Gill, S.S., Raj, B., Comparative Analysis of OFETs Materials and Devices for Sensor Applications. *J. Silicon*, Springer, 14, 4463–4471, 2022.
69. Sharma, S.K., Kumar, P., Raj, B., Raj, B., $In_{1-x}Ga_xAs$ Double Metal Gate-Stacking Cylindrical Nanowire MOSFET for highly sensitive Photo detector. *J. Silicon*, Springer, 14, 3535–3541, 2022.
70. Raj, B., Saxena, A.K., Dasgupta, S., Analytical Modeling of Quasi Planar Nanoscale Double Gate FinFET with Source/Drain Resistance and field dependent carrier mobility: A quantum mechanical study. *J. Comput. (JCP)*, Academy Publisher, FINLAND, 4, 9, 1–8, 2009.

71. Bhushan, S., Khandelwal, S., Raj, B., Analyzing Different Mode FinFET based Memory Cell at Different Power Supply for Leakage Reduction, in: *Seventh International Conference on Bio-Inspired Computing: Theories and Application, (BIC-TA 2012)*, Advances in Intelligent Systems and Computing, vol. 202, pp. 89–100, 2013.
72. Singh, J. and Raj, B., Temperature Dependent Analytical Modeling and Simulations of Nanoscale Memristor. *Eng. Sci. Technol. Int. J.*, Elsevier's, 21, 862–868, Oct 2018.
73. Singh, S., Bala, S., Raj, B., Raj, B., Improved Sensitivity of Dielectric Modulated Junctionless Transistor for Nanoscale Biosensor Design. *Sens. Lett.*, ASP, 18, 328–333, Apl 2020.
74. Kumar, V., Vishvakarma, S.K., Raj, B., Design and Performance Analysis of ASIC for IoT Applications. *Sens. Lett.*, ASP, 18, 31–38, Jan 2020.
75. Jaiswal, A., Sarin, R.K., Raj, B., Sukhija, S., A Novel Circular Slotted Microstrip-fed Patch Antenna with Three Triangle Shape Defected Ground Structure for Multiband Applications. *Adv. Electromagn. (AEM)*, 7, 3, 56–63, Aug 2018.
76. Kaur, P., Pandey, V., Raj, B., Comparative Study of Efficient Design, Control and Monitoring of Solar Power using IoT. *Sens. Lett.*, ASP, 18, 419–426, May 2020.
77. Bhardwaj, A.K., Gupta, S., Raj, B., Development & Analysis of compact model for Double Gate Schottky Barrier CNTFET. *J. Nanoelectron. Optoelectron.*, ASP, 15, 1199–1208, Aug 2020.
78. Wadhwa, G., Kamboj, P., Singh, J., Raj, B., Design and Investigation of Junctionless DGTFET for Biological Molecule Recognition. *Trans. Electr. Electron. Mater.*, Springer, 22, 282–289, 2021.
79. Verma, S.K., Singh, S., Wadhwa, G., Raj, B., Detection of Biomolecules using Charge-Plasma based Gate Underlap Dielectric Modulated Dopingless TFET. *Trans. Electr. Electron. Mater. (TEEM)*, Springer, 21, 528–535, June 2020.
80. Jain, N. and Raj, B., Impact of underlap spacer region variation on Electrostatic and analog/RF performance of symmetrical high-k SOI FinFET at 20 nm channel length. *J. Semicond. (JoS)*, IOP Science, 38, 12, 122002, Dec 2017.
81. Singh, S. and Raj, B., Analytical Modeling and Simulation analysis of T-shaped III-V heterojunction Vertical T-FET. *Superlattices Microstruct.*, Elsevier, 147, 106717, Nov 2020.
82. Singh, G., Sarin, R.K., Raj, B., Design and Analysis of Area Efficient QCA Based Reversible Logic Gates. *J. Microprocess. Microsys.*, Elsevier, 52, 59–68, May 2017.
83. Singh, S. and Raj, B., Parametric variation analysis on hetero-junction Vertical t-shape TFET for Supressing ambipolar conduction. *Indian J. Pure Appl. Phys.*, 58, 478–485, June 2020.

84. Soniya, Singh, S., Wadhwa, G., Raj, B., Design and Analysis of Dual Source Vertical Tunnel Field Effect Transistor for high performance. *Trans. Electr. Electron. Mater.*, Springer, 21, 74–82, Oct 2019.
85. Kaur, M., Gupta, N., Kumar, S., Raj, B., Singh, A.K., RF Performance Analysis of Intercalated Graphene Nanoribbon Based Global Level Interconnects. *J. Comput. Electron.*, Springer, 19, 1002–1013, June 2020.
86. Wadhwa, G. and Raj, B., Design and Performance Analysis of Junctionless TFET Biosensor for high sensitivity. *IEEE Nanotechnol.*, 18, 567 – 574, 2019.
87. Singh, J. and Raj, B., Enhanced Nonlinear Memristor Model Encapsulating Stochastic Dopant Drift. *JNO*, ASP, 14, 958–963, 2019.
88. Bhardwaj, A., Kumar, P., Raj, B., Anand, S., Design and Performance Optimization of doping-less Vertical Nanowire TFET using Gate Stack Technique. *J. Electron. Mater. (JEMS)*, Springer, 41, 7, 4005–4013, 2022.
89. Singh, J. and Raj, B., Tunnel Current Model of Asymmetric MIM Structure Levying Various Image Forces to Analyze the Characteristics of Filamentary Memristor. *Appl. Phys. A*, Springer, 125, 3, 203.1 to 203.11, Feb 2019.
90. Goyal, C., Ubhi, J.S., Raj, B., Low Leakage Zero Ground Noise Nanoscale Full Adder using Source Biasing Technique. *J. Nanoelectron. Optoelectron.*, American Scientific Publishers, 14, 360–370, March 2019.
91. Singh, G., Sarin, R.K., Raj, B., A Novel Robust Exclusive-OR Function Implementation in QCA Nanotechnology with Energy Dissipation Analysis. *J. Comput. Electron.*, Springer, 15, 2, 455–465, June 2016.
92. Wadhwa, G., Kamboj, P., Raj, B., Design optimisation of junctionless TFET biosensor for high sensitivity. *Adv. Nat. Sci.: Nanosci. Nanotechnol.*, 10, 045001 (7), 2019.
93. Bansal, P. and Raj, B., MEMRISTOR MODELING AND ANALYSIS FOR LINEAR DOPANT DRIFT KINETICS. *J. Nanoeng. Nanomanuf.*, American Scientific Publishers, 6, 1–7, 2016.
94. Ashima, Vaithiyanathan, D., Raj, B., Performance Analysis of Charge Plasma induced Graded Channel Si nanotube. *J. Eng. Res. (JER)*, EMSME Special Issue, 146–154, Aug 2021.
95. Tomar, A.S., Magraiya, V.K., Raj, B., Scaling of Access and Data Transistor for High Performance DRAM Cell Design. *Quantum Matter*, 2, 412–416, Oct. 2013.
96. Jain, N. and Raj, B., Parasitic Capacitance and Resistance Model Development and Optimization of Raised Source/Drain SOI FinFET Structure for Analog Circuit Applications. *J. Nanoelectron. Optoelectron.*, ASP, USA, 13, 531–539, Apl 2018.
97. Singh, S., Vishvakarma, S.K., Raj, B., Analytical Modeling of Split-Gate Junction-Less Transistor for a Biosensor Application. *Sens. Bio-Sens.*, Elsevier, 18, 31–36, April 2018.
98. Gopal, M. and Raj, B., Low Power 8T SRAM Cell Design for High Stability Video Applications. *ITSI Trans. Electr. Electron. Eng.*, 1, 5, 91–97, 2013.

99. Raj, B., Mitra, J., Bihani, D.K., Rangharajan, V., Saxena, A.K., Dasgupta, S., Analysis of Noise Margin, Power and Process Variation for 32 nm FinFET Based 6T SRAM Cell. *J. Comput. (JCP)*, Academy Publisher, FINLAND, 5, 6, 1–8, 2010.
100. Sharma, D., Mehra, R., Raj, B., Comparative Analysis of Photovoltaic Technologies for High Efficiency Solar Cell Design. *Superlattices Microstruct.*, Elsevier, 153, 106861, May 2021.
101. Kaur, P., Buttar, A.S., Raj, B., A comprehensive Analysis of Nanoscale Transistor based Biosensor: A Review. *Indian J. Pure Appl. Phys.*, 59, 304–318, April 2021.
102. Yadav, D., Raj, B., Raj, B., Design and Simulation of Low Power Microcontroller for IoT Applications. *J. Sens. Lett.*, ASP, 18, 401–409, May 2020.
103. Singh, S. and Raj, B., A 2-D Analytical Surface Potential and Drain current Modeling of Double-Gate Vertical t-shaped Tunnel FET. *J. Comput. Electron.*, Springer, 19, 1154–1163, Apl 2020.
104. Singh, J. and Raj, B., An Accurate and Generic Window function for Non-linear Memristor Model. *J. Comput. Electron.*, Springer, 18, 2, 640–647, June 2019.
105. Kaur, M., Gupta, N., Kumar, S., Raj, B., Singh, A.K., Comparative RF and Crosstalk Analysis of Carbon Based Nano Interconnects. *IET Circuits Devices Syst.*, 15, 6, 493–503, Feb 2021.
106. Kandasamy, N., Ahmad, F., Ajitha, D., Raj, B., Telagam, N., Quantum Dot Cellular Automata based Scan Flip Flop and Boundary Scan Register. *IETE J. Res.*, 66, 535–548, 2020.
107. Singh, J. and Raj, B., Design and Investigation of 7T2M NVSARM with Enhanced Stability and Temperature Impact on Store/Restore Energy. *IEEE Trans. Very Large Scale Integr. Syst.*, 27, 6, 1322 – 1328, June 2019.
108. Bhardwaj, A.K., Gupta, S., Raj, B., Singh, A., Impact of Double Gate Geometry on the Performance of Carbon Nanotube Field Effect Transistor Structures for Low Power Digital Design. *Comput. Theor. Nanosci.*, ASP, 16, 1813–1820, 2019.
109. Jain, N. and Raj, B., Thermal Stability Analysis and Performance Exploration of Asymmetrical Dual-k underlap Spacer (ADKUS) SOI FinFET for Security and Privacy Applications. *Indian J. Pure Appl. Phys. (IJPAP)*, 57, 352–360, May 2019.
110. Jain, N. and Raj, B., Dual-k Spacer Region Variation at the Drain Side of Asymmetric SOI FinFET Structure: Performance Analysis towards the Analog/RF Design Applications. *J. Nanoelectron. Optoelectron.*, American Scientific Publishers, 14, 349–359, March 2019.

111. Singh, J., Sharma, S., Raj, B., Khosla, M., Analysis of barrier layer thickness on performance of In1-xGaxAs based Gate Stack Cylindrical Gate Nanowire MOSFET. *JNO*, ASP, 13, 1473–1477, Oct 2018.
112. Jain, N. and Raj, B., Analysis and Performance Exploration of High-k SOI FinFETs Over the Conventional Low-k SOI FinFET toward Analog/RF Design. *J. Semicond. (JoS)*, IOP Science, 39, 12, 124002-1–7, Dec 2018.

6

Ferroelectric Tunnel FET: Next Generation of Classical Low Power CMOS Technology

Naima Guenifi[1], Shiromani Balmukund Rahi[2*], Houda Chabane[1] and Khadidja Dibi[1]

[1]LEA Electronics Department, University Mostefa Benboulaid of Batna 2, Batna, Algeria
[2]University School of Information and Communication Technology, Gautam Buddha University, Greater Noida, Uttar Pradesh, India

Abstract
After identification and experimental verification of the negative capacitance (NC) in ferroelectric material, significant research efforts have been continued with classical FET structures like bulk MOSFET, FinFET, gate all around(GAA) and nanosheet FET for/achieving improvement of device figures of merits(FoMs) such as I_{ON}, I_{OFF}, I_{ON}/I_{OFF} subthreshold swing(SS) for various applications. Low power use has become a serious issue for semiconductor players. Due to the large values of subthreshold swing (>60 mV/decade) conventional FETs structures such as bulk MOSFET, FinFET, gate all around (GAA), and nanosheet FET have failed to fulfill the requirements of ultra-low power uses. The scientific field behind classical MOS structures has not allowed the use of <0.5 V power supply. The tunnel FET was found to be the most suitable FET device, followed by band-to-band tunnel tunneling(BTBT) for ultra-low power uses owing to its smaller values of SS <60 mV/decade. However, a low on-current (I_{ON}) limitation forces more scientific exercise to achieve improved circuit design parameters with tunnel FET, such as I_{ON}, I_{OFF}, I_{ON}/I_{OFF}, and SS. The ferroelectric material was found to be suitable for reducing the applied power supply (V_{DD}). In this chapter, the authors focused on investigating the impact of ferroelectric gate materials in conventional tunnel FET. We used a double-gate tunnel FET (DG-TFET) structure. In conventional DGTFET structures, composite gate material engineering has been implemented with the help of a ferroelectric material and HFO_2. The negative capacitance (NC) in ferroelectric materials enables internal voltage amplification inside device, potentially reduce

**Corresponding author*: sbrahi@gmail.com

Balwinder Raj, Shiromani Balmukund Rahi and Nandakishor Yadav (eds.) FeFET Devices, Trends, Technology and Applications, (131–154) © 2025 Scrivener Publishing LLC

power (V_{DD}) without any compromise the devices figure of merits(FoMs) such as I_{ON}, I_{OFF} and I_{ON}/I_{OFF} ratio. The obtained results show a low I_{OFF} ($\sim 10^{-20}$ $A/\mu m$) and I_{ON} ($\sim 10^{-5}$ $A/\mu m$), while the SS was also reduced significantly.

Keywords: Tunnel FET, ferroelectric, low power, voltage amplification

6.1 Introduction

Tunnel field-effect transistors (TFETs) have emerged as promising candidates for ultra-low power uses due to their low off-state current (I_{OFF}), measured at $V_{GS} = 0.0V$ and steep subthreshold swing (SS <60 mV/decade) compared to conventional FETs structures such as bulk MOSFET, FinFET, gate all around(GAA), and nanosheet FET help to reduce power supply, V_{DD} <0.5 V. Figure 6.1 shows the prediction of the ferroelectric market for future generations. Tunnel FETs (TFETs) offer potential for future low-power logic circuits with sub-0.5 V operation due to their lower subthreshold swing (SS) compared to conventional MOSFETs [1–5]. Their integration with the existing CMOS technology presents challenges at the process level, particularly when combined with ferroelectric materials [6–10]. This is because TFETs rely on quantum-mechanical band-to-band tunneling (BTBT) in gated reverse-biased p–i–n junctions, which is fundamentally different from the operation of MOSFETs [11–16].

In addition to utilizing smaller-bandgap materials to enhance tunneling efficiency, the implementation of a type-II staggered heterojunction at the tunneling junction can effectively reduce the bandgap. This reduction in the effective bandgap, compared to that of bulk materials, leads to a

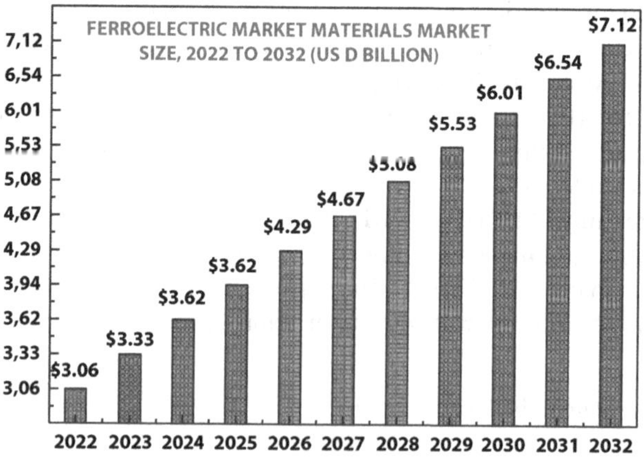

Figure 6.1 Ferroelectric market.

significant increase in the magnitude of the on-state tunneling current. In this chapter, we used pseudomorphically grown Si/$Si_{1-x}Ge_x$ layered on relaxed $Si_{1-x}Ge_x$, particularly for x = 0.4 biaxial tensile strain in the Si layer and compressive strain SiGe. High-quality Si/SiGe is used in this study. Overall, the mismatched between Si/SiGe causes a biaxial strain at the interface, resulting in a decrease in the effective bandgap. In this chapter, we attempt to identify the impact of ferroelectric materials on conventional double-gate TFET structures. The following section is devoted to an in-depth investigation of the impact of ferroelectrics in conventional DGTFET [17–24].

6.2 Implementation of Ferroelectric Material in Tunnel FET

6.2.1 Device Structure

In this section, the details of a tunnel FET having a double-gate structure with SiGe, a mole-dependent band gap [48] material in the source region, and Si materials in the drain and channel regions are discussed [25–32]. The combination of two different bandgap(E_G) materials fulfill the heterostructure interface in the tunnel region of the device. This emerging concept provides a relatively lower tunneling region than the homostructure. It improves the I_{ON} current [33–37] of the device governed by band-to-band tunneling (BTBT), followed by the WKB approximation theory for

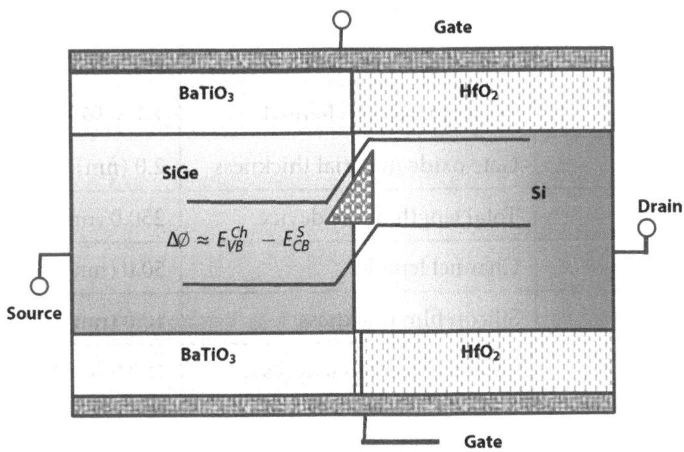

Figure 6.2 Schematic diagram of ferroelectric-based DGTFET.

tunneling measured with the help of Equation (6.1). Figure 6.2 shows a schematic view of a double-gate tunnel FET with a composite gate material ferroelectric material on the source side and HfO_2 on the drain side. Table 6.1 summarizes the key device design parameters, offering valuable information for device design engineers.

The gate oxide thickness (tox), an important device parameter for performance optimization, was set to 2.0 nm. The silicon source channel thickness is 10.0 nm, and the entire channel length from the source to the drain region is 50.0 nm [38–43]. Uniform doping concentrations of 2.5×10^{20} cm^{-3}, 1.0×10^{18} cm^{-3}, and 1.1×10^{15} cm^{-3} have been used for the source (Ns), drain (N_D), and channel (N_C) regions, respectively. The work function for the gate material in this region was chosen to be 5.2 eV. The mesh size was set to 5.0×10^{-4} μm at the source/channel interface and 10^{-3} μm in the rest of the structure [44–49]. Newton's iterative numerical method was employed to achieve the best convergence and reduce computation time.

6.2.2 Device Lay-Out Information of Fe Tunnel

Table 6.1 provides the device lay-out information used in schematic shown in Figure 6.1.

Table 6.1 Device design parameters for DG tunnel FET.

S.n.	Symbol	Physical parameters	Numeric value
1	φ_M	Work function	5.2 (eV)
2	N_S	Doping levels for source	2.5×10^{20} (cm^{-3})
3	N_D	Doping level for Drain	1.0×10^{18} (cm^{-3})
4	N_C	Doping level for channel	1.1×10^{15} (cm^{-3})
5	t_{ox}	Gate oxide material thickness	2.0 (nm)
6	L_t	Total length of the device	250.0 (nm)
7	L_{ch}	Channel length	50.0 (nm)
8	t_{Si}	Silicon film thickness	10.0 (nm)
9	L_S/L_D	Source and drain lengths	100.0 (nm)

6.3 Results and Analysis

This section has been devoted to the depth analysis of the Tunnel FET shown in Figure 6.2 which has a ferroelectric gate material inside the source and channel regions. The device follows band to band tunneling instead of drift diffusion transport phenomena of change carriers inside it [50–57]. This work employs the Wentzel–Kramers–Brillouin (WKB) approximation with a triangular potential and displacement vector continuity method to analyze charge carrier transport in Fe-DGTFETs [1, 3, 8]. In contrast, conventional tunnel FETs (TFETs) rely on band-to-band tunneling (BTBT) governed by an applied electric field within the tunneling region and channel material [24, 25, 32]. Equation (6.1) describes the tunneling mechanism in Fe-DGTFETs, which leverages the WKB approximation with a triangular potential [58–64].

$$T(E) \propto \left(-\frac{4\sqrt{2m * E_{g\text{-}effective}^{\frac{3}{2}}}}{3|q|h(E_{g\text{-}effective} + \Delta\phi)}\sqrt{\frac{\varepsilon_{si}}{\varepsilon_{ox}}t_{ox}t_{si}}\right)\Delta\phi \tag{6.1}$$

In Equation (6.1), m* is effective mass, $E_{g\text{-effective}}$ is effective band gap, $\Delta\phi$ is tunneling window, ε_{si}, ε_{ox} is dielectrics constants of Si and oxide material respectively, and t_{si}, t_{ox} is thickness of Si and oxide material, respectively. The numerical values of the device design parameters are listed in Table 6.1. The tunneling window can be expressed as: $E_{VB}^{ch} - E_{CB}^{S}$, as shown in Figure 6.2. As Equation (6.1) indicates, it has a sufficient design domain for device design engineers. As explained in References [32, 47–49], in a conventional high-k tunnel FET, the electric field (E) and permittivity (ε_{ox}) follow the displacement vector (D) with Equation (6.2), whereas in a ferroelectric material, it follows Equation (6.3). The additional parameter P causes more electric field in the Fe devices than in the conventional high-k configuration, depending on the thickness T_{FE} [32, 49].

$$D_{high\text{-}k} = \varepsilon E \tag{6.2}$$

$$D_{Ferro\text{-}k} = \varepsilon E + P \tag{6.3}$$

The increment in electric filed inside device has been realized by displacement vector continuity followed condition followed by Equation (6.4).

$$D_{Ferro-k} = D_{high-k} \tag{6.4}$$

By using, Equations (6.2), (6.3), and (6.4), we have, Equations (6.5) and (6.6).

$$\epsilon_1 E_1 + P = \epsilon_2 E_2 \tag{6.5}$$

or

$$\Rightarrow P = \epsilon_2 E_2 \sim \epsilon_1 E_1 \tag{6.6}$$

The simplified Equation (6.6) suggests an amplified electric field within the device due to the presence of a ferroelectric material with polarization P. The polarization of the ferroelectric layer was assumed to result from a superposition of individual, non-interacting dipoles [65–73]. To reduce the simulation time, the dipole density function is assumed to be Gaussian, simplifying the polarization P, as shown in Equation (6.7).

$$P = k \cdot P_s \cdot \tanh\left(\frac{E_{eff} + E_c}{2\delta}\right) + P_{off} \tag{6.7}$$

And δ is defined as Equation (6.8)

$$\delta - E_c\left[\ln\left(\frac{1+\frac{P_r}{P_s}}{1-\frac{P_r}{P_s}}\right)\right]^{-1} \tag{6.8}$$

The scaling factor k and offset parameter P_{off} enable the creation of unsaturated polarization loops, i.e., subloops. The + or − sign in (Equation (6.7)) introduces nonvolatility by separating the polarization states depending on the sweep direction of the applied electric field. The Parameters E_{eff}, Ec, Pr, and Ps are the effective electric field, coercive field, remnant polarization, and saturation polarization, respectively [22, 23, 25, 29, 30]. Figure 6.3 shows the improvement in electric field inside the device, shown in Figure 6.2.

Figure 6.4a shows the 2-D energy band diagram for the Fe Tunnel FET shown in Figure 6.2. Figure 6.4b show the simulated 3-D energy band diagram. Figures 6.4(a,b) show the comparison of the on- and off- states for the device [74–82]. It has been observed that in off-sate tunneling region inside device is too large, causes no tunneling between source and channel inside tunnel region but is on-state this bandgap has ban practically reduced much more due to applied gate electric filed, causes desirable band to band (BTBT) tunnel inside tunnel region.

Figure 6.3 compares the electric field with a conventional double-gate TFET structure with HfO_2, a high-k gate with an Fe–DG TFET, as shown in Figure 6.2. The amplified electric held in the Fe-DG TFET is shown in Figure 6.3. In the case of the Fe Tunnel inside the device, the electric field was improved without any increment in the external applied voltage (V_{DD}), as shown in Figure 6.3. This improvement in the electric field (Figure 6.2) is due to the ferroelectric material inside the source region, which causes more tunneling of charge than the conventional high-k DGTFET. The incremental change in the electric field inside the Fe-based FET structures causes voltage amplification inside the device [3, 8, 51–55].

For clear picturization of the tunneling phenomena in the device shown in Figure 6.4b. By examining the alignment of the energy bands, as shown in Figure 6.4b for various V_{GS} values, one can identify regions where the

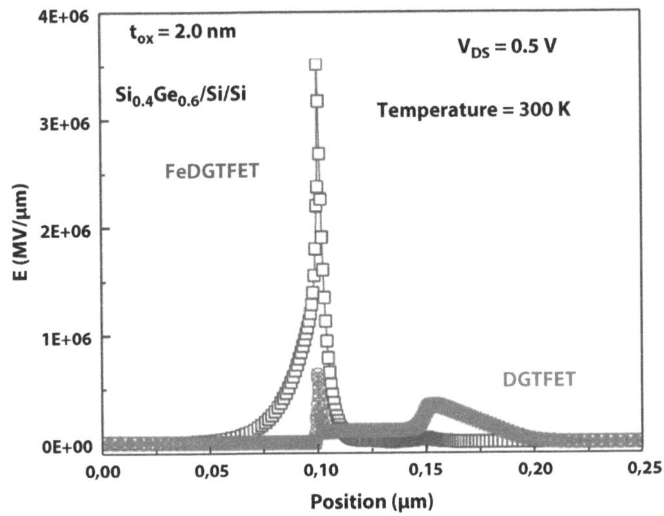

Figure 6.3 Impact of electric field due to application ferroelectric material.

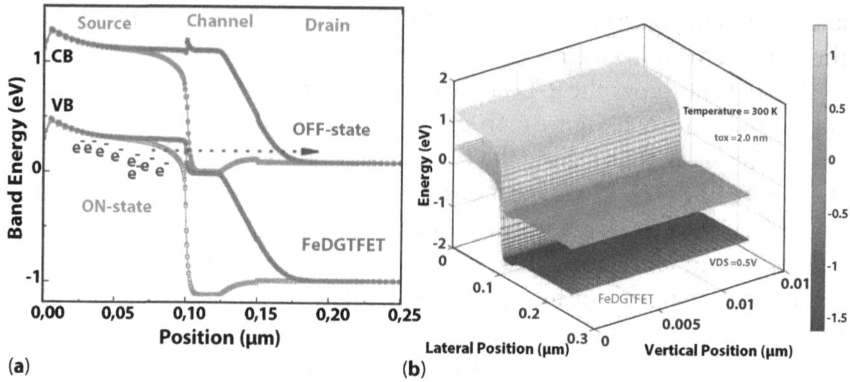

Figure 6.4 (a) Energy band diagram of Fe-DGTFET, (b): Energy band diagram of Fe- DGTFET.

valence bands of the source and drain overlap in the band space, indicating potential tunneling current regions. It has been observed that how the alignment of energy bands changes with the applied V_{GS} enables an understanding of how the tunneling current is modulated by the applied voltage. The oxide thickness (t_{ox}) influences the electron tunneling probability, with thinner oxides (2.0 nm) allowing for higher tunneling currents [83–87].

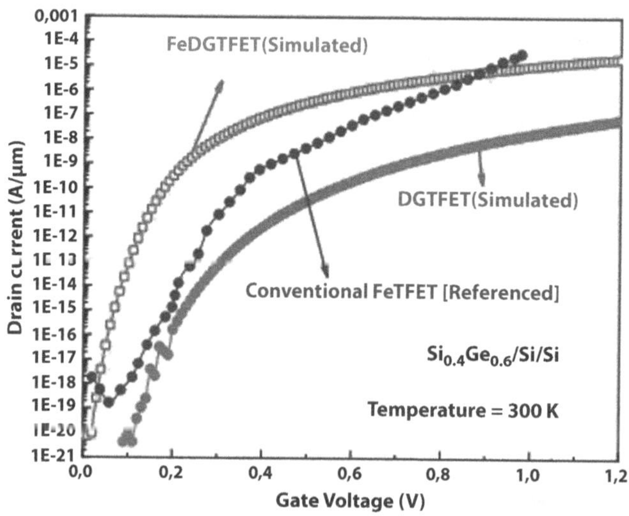

Figure 6.5 Comparison of ID-VG characteristic of Fe Tunnel with conventional TFET structures.

Figure 6.5 shows a quantitative comparison of tunnel FET structures. Here, we have considered three cases: in case 1, a conventional double-gate tunnel FET; in case 2, a conventional FeTunnel; and in case 3, a double-gate tunnel FET with an Fe gate material with HFO_2. For case 3, as shown in Figure 6.5, the I–V characteristics improved with a lower I_{OFF} current and steep subthreshold swing (SS). The I_{ON} currents in cases 2 and 3 were almost the same.

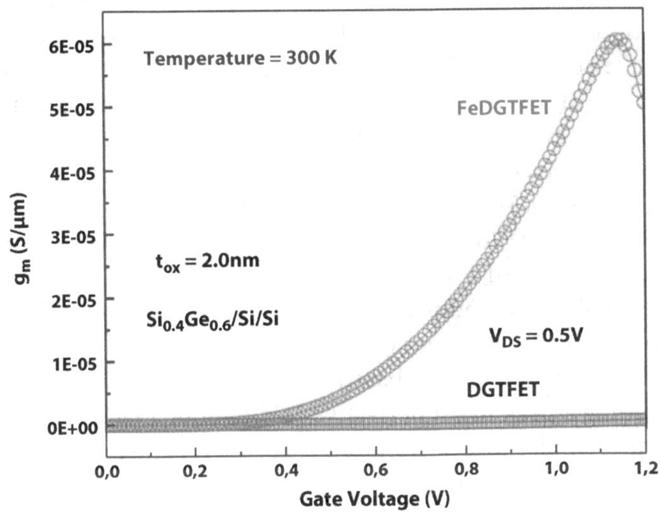

Figure 6.6 Comparison of first derivative of ID-VGS of conventional double gate tunnel FET with Fe based DGTFET.

Table 6.2 Comparison between DGTFET homojunction heterojunction and FeDGTFET.

Design parameter	High-k dielectric ($HfO_2 \approx 25$)		FeDGTFET
	Homo-DG-TFET (Si/Si/Si)	Hetero-DG-TFET (Si/GaAs/G aAs)	
I_{ON} (A/μm)	4.0×10^{-6}	4.0×10^{-6}	1.0×10^{-4}
I_{OFF} (A/μm)	1.0×10^{-18}	1.0×10^{-20}	1.0×10^{-20}
I_{ON}/I_{OFF}	4.0×10^{12}	4.0×10^{14}	1.0×10^{16}
SS (mV/dec)	39.68	34.25	31.25
I_{amb}(A/μm)	5.0×10^{-11}	1.0×10^{-19}	1.0×10^{-19}

A comparison of the electrical parameters of the different structures studied in a previous work [reference] and this work. According to the results, an improvement in the parameters was observed in the case of the FeDGTFET (Table 6.2).

To explore the impact of the work function (WF) and source doping concentration (Ns) on device performance, we conducted over 400 simulations using a MATLAB script interfaced with Silvaco. All other device parameters remained constant [88–94]. The simulations were swept through 20 work function values (4.9 eV to 5.2 eV) and 20 source doping concentrations (1.0×10^{20} cm^{-3} to 1.0×10^{22} cm^{-3}). Key device parameters, including the subthreshold slope (SS), on-current (I_{ON}), off-current (I_{OFF}), their ratio (I_{ON}/I_{OFF}), and threshold voltage (Vth), were extracted from each simulation and saved for analysis [95–101]. Figure 6.7 illustrates the combined influence of the source doping and work function on the subthreshold slope (SS). We observed a significant decrease in the SS, reaching values close to 47 mV/decade, before plateauing once the source doping concentration exceeded 2×10^{20} cm^{-3}.

Figure 6.5 shows a comparison of the first derivative of I_D–V_{GS} of a conventional double-gate tunnel FET with an Fe-based DGTFET. The improvement in g_m for the Fe-based Tunnel FET is shown in Figure 6.6. After obtaining a peak in 1st derivative I–V, a sudden fall indicates a non-linear feature, similar to a conventional MOSFET. Figure 6.8 illustrates the gain versus cut-off frequency (f_T) and gate voltage (V_{GS}) for an Fe-DGTFET in a three-dimensional representation. The component parameters under

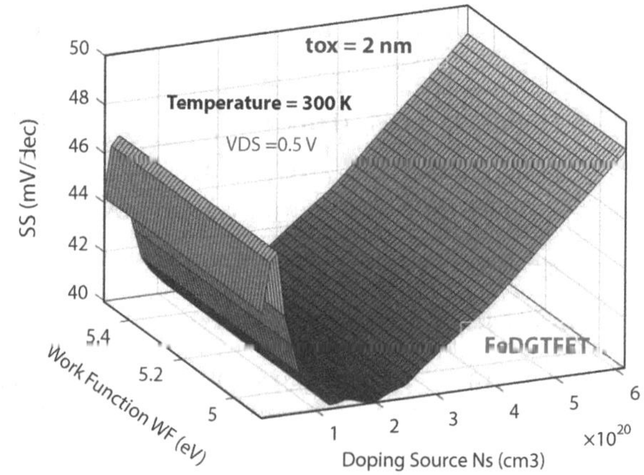

Figure 6.7 SS versus work function and doping source.

investigation include V_{DS} = 0.5 V, a temperature of 300 K, and an oxide thickness (tox) of 2.0 nm [102–107].

From Figure 6.8, it is evident that the gain reached 60 GHz with a broad bandwidth. Notably, the component exhibited high linearity, ensuring that the signal under study remained undistorted. The results were obtained through simulations using Silvaco and further analyzed using MATLAB, leveraging the mathematical model shown in Equation (6.9). The gain bandwidth of the Fe-DGTFET was measured with an input voltage with a cutoff frequency of the device.

$$f_r = \frac{g_m}{2\pi(C_{gs} + C_{gd})} = \frac{g_m}{2\pi C_{gg}} \tag{6.9}$$

Equation (6.10) has been used to determine the gain band width of Fe-DGTFET

$$GBW = \frac{g_m}{2\pi 10 C_{gd})} \tag{6.10}$$

Figure 6.9 shows the transductance (g_m) versus drain current (I_{DS}) of the FeDGTFET; an initial increase in gm with I_{DS} was observed [108–120]. This indicated an improvement in the device's current gain. This initial growth phase of g_m signifies that the transistor possesses high electrostatic control, where small variations in the gate voltage lead to significant

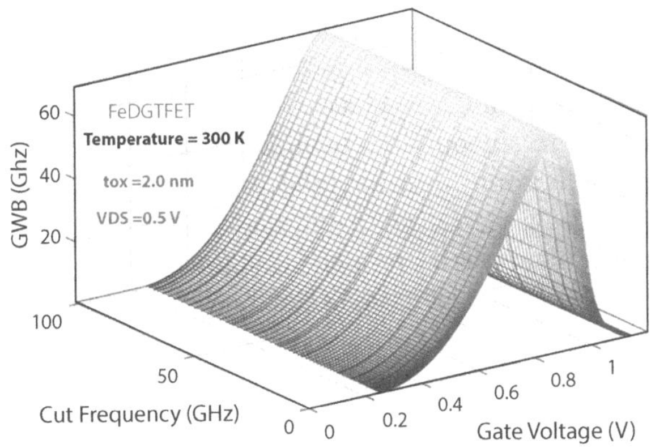

Figure 6.8 Gain versus cut- frequency (fT) and gate voltage (VGS).

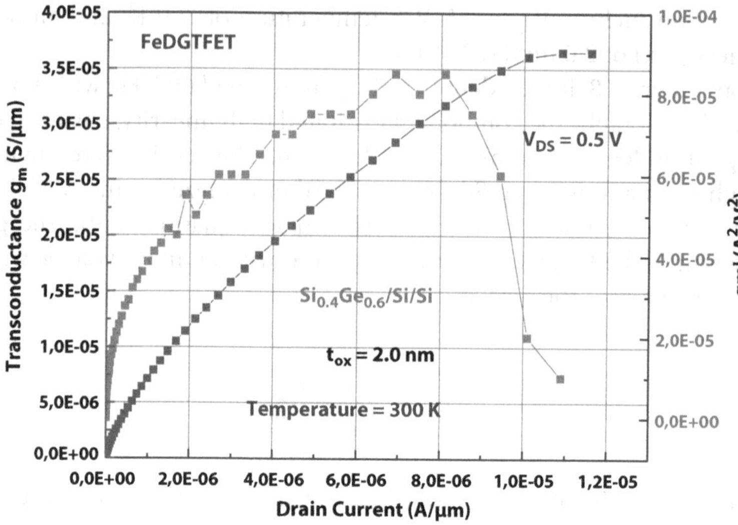

Figure 6.9 Variation of transconductance and 1st derivative with drain current.

increases in the drain current, thus favoring efficient amplification [121–132]. As I_{DS} continues to increase and reaches its maximum, this stabilization of gm implies that the gain no longer improves despite the increase in drain current, marking an optimal but reached amplifier performance. Simultaneously, the 1st-order derivative of the transconductance (g_m') initially increases, signaling a rapid growth of g_m, and then decreases after reaching a maximum, indicating that the growth of gm slows down and stabilizes [133–146].

6.4 Conclusion

Progress in conventional CMOS-based applications is rapidly increasing. With the progress in silicon circuit miniaturization, lowering the power consumption has become a major objective for semiconductor players. The supply voltage(V_{DD}) scaling in ultralarge-scale integration (ULSI) is limited by the physical barrier termed "Boltzmann Tyranny". To solve the existing problems of classical CMOS technology, a ferroelectric negative capacitance Tunnel FET (Fe-TFET) is proposed in this chapter. To reduce the subthreshold swing (SS) through an internal voltage amplification mechanism, the supply voltage can be effectively scaled, and the power dissipation of the ULSI is significantly lowered. Various influencing factors of DG Tunnel, such as I_{ON}, I_{OFF} SS, I_{ON}/I_{OFF} ratio, and SS, were also theoretically analyzed and varied with the help of a standard device simulation technique. The obtained results of the Fe-DGTFET show an amplification of the internal voltage inside the device. The improved electric field inside the device, without increasing the applied voltage (V_{DD}), also increased the tunneling current. The transconductance (g_m), cutoff frequency (f_T), and gain bandwidth (GBW) of the proposed ferroelectric device were also determined. The obtained results indicate that ferroelectric-based FET can be used in power saving applications.

References

1. Ionescu, A.M., Lattanzio, L., Salvatore, G.A., De Michielis, L., Boucart, K., Bouvet, D., The hysteretic ferroelectric tunnel FET. *IEEE Trans. Electron Devices*, 57, 12, 3518–3524, 2010.
2. Tu, L., Wang, X., Wang, J., Meng, X., Chu, J., Ferroelectric negative capacitance field effect transistor. *Adv. Electron. Mater.*, 4, 11, 1800231, 2018.
3. Kobayashi, M., Jang, K., Ueyama, N., Hiramoto, T., Negative capacitance for boosting tunnel FET performance. *IEEE Trans. Nanotechnol.*, 16, 2, 253–258, 2017.
4. Rollo, T. and Esseni, D., New design perspective for ferroelectric NC- FETs. *IEEE Electron Device Lett.*, 39, 4, 603–606, 2018.
5. Jung, H., Impact of Hysteresis Curve on Subthreshold Swing in Ferroelectric FET. *Vietnam J. Sci. Technol.*, 62, 1, 156–169, 2024.
6. Choe, G. and Yu, S., Variability study of ferroelectric field-effect transistors towards 7nm technology node. *IEEE J. Electron Devices Soc.*, 9, 1131–1136, 2021.
7. Tian, G., Xu, G., Yin, H., Li, L., Yan, G., Luo, Y., Sun, X., Negative capacitance double-gate vertical tunnel FET with improvedsubthreshold characteristics, in: *2022 China Semiconductor Technology International Conference (CSTIC)*, IEEE, pp. 1–3, 2022.
8. Rahi, S.B., Tayal, S., Upadhyay, A.K., A review on emerging negative capacitance field effect transistor for low power electronics. *Microelectron. J.*, 116, 105242, 2021.
9. Gupta, R., Beg, S., Singh, S., Evaluation of Design and Performance of Biosensor Utilizing Ferroelectric Vertical Tunnel Field-Effect Transistor (V-TFET). *Silicon*, 16, 1–12, 2024.
10. Lin, H.-H. and Hu, V.P.-H., Device designs and analog performance analysis for negative-capacitance vertical-tunnel FET, in: *20th International Symposium on Quality Electronic Design (ISQED)*, IEEE, pp. 241–246, 2019.

11. Lee, M.-H., Wei, Y.-T., Lin, J.-C., Chen, C.-W., Tu, W.-H., Tang, M.J.A.A., Ferroelectric gate tunnel field-effect transistors with low-power steep turn-on. *AIP Adv.*, 4, 10, 1–7, 2014.
12. Kobayashi, M., Jang, K., Ueyama, N., Hiramoto, T., Negative capacitance as a performance booster for tunnel FET, in: *2016 IEEE Silicon Nanoelectronics Workshop (SNW)*, IEEE, pp. 150–151, 2016.
13. Xu, H., Two dimensional analytical model for a negative capacitance double gate tunnel field effect transistor with ferroelectric gate dielectric. *J. Semicond.*, 39, 10, 104004, 2018.
14. Upadhyay, A.K., Rahi, S.B., Tayal, S., Song, Y.S., Recent progress on negative capacitance tunnel FET for low-power applications: Device perspective. *Microelectron. J.*, 129, 105583, 2022.
15. Harikumar, K.R., Midhun Das, P., Shikha, U.S., James, R.K., Pradeep, A., Performance Analysis of Dielectric Engineered Negative Capacitance Tunnel FETs, in: *2022 IEEE 3rd International Conference on VLSI Systems, Architecture, Technology and Applications (VLSI SATA)*, IEEE, pp. 1–4, 2022.
16. Tayal, S., Upadhyay, A.K., Kumar, D., Rahi, S.B. (Eds.), *Emerging low-power semiconductor devices: Applications for future technology nodes*, CRC Press, Boca Raton, FL, USA, 2022.
17. Naima, G. and Rahi, S.B., Low power circuit and system design hierarchy and thermal reliability of tunnel field effect transistor. *Silicon*, 14, 2021.
18. Thoti, N. and Li, Y., Design of GAA nanosheet ferroelectric area tunneling FET and its significance with DC/RF characteristics including linearity analyses. *Nanoscale Res. Lett.*, 17, 1, 53, 2022.
19. Guenifi, N., Rahi, S.B., Benmahdi, F., Chaabane, H., Optimization for device figure of merit of ferroelectric tunnel FET using genetic algorithm. *ECS J. Solid State Sci. Technol.*, 12, 2, 023001, 2023.
20. Wahab, M.A. and Alam, M.A., A verilog-A compact model for negative capacitance FET. Version 1.1. 3, 2017.
21. Chen, Z., Li, W., Fan, Z., Dong, S., Chen, Y., Qin, M., Zeng, M., *et al.*, All-ferroelectric implementation of reservoir computing. *Nat. Commun.*, 14, 1, 3585, 2023.
22. Zhao, Y., Liang, Z., Huang, Q., Wang, H., Peng, Y., Han, G., Huang, R., Experimental study on the transient response of negative capacitance tunnel FET, in: *2019 Electron Devices Technology and Manufacturing Conference (EDTM)*, IEEE, pp. 88–90, 2019.
23. Paras, N., Rahi, S.B., Upadhyay, A.K., Bharti, M., Song, Y.S., Design and analysis of novel La: HfO2 gate stacked ferroelectrictunnel FET for non-volatile memory applications. *Memories - Mater. Devices Circuits Syst.*, 7, 100101, 2024.
24. Yadav, D.S., Rahi, S.B., Tirkey, S. (Eds.), *Advanced Field-Effect Transistors: Theory and Applications*, CRC Press, Boca Raton, FL, USA, 2023.

25. Guenifi, N., Rahi, S.B., Benmahdi, F., Chabane, H., Optimization of tunneling current in ferroelectric tunnel FET using genetic algorithm. *J. Supercomput.*, 79, 14, 15773–15789, 2023.
26. Song, Y.S., Tayal, S., Rahi, S.B., Upadhyay, A.K. (Eds.), *Negative Capacitance Field Effect Transistors: Physics, Design, Modeling and Applications*, CRC Press, Boca Raton, FL, USA, 2023.
27. Thoti, N. and Li, Y., Promised design of energy-efficient negative- capacitance vertical tunneling FET. *ECS J. Solid State Sci. Technol.*, 10, 7, 075002, 2021.
28. Samuel, T.S.A., Song, Y.S., Tayal, S., Vimala, P., Rahi, S.B. (Eds.), *Tunneling Field Effect Transistors: Design, Modeling and Applications*, CRC Press, Boca Raton, FL, USA, 2023.
29. Go, S., Kim, S., Park, J.Y., Lee, D.K., Noh, H.J., Park, S.R., Kim, Y., Kim, D.H., Kim, S., Impact of sidewall spacer materials and gate underlap length on negative capacitance double-gate tunnel field- effect transistor (NCDG-TFET). *Solid State Electron.*, 198, 108483, 2022.
30. Bind, U.C., Rahi, S.B., Basic Operational Principle of Anti-ferroelectric and Ferroelectric Materials, in: *Negative Capacitance Field Effect Transistors*, pp. 62–82, CRC Press, Boca Raton, FL, USA, 31 October 2023.
31. Guha, S. and Pachal, P., Heterojunction negative-capacitance tunnel-FET as a promising candidate for sub-0.4 VV DD digital logic circuits. *IEEE Trans. Nanotechnol.*, 20, 576–583, 2021.
32. Rahi, S.B., Asthana, P., Gupta, S., Heterogate junctionless tunnel field-effect transistor: future of low-power devices. *J. Comput. Electron.*, 16, 30–38, 2017.
33. Kim, S., Go, S., Kim, S., Simulation study about negative capacitance effects on recessed channel tunnel FET. *Jpn. J. Appl. Phys.*, 60, SC, SCCE07, 2021.
34. Hu, V.P.-H., Lin, H.-H., Lin, Y.-K., Hu, C., Optimization of negative-capacitance vertical-tunnel FET (NCVT-FET). *IEEE Trans. Electron Devices*, 67, 6, 2593–2599, 2020.
35. Liu, C., Chen, P.-G., Xie, M.-J., Liu, S.-N., Lee, J.-W., Huang, S.-J., Liu, S., *et al.*, Simulation-based study of negative-capacitance double-gate tunnel field-effect transistor with ferroelectric gate stack. *Jpn. J. Appl. Phys.*, 55, 4S, 04EB08, 2016.
36. Saeidi, A., Jazaeri, F., Bellando, F., Stolichnov, I., Luong, G.V., Zhao, Q.-T., Mantl, S., Enz, C.C., Ionescu, A.M., Negative capacitance as performance booster for tunnel FETs and MOSFETs: an experimental study. *IEEE Electron Device Lett.*, 38, 10, 1485–1488, 2017.
37. Zhao, Y., Liang, Z., Huang, Q., Chen, C., Yang, M., Sun, Z., Zhu, K., *et al.*, A novel negative capacitance tunnel FET with improved subthreshold swing and nearly non-hysteresis through hybrid modulation. *IEEE Electron Device Lett.*, 40, 6, 989–992, 2019.
38. Saeidi, A., Jazaeri, F., Stolichnov, I., Luong, G.V., Zhao, Q.-T., Mantl, S., Ionescu, A.M., Effect of hysteretic and non-hysteretic negative capacitance on tunnel FETs DC performance. *Nanotechnology*, 29, 9, 095202, 2018.

39. Cao, W. and Banerjee, K., Is negative capacitance FET a steep-slope logic switch? *Nat. Commun.*, 11, 1, 196, 2020.
40. Kim, H.W. and Kwon, D., Gate-normal negative capacitance tunnel field-effect transistor (TFET) with channel doping engineering. *IEEE Trans. Nanotechnol.*, 20, 278–281, 2021.
41. Li, W., Jia, Q., Pan, Y., Chen, X., Yin, Y., Wu, Y., Wang, Y., Wen, Y., Wang, C., Wang, S., A T-shaped gate tunneling field effect transistor with negative capacitance, super-steep subthreshold swing. *Nanotechnology*, 32, 39, 395202, 2021.
42. Rollo, T., Ferroelectric Negative Capacitance Transistors as Beyond Tunnel-FETs, Steep-Slope Devices: a Modeling, Simulation and Design Study, 2019.
43. Rooyackers, R., Trends and challenges in Tunnel-FETs for low power electronics, in: *2019 34th Symposium on Microelectronics Technology and Devices (SBMicro)*, IEEE, pp. 1–6, 2019.
44. Guha, S. and Pachal, P., Exploring the Feasibility of Implementing Negative-Capacitance Tunnel-FET (NC-TFET) in Low-Power Digital Circuits, in: *2022 IEEE VLSI Device Circuit and System (VLSI DCS)*, IEEE, pp. 1–4, 2022.
45. Lone, M.A., Solay, L.R., Singh, A., Intekhab Amin, S., Anand, S., Design and Implementation of Negative Capacitance Based Electrostatic Doped Double Gate Tunnel Field Effect Transistor. *Silicon*, 14, 18, 12293–12301, 2022.
46. Alam, M.A., Si, M., Ye, P.D., A critical review of recent progress on negative capacitance field-effect transistors. *Appl. Phys. Lett.*, 114, 1–7, 9, 2019.
47. Samuel, T.S.A., Suh Song, Y., Tayal, S., Vimala, P., Rahi, S.B. (Eds.), *Tunneling Field Effect Transistors: Design, Modeling and Applications*, 1st ed., CRC Press, Boca Raton, FL, USA, 2023, https://doi.org/10.1201/9781003327035.
48. Rahi, S.B. and Ghosh, B., High-k double gate junctionless tunnel FET with a tunable bandgap. *RSC Adv.*, 5, 67, 54544–54550, 2015.
49. Ilatikhameneh, H., *et al.*, Dielectric Engineered Tunnel Field-Effect Transistor. *IEEE Electron Device Lett.*, 36, 10, 1097–1100, October 2015.
50. Boucart, K. and Ionescu, A.M., Double Gate Tunnel FET with ultrathin silicon body and high-k gate dielectric. *IEEE Electron Devices*, 54, 7, 1725–1733, 2007.
51. Krivokapic, Z., Aziz, A., Song, D., Rana, U., Galatage, R., Banna, S., NCFET: Opportunities & challenges for advanced technology nodes, in: *2017 Fifth Berkeley Symposium on Energy Efficient Electronic Systems & Steep Transistors Workshop (E3S)*, IEEE, pp. 1–3, 2017.
52. Salamin, S., Rapp, M., Amrouch, H., Pahwa, G., Chauhan, Y., Henkel, J., NCFET-aware voltage scaling, in: *2019 IEEE/ACM International Symposium on Low Power Electronics and Design (ISLPED)*, IEEE, pp. 1–6, 2019.
53. Liang, Y., Li, X., Gupta, S.K., Datta, S., Narayanan, V., Analysis of DIBL effect and negative resistance performance for NCFET based on a compact SPICE model. *IEEE Trans. Electron Devices*, 65, 12, 5525–5529, 2018.
54. Liang, Y., Li, X., George, S., Srinivasa, S., Zhu, Z., Gupta, S.K., Datta, S., Narayanan, V., Influence of body effect on sample-and-hold circuit design

using negative capacitance FET. *IEEE Trans. Electron Devices*, 65, 9, 3909–3914, 2018.

55. Datta, S., Liu, H., Narayanan, V., Tunnel FET technology: A reliability perspective. *Microelectron. Reliab.*, 54, 5, 861–874, 2014.
56. Guenifi, N. and Rahi, S.B., Design and Optimization of Heterostructure Double Gate Tunneling Field Effect Transistor for Ultra Low Power Circuit and System, in: *Electrical and Electronic Devices, Circuits and Materials Technological Challenges and Solutions*, 2021, DOI:10.1002/9781119755104.
57. Yadav, D., Chouhan, S.S., Vishvakarma, S.K., Raj, B., Application Specific Microcontroller Design for IoT based WSN. *Sens. Lett.*, ASP, 16, 374–385, May 2018.
58. Singh, G., Sarin, R.K., Raj, B., Fault-Tolerant Design and Analysis of Quantum-Dot Cellular Automata Based Circuits. *IEEE/IET Circuits Devices Syst.*, 12, 638–644, 2018.
59. Singh, J. and Raj, B., Modeling of Mean Barrier Height Levying Various Image Forces of Metal Insulator Metal Structure to Enhance the Performance of Conductive Filament Based Memristor Model. *IEEE Nanotechnol.*, 17, 2, 268–267, March 2018.
60. Jain, A., Sharma, S., Raj, B., Analysis of Triple Metal Surrounding Gate (TM-SG) III-V Nanowire MOSFET for Photosensing Application. *Opto-Electron. J.*, Elsevier, 26, 2, 141–148, May 2018.
61. Jain, A., Sharma, S., Raj, B., Design and Analysis of High Sensitivity Photosensor Using Cylindrical Surrounding Gate MOSFET For Low Power Sensor Applications. *Eng. Sci. Technol. Int. J.*, Elsevier's, 19, 4, 1864–1870, December 2016.
62. Saiphani Kumar, G., Singh, A., Raj, B., Design and Analysis of Gate All Around CNTFET based SRAM cell Design. *J. Comput. Electron.*, Springer, 17, 1, 138–145, March 2018.
63. Singh, G.P., Sohi, B.S., Raj, B., Material Properties Analysis of Graphene Base Transistor (GBT) for VLSI Analog Circuits. *Indian J. Pure Appl. Phys. (IJPAP)*, 55, 896–902, Dec-2017.
64. Kumar, S. and Raj, B., Estimation of Stability and Performance metric for Inward Access Transistor based 6T SRAM Cell Design using n-type/p-type DMDG-GDOV TFET. *IEEE VLSI Circuits Syst. Lett.*, 3, 2, 25–39, June 2017.
65. Sharma, S., Kumar, A., Pattanaik, M., Raj, B., Forward Body Biased Multimode Multi-Threshold CMOS Technique for Ground Bounce Noise Reduction in Static CMOS Adders. *Int. J. Inf. Electron. Eng.*, 3, 3, 567–572, 2013.
66. Guenifi N., Rahi, SB., Design and optimization of heterostructure double gate tunneling field effect transistor for ultra low power circuit and system. *Electrical and Electronic Devices, Circuits, and Materials: Technological Challenges and Solutions*, 19–36, 2021
67. Bhardwaj, A.K., Gupta, S., Raj, B., Investigation of Parameters for Schottky Barrier (SB) Height for Schottky Barrier Based Carbon Nanotube Field

Effect Transistor Device. *J. Nanoelectron. Optoelectron.*, ASP, 15, 783–791, July 2020.

68. Bansal, P. and Raj, B., Memristor: A Versatile Nonlinear Model for Dopant Drift and Boundary Issues. *JCTN*, American Scientific Publishers, 14, 5, 2319–2325, May 2017.
69. Jain, N. and Raj, B., An Analog and Digital Design Perspective Comprehensive Approach on Fin-FET (Fin-Field Effect transistor) Technology - A Review. *Rev. Adv. Sci. Eng. (RASE)*, ASP, 5, 1–14, 2016.
70. Raj, B., Saxena, A.K., Dasgupta, S., Analytical Modeling for the Estimation of Leakage Current and Subthreshold Swing Factor of Nanoscale Double Gate FinFET Device. *Microelectron. Int.*, UK, 26, 53–63, 2009.
71. Soniya, Singh, S., Wadhwa, G., Raj, B., An Analytical Modeling for Dual Source Vertical Tunnel Field Effect Transistor. *Int. J. Recent Technol. Eng. (IJRTE)*, 8, 2, 603–608, July 2019.
72. Singh, S. and Raj, B., Design and analysis of hetrojunction Vertical T-shaped Tunnel Field Effect Transistor. *J. Electron. Mater.*, Springer, 48, 10, 6253–6260, October 2019.
73. Goyal, C., Ubhi, J.S., Raj, B., A Low Leakage CNTFET based Inexact Full Adder for Low Power Image Processing Applications. *Int. J. Circuit Theory Appl.*, Wiley, 47, 9, 1446–1458, September 2019.
74. Raj, B., Saxena, A.K., Dasgupta, S., A Compact Drain Current and Threshold Voltage Quantum Mechanical Analytical Modeling for FinFETs. *J. Nanoelectron. Optoelectron. (JNO)*, USA, 3, 2, 163–170, 2008.
75. Wadhwa, G. and Raj, B., An Analytical Modeling of Charge Plasma based Tunnel Field Effect Transistor with Impacts of Gate underlap Region. *Superlattices Microstruct.*, Elsevier, 142, 106512, June 2020.
76. Singh, S. and Raj, B., Modeling and Simulation analysis of SiGe hetrojunction Double GateVertical t-shaped Tunnel FET. *Superlattices Microstruct.*, Elsevier, 142, 106496, June 2020.
77. Jain, N. and Raj, B., Device and Circuit Co-Design Perspective Comprehensive Approach on FinFET Technology - A Review. *J. Electron Devices*, 23, 1, 1890–1901, 2016.
78. Kumar, S. and Raj, B., Analysis of I_{ON} and Ambipolar Current for Dual-Material Gate-drain Overlapped DG-TFET. *J. Nanoelectron. Optoelectron.*, American Scientific Publishers, USA, 11, 323–333, June 2016.
79. Anjum, N., Bali, T., Raj, B., Design And Simulation Of Handwritten Multiscript Character Recognition. *Int. J. Adv. Res. Comput. Commun. Eng.*, 2, 7, 2544–2549, July 2013.
80. Singh, K. and Raj, B., Performance and analysis of temperature dependent Multi-walled Carbon Nanotubes as Global Interconnects at different technology nodes. *J. Comput. Electron.*, Springer, 14, 2, 469–476, June 2015.
81. Kumar, S. and Raj, B., Compact channel potential analytical Modeling of DG-TFET based on Evanescent–mode Approach. *J. Comput. Electron.*, Springer, 14, 2, 820–827, July 2015.

82. Singh, K. and Raj, B., Temperature Dependent Modeling and Performance Evaluation of Multi-Walled CNT and Single-Walled CNT as Global Interconnects. *J. Electron. Mater.*, Springer, 44, 12, 4825–4835, Dec 2015.
83. Sharma, V.K., Pattanaik, M., Raj, B., INDEP approach for leakage reduction in nanoscale CMOS circuits. *Int. J. Electron.*, Taylor & Francis, 102, 2, 200–215, 2014.
84. Singh, K. and Raj, B., Influence of Temperature on MWCNT bundle, SWCNT bundle and Copper interconnects for Nanoscaled Technology nodes. *J. Mater. Sci.: Mater. Electron.*, Springer, 26, 8, 6134–6142, 2015.
85. Anjum, N., Bali, T., Raj, B., Design and Simulation of Handwritten Gurumukhi and Devanagri Numerical Recognition. *Int. J. Comput. Appl.*, Foundation of Computer Science, New York, USA, 73, 12, 16–21, 2013.
86. Khandelwal, S., Gupta, V., Raj, B., Gupta, R.D., Process Variability Aware Low Leakage Reliable Nano Scale DG-FinFET SRAM Cell Design Technique. *J. Nanoelectron. Optoelectron.*, 10, 6, 810–817, December 2015.
87. Sharma, V.K., Pattanaik, M., Raj, B., ONOFIC Approach: Low Power High Speed Nanoscale VLSI Circuits Design. *Int. J. Electron.*, Taylor & Francis, 101, 1, 61–73, 2014.
88. Khandelwal, S., Raj, B., Gupta, R.D., FinFET based 6T SRAM Cell Design: Analysis of Performance Metric, Process Variation and Temperature Effect. *J. Comput. Theor. Nanosci.*, ASP, USA, 12, 2500–2506, 2015.
89. Singh, S., Yadav, S., Rahul, J., Srivastava, A., Raj, B., Impact of HfO_2 in Graded Channel Dual Insulator Double Gate MOSFET. *J. Comput. Theor. Nanosci.*, American Scientific Publishers, 12, 6, 950–953, April 2015.
90. Sharma, V.K., Pattanaik, M., Raj, B., PVT variations aware low leakage INDEP approach for nanoscale CMOS Circuits. *Microelectron. Reliab.*, Elsevier, 54, 90–99, 2014.
91. Raj, B., Saxena, A.K., Dasgupta, S., Quantum Mechanical Analytical Modeling of Nanoscale DG FinFET: Evaluation of Potential, Threshold Voltage and Source/Drain Resistance. *Elsevier's J. Mater. Sci. Semicond. Process.*, Elsevier, 16, 4, 1131–1137, 2013.
92. Gopal, M., Prasad, S.S.D., Raj, B., 8T SRAM Cell Design for Dynamic and Leakage Power Reduction. *Int. J. Comput. Appl.*, Foundation of Computer Science, New York, USA, 71, 9, 43–48, June 2013.
93. Pattanaik, M., Raj, B., Sharma, S., Kumar, A., Diode Based Trimode Multi-Threshold CMOS Technique for Ground Bounce Noise Reduction in Static CMOS Adders. *Adv. Mater. Res.*, Trans Tech Publications, Switzerland, 548, 885–889, 2012.
94. Raj, B., Saxena, A.K., Dasgupta, S., Nanoscale FinFET Based SRAM Cell Design: Analysis of Performance metric, Process variation, Underlapped FinFET and Temperature effect. *IEEE Circuits Syst. Mag.*, 11, 2, 38–50, 2011.
95. Sharma, V.K., Pattanaik, M., Raj, B., Leakage Current ONOFIC Approach for Deep Submicron VLSI Circuit Design. *Int. J. Electr. Comput. Electron.*

Commun. Eng., World Academy of Sciences, Engineering and Technology, 7, 4, 239–244, 2013.
96. Kaur, P., Gill, S.S., Raj, B., Comparative Analysis of OFETs Materials and Devices for Sensor Applications. *J. Silicon*, Springer, 14, 4463–4471, 2022.
97. Sharma, S.K., Kumar, P., Raj, B., Raj, B., $In_{1-x}Ga_xAs$ Double Metal Gate-Stacking Cylindrical Nanowire MOSFET for highly sensitive Photo detector. *J. Silicon*, Springer, 14, 3535–3541, 2022.
98. Raj, B., Saxena, A.K., Dasgupta, S., Analytical Modeling of Quasi Planar Nanoscale Double Gate FinFET with Source/Drain Resistance and field dependent carrier mobility: A quantum mechanical study. *J. Comput. (JCP)*, Academy Publisher, Finland, 4, 9, 1–8, 2009.
99. Bhushan, S., Khandelwal, S., Raj, B., Analyzing Different Mode FinFET based Memory Cell at Different Power Supply for Leakage Reduction, in: *Seventh International Conference on Bio-Inspired Computing: Theories and Application, (BIC-TA 2012)*, vol. 2, pp. 89–100, Springer India, 2013.
100. Singh, J. and Raj, B., Temperature Dependent Analytical Modeling and Simulations of Nanoscale Memristor. *Eng. Sci. Technol. Int. J.*, Elsevier's, 21, 862–868, Oct 2018.
101. Singh, S., Bala, S., Raj, B., Raj, B., Improved Sensitivity of Dielectric Modulated Junctionless Transistor for Nanoscale Biosensor Design. *Sens. Lett.*, ASP, 18, 328–333, Apl 2020.
102. Kumar, V., Vishvakarma, S.K., Raj, B., Design and Performance Analysis of ASIC for IoT Applications. *Sens. Lett.*, ASP, 18, 31–38, Jan 2020.
103. Jaiswal, A., Sarin, R.K., Raj, B., Sukhija, S., A Novel Circular Slotted Microstrip-fed Patch Antenna with Three Triangle Shape Defected Ground Structure for Multiband Applications. *Adv. Electromagn. (AEM)*, 7, 3, 56–63, Aug 2018.
104. Wadhwa, G. and Raj, B., Label Free Detection of Biomolecules using Charge-Plasma-Based Gate Underlap Dielectric Modulated Junctionless TFET. *J. Electron. Mater. (JEMS)*, Springer, 47, 8, 4683–4693, August 2018.
105. Singh, G., Sarin, R.K., Raj, B., Design and Performance Analysis of a New Efficient Coplanar Quantum-Dot Cellular Automata Adder. *Indian J. Pure Appl. Phys. (IJPAP)*, 55, 97–103, February 2017.
106. Kaur, P., Pandey, V., Raj, B., Comparative Study of Efficient Design, Control and Monitoring of Solar Power using IoT. *Sens. Lett.*, ASP, 18, 419–426, May 2020.
107. Bhardwaj, A.K., Gupta, S., Raj, B., Development & Analysis of compact model for Double Gate Schottky Barrier CNTFET. *J. Nanoelectron. Optoelectron.*, ASP, 15, 1199–1208, Aug 2020.
108. Wadhwa, G., Kamboj, P., Singh, J., Raj, B., Design and Investigation of Junctionless DGTFET for Biological Molecule Recognition. *Trans. Electr. Electron. Mater.*, Springer, 22, 282–289, 2021.
109. Verma, S.K., Singh, S., Wadhwa, G., Raj, B., Detection of Biomolecules using Charge-Plasma based Gate Underlap Dielectric Modulated Dopingless

TFET. *Trans. Electr. Electron. Mater. (TEEM)*, Springer, 21, 528–535, June 2020.
110. Jain, N. and Raj, B., Impact of underlap spacer region variation on Electrostatic and analog/RF performance of symmetrical high-k SOI FinFET at 20 nm channel length. *J. Semicond. (JoS)*, IOP Science, 38, 12, 122002, Dec 2017.
111. Singh, S. and Raj, B., Analytical Modeling and Simulation analysis of T-shaped III-V heterojunction Vertical T-FET. *Superlattices Microstruct.*, Elsevier, 147, 106717, Nov 2020.
112. Singh, G., Sarin, R.K., Raj, B., Design and Analysis of Area Efficient QCA Based Reversible Logic Gates. *J. Microprocess. Microsys.*, Elsevier, 52, 59–68, May 2017.
113. Singh, S. and Raj, B., Parametric variation analysis on hetero-junction Vertical t-shape TFET for Supressing ambipolar conduction. *Indian J. Pure Appl. Phys.*, 58, 478–485, June 2020.
114. Soniya, Singh, S., Wadhwa, G., Raj, B., Design and Analysis of Dual Source Vertical Tunnel Field Effect Transistor for high performance. *Trans. Electr. Electron. Mater.*, Springer, 21, 74–82, Oct 2019.
115. Kaur, M., Gupta, N., Kumar, S., Raj, B., Singh, A.K., RF Performance Analysis of Intercalated Graphene Nanoribbon Based Global Level Interconnects. *J. Comput. Electron.*, Springer, 19, 1002–1013, June 2020.
116. Wadhwa, G. and Raj, B., Design and Performance Analysis of Junctionless TFET Biosensor for high sensitivity. *IEEE Nanotechnol.*, 18, 567 – 574, 2019.
117. Singh, J. and Raj, B., Enhanced Nonlinear Memristor Model Encapsulating Stochastic Dopant Drift. *JNO*, ASP, 14, 958–963, 2019.
118. Bhardwaj, A., Kumar, P., Raj, B., Anand, S., Design and Performance Optimization of doping-less Vertical Nanowire TFET using Gate Stack Technique. *J. Electron. Mater. (JEMS)*, Springer, 41, 7, 4005–4013, 2022.
119. Singh, J. and Raj, B., Tunnel Current Model of Asymmetric MIM Structure Levying Various Image Forces to Analyze the Characteristics of Filamentary Memristor. *Appl. Phys. A*, Springer, 125, 3, 203.1 to 203.11, Feb 2019.
120. Goyal, C., Ubhi, J.S., Raj, B., Low Leakage Zero Ground Noise Nanoscale Full Adder using Source Biasing Technique. *J. Nanoelectron. Optoelectron.*, American Scientific Publishers, 14, 360–370, March 2019.
121. Tomar, A.S., Magraiya, V.K., Raj, B., Scaling of Access and Data Transistor for High Performance DRAM Cell Design. *Quantum Matter*, 2, 412–416, Oct. 2013.
122. Jain, N. and Raj, B., Parasitic Capacitance and Resistance Model Development and Optimization of Raised Source/Drain SOI FinFET Structure for Analog Circuit Applications. *J. Nanoelectron. Optoelectron.*, ASP, USA, 13, 531–539, Apl 2018.
123. Singh, S., Vishvakarma, S.K., Raj, B., Analytical Modeling of Split-Gate Junction-Less Transistor for a Biosensor Application. *Sens. Bio-Sens.*, Elsevier, 18, 31–36, April 2018.

124. Gopal, M. and Raj, B., Low Power 8T SRAM Cell Design for High Stability Video Applications. *ITSI Trans. Electr. Electron. Eng.*, 1, 5, 91–97, 2013.
125. Raj, B., Mitra, J., Bihani, D.K., Rangharajan, V., Saxena, A.K., Dasgupta, S., Analysis of Noise Margin, Power and Process Variation for 32 nm FinFET Based 6T SRAM Cell. *J. Comput. (JCP)*, Academy Publisher, Finland, 5, 6, 1–8, 2010.
126. Sharma, D., Mehra, R., Raj, B., Comparative Analysis of Photovoltaic Technologies for High Efficiency Solar Cell Design. *Superlattices Microstruct.*, Elsevier, 153, 106861, May 2021.
127. Kaur, P., Buttar, A.S., Raj, B., A comprehensive Analysis of Nanoscale Transistor based Biosensor: A Review. *Indian J. Pure Appl. Phys.*, 59, 304–318, April 2021.
128. Yadav, D., Raj, B., Raj, B., Design and Simulation of Low Power Microcontroller for IoT Applications. *J. Sens. Lett.*, ASP, 18, 401–409, May 2020.
129. Singh, S. and Raj, B., A 2-D Analytical Surface Potential and Drain current Modeling of Double-Gate Vertical t-shaped Tunnel FET. *J. Comput. Electron.*, Springer, 19, 1154–1163, Apl 2020.
130. Singh, J. and Raj, B., An Accurate and Generic Window function for Non-linear Memristor Model. *J. Comput. Electron.*, Springer, 18, 2, 640–647, June 2019.
131. Kaur, M., Gupta, N., Kumar, S., Raj, B., Singh, A.K., Comparative RF and Crosstalk Analysis of Carbon Based Nano Interconnects. *IET Circuits Devices Syst.*, 15, 6, 493–503, Feb 2021.
132. Kandasamy, N., Ahmad, F., Ajitha, D., Raj, B., Telagam, N., Quantum Dot Cellular Automata based Scan Flip Flop and Boundary Scan Register. *IETE J. Res.*, 66, 535–548, 2020.
133. Singh, J. and Raj, B., Design and Investigation of 7T2M NVSARM with Enhanced Stability and Temperature Impact on Store/Restore Energy. *IEEE Trans. Very Large Scale Integr. Syst.*, 27, 6, 1322 – 1328, June 2019.
134. Bhardwaj, A.K., Gupta, S., Raj, B., Singh, A., Impact of Double Gate Geometry on the Performance of Carbon Nanotube Field Effect Transistor Structures for Low Power Digital Design. *Comput. Theor. Nanosci.*, ASP, 16, 1813–1820, 2019.
135. Jain, N. and Raj, B., Thermal Stability Analysis and Performance Exploration of Asymmetrical Dual-k underlap Spacer (ADKUS) SOI FinFET for Security and Privacy Applications. *Indian J. Pure Appl. Phys. (IJPAP)*, 57, 352–360, May 2019.
136. Jain, N. and Raj, B., Dual-k Spacer Region Variation at the Drain Side of Asymmetric SOI FinFET Structure: Performance Analysis towards the Analog/RF Design Applications. *J. Nanoelectron. Optoelectron.*, American Scientific Publishers, 14, 349–359, March 2019.
137. Singh, J., Sharma, S., Raj, B., Khosla, M., Analysis of barrier layer thickness on performance of In1-xGaxAs based Gate Stack Cylindrical Gate Nanowire MOSFET. *JNO*, ASP, 13, 1473–1477, Oct 2018.

138. Jain, N. and Raj, B., Analysis and Performance Exploration of High-k SOI FinFETs Over the Conventional Low-k SOI FinFET toward Analog/RF Design. *J. Semicond. (JoS)*, IOP Science, 39, 12, 124002-1–7, Dec 2018.
139. Goyal, C., Ubhi, J.S., Raj, B., A reliable leakage reduction technique for approximate full adder with reduced ground bounce noise. *J. Math. Probl. Eng.*, Hindawi, 2018, Article ID 3501041, 16, 15 Oct 2018.
140. Anuradha, Singh, J., Raj, B., Khosla, M., Design and Performance Analysis of Nano-scale Memristor-based Nonvolatile SRAM. *J. Sens. Lett.*, American Scientific Publishers, 16, 798–805, Oct 2018.
141. Wadhwa, G. and Raj, B., Parametric Variation Analysis of Charge-Plasma-based Dielectric Modulated JLTFET for Biosensor Application. *IEEE Sens. J.*, 18, 15, 6070 - 6077, AUGUST 1, 2018, (SCI).
142. Singh, J. and Raj, B., Comparative Analysis of Memristor Models for Memories Design. *JoS*, IoP, 39, 7, 074006-1–12, July 2018.
143. Singh, G., Sarin, R.K., Raj, B., A Novel Robust Exclusive-OR Function Implementation in QCA Nanotechnology with Energy Dissipation Analysis. *J. Comput. Electron.*, Springer, 15, 2, 455–465, June 2016.
144. Wadhwa, G., Kamboj, P., Raj, B., Design optimisation of junctionless TFET biosensor for high sensitivity. *Adv. Nat. Sci.: Nanosci. Nanotechnol.*, 10, 045001 (7), 2019.
145. Bansal, P. and Raj, B., Memristor Modeling and Analysis for Linear Dopant Drift Kinetics. *J. Nanoeng. Nanomanuf.*, American Scientific Publishers, 6, 1–7, 2016.
146. Ashima, Vaithiyanathan, D., Raj, B., Performance Analysis of Charge Plasma induced Graded Channel Si nanotube. *J. Eng. Res. (JER)*, EMSME Special Issue, 9, 146–154, Aug 2021.

7

Identification of Negative Capacitance in Ferroelectric in FET Devices

Umesh Chandra Bind[1]*, Shiromani Balmukund Rahi[2] and Keshav Kumar[3]

[1]Department of Applied Science and Humanities, Institute of Engineering and Technology, Dr APJ Abdul Kalam Technical University, Lucknow, UP, India
[2]University School of Information and Communication Technology, Gautam Buddha University, Greater Noida, UP, India
[3]New Chemistry Unit, International Centre for Material Science and School of Advanced Materials, Jawaharlal Nehru Centre for Advanced Scientific Research, Jakkur, Bengaluru, India

Abstract
Internal heating in portable devices is one of the most challenging tasks for researchers owing to its excessive utilization in the daily lives of people worldwide. Therefore, these devices are an ingenuous part of our lives owing to their capability to have an excess amount of information and their utilization. These devices are smaller in size (nanosize) and are constituents that have better quality from almost all perspectives than their older counterparts. Currently, these devices must consume less power to improve their applicability and utilization for a longer duration. The Negative Capacitance (NC) features developed in the field effect transistor (FET) owing to the presence of ferroelectric materials in the gate stack reduce the subthreshold limit and leakage current, which makes NCFET an important part of portable as well as advanced devices. This chapter describes the evidence of NC in a ferroelectric material sandwiched within a gate stack and its applicability in NCFET transistors. It also covers the present status of evidence of NC in hafnia, AlScN, and fluorite-based advanced and futuristic ferroelectric transistors and their futuristic perspective role in NCFET devices.

***Keywords*:** Negative capacitance, promising ferroelectrics, subthreshold slope, low power devices, sensing, fluorite-based ferroelectrics, AlScN, novel devices

**Corresponding author*: umesh.bind@gmail.com

Balwinder Raj, Shiromani Balmukund Rahi and Nandakishor Yadav (eds.) FeFET Devices, Trends, Technology and Applications, (155–184) © 2025 Scrivener Publishing LLC

7.1 Introduction

The negative capacitance behavior in ferroelectrics has opened up a new era in nanoelectronics owing to voltage amplification and lowering of the power dissipation in logic integrated circuits in integrated chips [1–5]. The negative capacitance was first proposed by Landauer in 1976, followed by some novel research reports by Bratkovsky and Levanyuk, and Salahuddin and Data to express the characteristic features of ferroelectrics in capacitive mode [6–9]. Negative capacitance is as a hallmark of the ferroelectric domain structure with partial screening and voltage amplification components in the ferroelectronics owing to overcoming Boltzmann tyranny, which limits the performance of conventional FET transistors. The NC feature of ferroelectrics in FET enables ultralow power nanoelectronic potential applications to resolve the energy challenge present in VLSI technology over the past decade. Recently, some research reports have been very impressive and encouraging on NC in hafnia-based FinFETs that exhibit an enhancement in the ON-state current and transconductance along with a short-channel effect [10–13]. The present scenario of research on NC is very versatile, with the discovery of ferroelectric behavior in hafnia, AlScN, and fluorite-based structures [14–18]. The NC behavior in ferroelectrics has been rigorously applied to the concept of proof, modeling, material optimization, and device fabrication [19–22].

The transition between two polarizing states in the ferroelectric is associated with external energy that could be due to the applied field, strain, stress, temperature, or other parameters; therefore, all ferroelectrics are pyroelectric and piezoelectric in nature owing to their wide range of applications, such as memory devices, sensors, actuators, and energy storage devices. Ferroelectrics in integrated chips are generally used for information storage and switchable polarization [23–25]. Ferroelectric materials at the atomistic scale are non-centrosymmetric in structure with single atoms or groups of atoms at the lattice sites that have the ability to switch their position within the structure on the application of any external disturbance of some short of sufficient energy but still very low order of power that could easily disturb the lattice vibration [26, 27]. Such materials (e.g., hafnia, AlScN, and PZT) were difficult to integrate on semiconductor chips a few decades ago. However, presently, various techniques, such as atomic layer deposition, pulsed laser deposition, and other epitaxial growth techniques, depend on the ferroelectric material lattice mismatch [28–31]. To estimate the use of ferroelectrics in an integrated chip, the performance of the ferroelectric must be verified or observed from the device output. This can be implemented by observing the ferroelectric negative capacitance behavior in the device output, which can be identified with the subthreshold swing slope of the I_D vs V_{GS} curve, the thickness of the dielectric and ferroelectric layers and their capacitance, and the permittivity of the ferroelectric in the device, which can be determined via the capacitance and slope feature of hysteresis. In this chapter, the evidence of NC in ferroelectric devices is discussed briefly. It includes also a perspective scenario of ferroelectrics in future devices.

7.2 Negative Capacitance

One of the vital components of all electronic devices is the capacitor, which relates the stored charges to the voltage between plates and is characterized by its capacitance (C).

$$Q = CV \tag{7.1}$$

where, Q and V represent charge and applied voltage, respectively. The C also relates the stored electrical energy (U) within it with its charge.

$$U = \frac{Q^2}{2C} \tag{7.2}$$

Energy U is always positive [32]. This implies that capacitance cannot be negative. The fundamental principle of the minimum energy for the capacitor system also states that the capacitance cannot be negative. However, if the capacitor is filled with two inhomogeneous dielectrics between the plates, they are connected in series and reflect positive capacitance [33–37]. This does not mean that the capacitance of any one capacitor cannot be negative, as long as the total capacitance is positive to fulfill the thermodynamic stability condition. In other words, suppose an inhomogeneous dielectric having its two components, which have uniform characteristics throughout the material having C_1 and C_2 capacitances of each component, are placed between the capacitor's plates. The total capacitance C can be estimated as:

$$\frac{1}{C} = \frac{1}{C_1} + \frac{1}{C_2} \tag{7.3}$$

C must be positive to maintain the thermodynamic stability according to the principle of minimum energy. However, this relation is also true if C_1 or C_2 are negative and the thermodynamic stability condition $C \geq 0$ is satisfied. It means there is a possibility to observe negative capacitance in one of the components [38–45].

Mathematically, capacitance can be defined as:

$$C = \frac{Q}{V} \tag{7.4}$$

This relation in Eq. (7.4) holds good for the linear dielectric capacitors, where applied positive voltage charges the capacitor plates (Q >0) that create an electric field (E) in the dielectric as

$$E = \frac{V}{d} \tag{7.5}$$

where d is the separation between the plates. The capacitance equivalent to differential capacitance can be written as:

$$C = \frac{dQ}{dV} \tag{7.6}$$

However, differential capacitance is also applicable to nonlinear dielectrics such as ferroelectrics. The ferroelectrics did not behave similarly to the dielectric in the application of positive voltage. It fundamentally differs from the response to external capacitive charge (Q) at the interface of ferroelectrics to constrain displacement field (D)

$$D = \frac{Q}{A} \tag{7.7}$$

where A is the surface area of the capacitor plate. That means there is an important role of dielectric permittivity (ε) in the ferroelectrics capacitance behavior owing to electric field relation to displacement field

$$D = \varepsilon E \tag{7.8}$$

A positive total capacitance of the specimen suggest that the apllied external electric field (E_{ext}) is associated with displacement field (D) in such a manner so that

$$\frac{\Delta D}{\Delta E_{ext}} > 0 \tag{7.9}$$

The permittivity ε (r) at any point r can be defined as the ratio of the change in displacement field D(r) to change in the local electric field E(r).

$$\varepsilon(r) = \frac{\delta D(r)}{\varepsilon_0 \delta E(r)} \tag{7.10}$$

where ε_0 is the permittivity in a vacuum. Since Coulomb interaction is a long-range interaction, the permittivity ε(r) at any point in the materials depends on the material's response at r, which makes its character less obvious [15, 46–52]. In exacting, nothing stops it from being locally negative, such that the change in the displacement field D(r) responds to the variation in E(r). In other words, one can say that D(r) field opposes any change in E(r) in such a way that

$$\frac{\Delta D}{\Delta E_{ext}} > 0 \tag{7.11}$$

The capacitance of the local part of the materials is basically equal to the local permittivity ε(r), apart from the geometrical factors, which means that it can be negative. Remember that it is the capacitance of very small part of the ferroelectric slab specimen, in common; the capacitance of complete ferroelectric specimen is related to the total permittivity [53–59]. Therefore, one can say the permittivity at a local point is a static equilibrium quantity and is not directly related to the complete specimen.

The intrinsic response of ferroelectrics to the application of a positive voltage has transformed from a non-polar paraelectric phase to a switchable spontaneous polarization phase. It creates electric field E in the ferroelectric specimen, given by

$$E = \frac{V}{d} \tag{7.12}$$

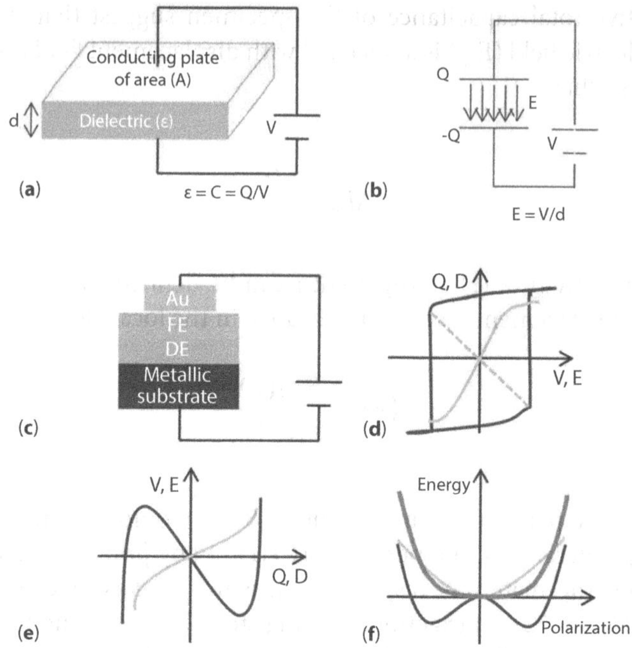

Figure 7.1 Schematic representation of (a) parallel plate capacitor filled with dielectrics, (b) polarization direction within the dielectric, (c) ferroelectric (violet line) and dielectric (brown line) layers in series capacitance (device), (d) charge or displacement field in the ferroelectric and dielectric as a function of the applied voltage or electric field, (e) voltage or electric field across the ferroelectric and dielectric superlattices as a function of the charge or displacement across it, and (f) free energy of the ferroelectric and dielectric as a function of polarization, respectively. The green line shows the resultant energy as a combination of both ferroelectric and dielectric versus polarization, which reflects the device's working stability with these ferroelectric and dielectric layered materials.

where d is the thickness of the specimen. This means that the intrinsic behavior of the ferroelectric upon the application of a positive voltage is fundamentally different from the response to an external capacitive charge (Q) at the interface of the ferroelectric. It creates a displacement field (D) that equivalently develops constraints in the ferroelectrics which is given by

$$D = \frac{Q}{A} \tag{7.13}$$

where A is the surface area of the ferroelectric specimen. Accordingly, the constitutive relations Q(V) (or D(E)) and V(Q) (or E(D)) reflect different behaviors, as shown in Figure 7.1. These constitutive relations determine the non-linear capacitive performance i.e., differential capacitance C at the operating voltage V_0 by

$$C(V_o) = \frac{dQ}{dV} \tag{7.14}$$

and its inverse differential capacitance C at operating charge Q_0 by

$$C^{-1}(Q_o) = \frac{dV}{dQ} \tag{7.15}$$

The multi-valued reflection of Q(V) from the negative capacitance (NC) region prevents the stabilization of negative capacitance state by controlling the voltage (V). In contrast, the constitutive equation V(Q) becomes a single-valued function of the charge Q [60–66]. This allows the realization of a static negative capacitance regime by changing the charge Q. Alternatively, one can say that instead of playing with voltage, just play with charge to obtain NC behavior in the ferroelectric specimen.

7.3 NC in Ferroelectrics

Ferroelectrics display hysteresis behavior of polarization upon application of an external electric field. In other words, charges on ferroelectric slabs as a function of the applied voltage are unstable owing to the constraints of the dipole orientation inside them [67–73]. This inherent stability is responsible for the negative capacitance (NC) behavior. Various ferroelectric systems have been reported [25, 33–35]. In the Landau theory of phase transitions, polarization is considered an order parameter. The free energy function can be calculated using first principle. Its derivative equation of polarization along with the Maxwell-Faraday equation and Gauss's law defines the electric response within ferroelectrics, which means that it determines the capacitance of the ferroelectric [74–82].

Let us consider a ferroelectric slab of a single domain (i.e., uniform polarization) to study its capacitance behavior. In a typical setup, the capacitor is obsessed by the voltage source and the total charge on the capacitor plates is determined by integrating the displacement current present in the circuit. In the ferroelectric phase, the constitutive relation Q(V) reflects a multivalued S-shaped function with a negative slope segment that facilitates the negative capacitance behavior [83–88]. However, this segment is unstable, and on the application of any voltage in this multivalued region, the actual charge of an adiabatic voltage-driven process corresponds to one of the two branches with a positive slope, which reflects the positive capacitance. This behavior resulted in the formation of the hysteresis-shaped Q(V) observations.

Furthermore, for a charge-driven ferroelectric capacitor, the multivalued S-shaped Q(V) constitutive relation is flipped and becomes an N-shaped Q(V) relation. This N-shaped Q(V) relation curve reflects the single-valued signature, which also includes a negative slope region and (V = 0, Q = 0) points (Figure 7.1e). The free energy of the dielectric and ferroelectric layers can jointly stabilize the device because with only dielectric, its polarization interval is very short, and with ferroelectric, it cannot be stabilized except at the center of the polarization zone (Figure 7.1f). Thus, negative capacitance can be achieved by controlling the charge Q. Now the question is. How can the amount of charge be controlled and estimated?

A very simple way to control and estimate the charges in incipient ferroelectrics is to connect with a regular dielectric capacitor in series [9]. For example, in the setup, the ferroelectric capacitance (C_1) can be connected to a dielectric capacitor (C_2, say, Cs) in series (Figure 7.2), which means that both capacitors must have the same charges on the capacitor plates and thus both are electrostatically coupled. Therefore, the accessible charge value can be estimated [89–95]. For a small value of Cs, the series capacitance will suppress the spontaneous polarization in the ferroelectrics and thus stabilize the negative slope, i.e., paraelectric state (V = 0, Q = 0), even if the ferroelectric is below its transition temperature. As we know that ferroelectric behavior also depends on temperature, it is very important to consider temperature when estimating the negative capacitance behavior [96–102]. In ferroelectrics below the nominal transition temperature, the internal voltage drop is opposite to the overall external bias, which results in an increase in the voltage at the ferroelectric and dielectric interfaces. In practice, it is conceptually observed that the effect of a series capacitor is indistinguishable from that of the well-studied finite screening lengths at the metal-ferroelectric interface and at the dielectric layers of dielectric ferroelectric superlattices and multilayers, as well as in the channel of the field-effect transistor [7, 36–40]. To understand the opposing mechanism of the internal voltage drop, one has to be aware of the crystal structure of the ferroelectrics, i.e., the location of the lattice point and the interaction

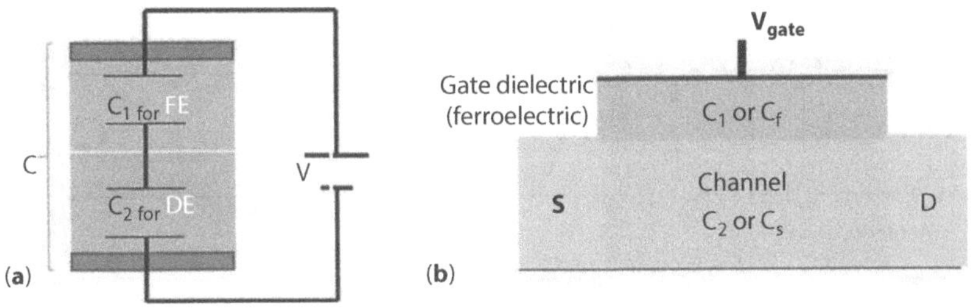

Figure 7.2 Schematic of (a) ferroelectric and dielectric stack as a layered structure, the capacitance of ferroelectric (C_1) and dielectric (C_2) connected in series within two metallic plates and reflect total capacitance (C), (b) FET with ferroelectric in gate stack reflect the non-uniform capacitance (C_1 or C_f) and channel capacitance (C_2 or C_s).

between the lattice atoms owing to different chemical characteristics of the lattice atom along with Coulombic interaction. Let us consider the one-dimensional instability in centrosymmetry [103–108]. From a stability point of view, every structure or system attempts to reduce its total energy to achieve stability in the system for a long period of time. Any disturbance produced in it owing to the addition of energy into the system makes it unstable, i.e., it receives some extra energy due to external energy disturbances such as temperature, stress, strain, charge, and field (electric or magnetic). This means that this system is sensitive to all of the above to meet its stability criteria. In other words, one can say that at its equilibrium position or stable condition, all parameters have a constant value, and tuning in it results in instability [109–115]. Here, we discuss a ferroelectric system, which is sensitive to temperature, electric field, and charges more dominantly than the rest of the parameters. This means that for 1-D stability to be achieved in a ferroelectric system, there must be a temperature T (i.e., T less than Tc as the critical temperature) and an electric field (E = 0) (Figure 7.3). As the temperature increases above T, the central atom reaches at the center and reflects the centrosymmetry.

When reduces, it goes to its lowest energy position, i.e., at a stable position. Under stable conditions, when an electric field or potential is applied across it, it turns its location along the direction of the electric field. This implies that it finds its lowest energy configuration by shifting its central atom along the electric field direction to reduce the total energy [116–123]. Thus, one can observe that even in a 1-D ferroelectric it has two stable positions. A similar characteristic is also observed in antiferrodistortivity, which means that lattice atoms have two stable configurations. Further, to understand this, let us consider a three-dimensional pictorial view of the perovskite structure of ferroelectrics, as shown in Figure 7.4. At a

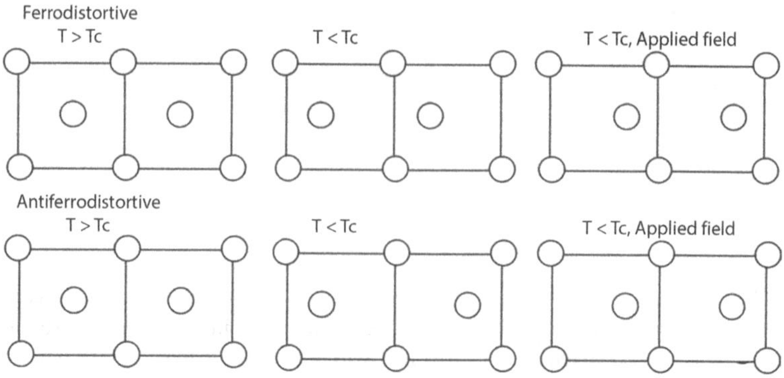

Figure 7.3 Schematic representation of non-centrosymmetric atom movement in ferroelectric 2-dimensional lattice structure as a function of external field.

temperature T higher than the critical temperature, the central atom is at a central position in the 3-D lattice structure, whereas as the temperature decreases below the critical temperature, the central atom become disordered along the vertical axis (i.e., along the z-axis) (Figures 7.4b, d). It is important point to understand that this central atom can move the lower or upper side along the vertical axis to lower its energy in order to achieve stability. This means that it has two points/positions to place its central atom to reduce the total energy (Figure 7.4e). The rhombohedral structural arrangement of the ferroelectric have lower energy stable states along the diagonal axis (i.e., polarized along a diagonal axis) (Figure 7.5a), where the monoclinic structure is polarized along the diagonal axis of the inclined surface (Figure 7.5b). These structures reflect the noncentrosymmetric crystalline behavior required for practical device applications [124–130].

7.4 Ferroelectric Materials in Practice for NC

Ferroelectricity can be observed in various types of noncentrosymmetric structures as discussed above.

They can be classified into different groups based on their ferroelectricity mechanism of reflecting negative capacitances, such as

i. Traditional ferroelectric
ii. Hafnium and zirconium-based ferroelectric
iii. Wurtzite aluminum scandium nitride

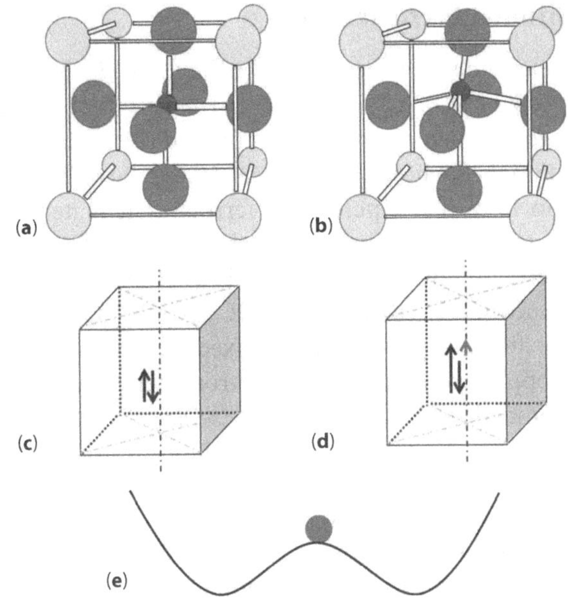

Figure 7.4 Schematic representation of ferroelectricity in perovskite ferroelectric materials in tetragonal structure (a) atom arrangement in tetragonal structure when the temperature is higher than the critical temperate (Tc) above which atomic arrangement occurs in such a way that there is symmetrical structure, (b) atomic arrangement of perovskite structure below the critical temperature (Tc) resulting in the development of a non-centrosymmetric structure, (c) equal charges or force field applied on the central atom, directional view of (a) structure, (d) total charges or force field applied on the central atom, directional view of (b) structure as result central atom shifted along c-direction or z-axis, (e) it reflects the position of a central atom in the lattice structure at (a) diagram position; it is at higher free energy leading to instability. To obtain a stable position, the central atom could move in the lower or upper direction along the c-axis or z-axis. This movement is attributed to the negative capacitance behavior of the perovskite structure [137–142].

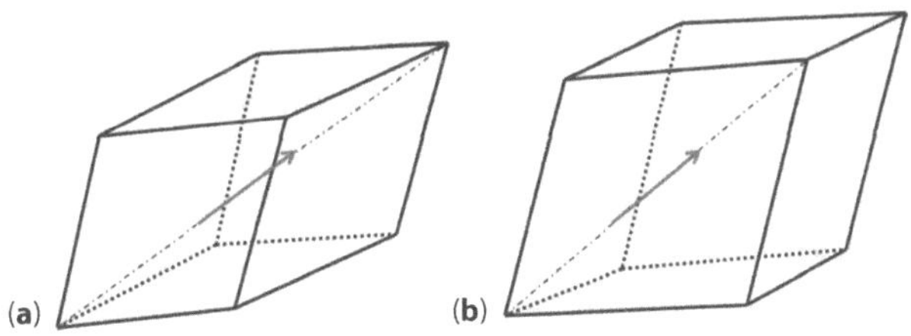

Figure 7.5 Schematic representation of central atom movement direction in (a) Rhombohedral ferroelectric structure, and (b) Monoclinic ferroelectric structures.

7.4.1 Traditional Ferroelectric

The first ferroelectric was observed as Rochelle salt in 1920, followed by the discovery of KH_2PO_4 in 1935 [41, 42]. The very first perovskite structure, $BaTiO_3$, was studied rigorously and proved the availability of various perovskite materials such as $Pb(Zr,Ti)O_3$ (PZT), $SrBi_2Ta_2O_9$ (SBT), and $BiFeO_3$(BFO) in 1940. These perovskite ferroelectric materials are used in several devices for different purposes such as memory storage, actuators, photovoltaics, etc., and have been reported to exhibit negative capacitance [43–46]. Kołtunowicz *et al.* reported NC in $(Fe_{0.45}Co_{0.45}Zr_{0.10})_x(PZT_{1-x})$ nanocomposite films deposited in a mixed argon–oxygen atmosphere owing to the hopping conductance of electrons between the fully oxidized nanoparticles embedded in the PZT matrix [43]. Jiang *et al.* simulated a negative-capacitance double-gate junctionless transistor (NCDG-JLT) with a ferroelectric (FE) gate dielectric and metal-ferroelectric-metal-insulator semiconductor (MFMIS) structure [131–136]. They observed an improvement in the on-state drain current and lower subthreshold swing simultaneously compared to its conventional DG JLT counterpart [44]. Hua *et al.* used a DC bias assistant model in which one of the layers of the bilayer capacitor has a high dielectric constant and the other layer has a low dielectric constant. They reported that NC can be obtained using an external DC bias electric field larger than a critical value, owing to the self-polarization effect in the ferroelectric layer [45].

7.4.2 Hafnium and Zirconium-Based Ferroelectric

Ferroelectricity in HfO_2 was first observed by Böscke *et al.* in 2011 for very first time [47, 48]. Generally, conventional ferroelectrics suffer from the degradation of ferroelectricity in ultra-thin films, but Böscke *et al.* successfully achieved a 10 nm thin film with robust remnant polarization (Pr) higher than 15 μC/cm² [47]. This report opens a new era in the research field of ferroelectricity of HfO_2 and ZrO_2 with various doping elements and deposition techniques [15, 49–54]. Sang *et al.* demonstrated the formation of an orthorhombic phase (space group $Pca2_1$) by convergent beam electron diffraction that reflects the crystallographic origin of the ferroelectricity and several polymorphs of HfO_2 and ZrO_2 [55]. Different crystallographic structures of HfO_2 and ZrO_2 thin films can be easily achieved by doping, thickness impurities, electric field cycling, and temperature. This nanoscale polymorphism can be attributed to the metastable configuration of the orthorhombic ferroelectric phase. Park *et al.* observed that there are dominant crystalline phase changes, as a function of the doping species and dopant concentration, to orthorhombic or tetragonal/cubic from monoclinic, and there is a specific concentration of dopant to which the orthorhombic crystalline phase is stable with high ferroelectricity [56]. They also observed remnant polarization Pr with different dopants and variation in the film thickness. Commonly, the orthorhombic phase and remnant polarization Pr vary with the film thickness. Materlik *et al.* reported that polymorphism in fluorite-structure-based ferroelectrics is strongly affected by surface energy [57]. These ferroelectrics are also influenced by electric-field cycling. The phenomenon for which the increase in remnant polarization with a low cycle count of electric field has termed as the "wake-up effect,' while the phenomenon in which the remanent polarization leads to reduction with an increase in electric field cycles is known as 'fatigue effect" [58–61]. Pešić and co-workers reported that the field-cycle-dependent remanent polarization in Gd:HfO_2 thin films at different temperatures and observed that the remanent polarization can be simulated for the redistribution identification of oxygen vacancies that are responsible for the variation in the local field, resulting in local phase transitions. They also found that the oxygen vacancy distribution is more homogeneous in the interfacial region. The cycle process resulted from the redistribution of oxygen vacancies during the wake-up and fatigue effects owing to the presence and absence of oxygen vacancies [59]. The phase transition has been tried to record with transmission electron microscopy (TEM) during the wake-up process by Grimley and coworkers but could not identify the amount of phase transition even by synchrotron-based X-ray diffraction studies [59].

7.4.3 Wurtzite Aluminum Scandium Nitride

Ferroelectricity in these types of structures originates from the destabilization of the wurtzite polar structure via the addition of other elements, such as nitrides, which tend to multifold coordination [29, 62, 63]. Generally, the coercive field of AlScN has a lower dielectric breakdown strength than that of ferroelectrics, which makes it ferroelectric by applying tensile stress on the wurtzite basal plane. The solid mixing of Sc with AlN flattens the energy landscape of the wurtzite structure, resulting in a reduced energy barrier between the N-polar and M-polar states, i.e., between the two polarizing states. The structural change from wurtzite to $P6_3/mmc$ lowers the energy barrier, which is suitable for M-polar (metal polar) to N-polar (nitrogen polar) switching along the c-axis or z-direction. This mechanism lowers the coercive field until it becomes smaller than the dielectric breakdown strength. This creates suitable conditions for an electric field to switch from $P6_3/mmc$ to $P6_3mc$, where AlScN reflects ferroelectric behavior. Such a mechanism can also be observed in other wurtzite semiconductor nanomaterials, such as GaN or ZnO synthesized with suitable elements that can be identified for solid solution formation to exhibit ferroelectricity [62, 64]. The simulation method can be used to identify possible methods for solid solutions with various elements. There are some reports of simulated research to identify solid solutions with various possibilities [65, 66].

The attraction of the scientific community to the ferroelectric behavior of AlScN is due to some unique features, such as: i) it is the first readily available nitride ferroelectric material and its fabrication is compatible with CMOS fabrication techniques such as PLD in the temperature range (lower than 400 °C) with good stability of polarization up to 600 °C, [67] ii) the spontaneous polarization and coercive field in the range of 80–110 $\mu C/cm^2$ and 1.8–5 mV/cm, respectively, which are rarely observed in other materials, and iii) spontaneous polarization and coercive field both can be adjusted according to the amount of ScN (typically in the range of 20--43%) and basal plane state [29].

7.5 Evidence of NC in Ferroelectrics

The experimental evidence available in the literature can be divided into two parts: (i) *the concept* and (ii) *device fabrication*. For the concept building of negative capacitances, research has usually focused on the various parameters of the ferroelectric layer connected in series with the dielectric layer for NC response observation [19, 20, 68, 69]. From the device fabrication point of view, several studies have reported a reduction in the subthreshold swing and leakage current with the implementation of a ferroelectric layer in series with a dielectric layer in the gate terminal of the FET, which influences the applied voltage by NC [67, 70–76]. Both approaches are used to find suitable ferroelectric materials that are compatible with the available commercial FET dielectric layers. This is feasible only by isolating the ferroelectric layer as a capacitor, which is externally accessible. This can be achieved in two ways: first, by introducing an internal metallization node in the transistor stack layer, and second, by connecting the ferroelectric capacitor to external resistors or capacitors in series [143–149].

7.5.1 For the Concept of NC

The first concept of negative capacitance was recognized by Bratkovsky and Levanyuk during the analysis of ultrathin high-quality $SrRuO_3/BaTiO_3/$

$SrRuO_3$ capacitors and the theoretical estimate of the metallic screening length to re-plot the actual voltage drop hysteresis across the ferroelectric. The actual voltage drop occurred across the ferroelectric layers and the interface capacitance, which arises due to the finite intrinsic screening length at the interface. The resulting plot reflects negative slopes attributed to the presence of multiple domains in the materials (Figure 7.6a).

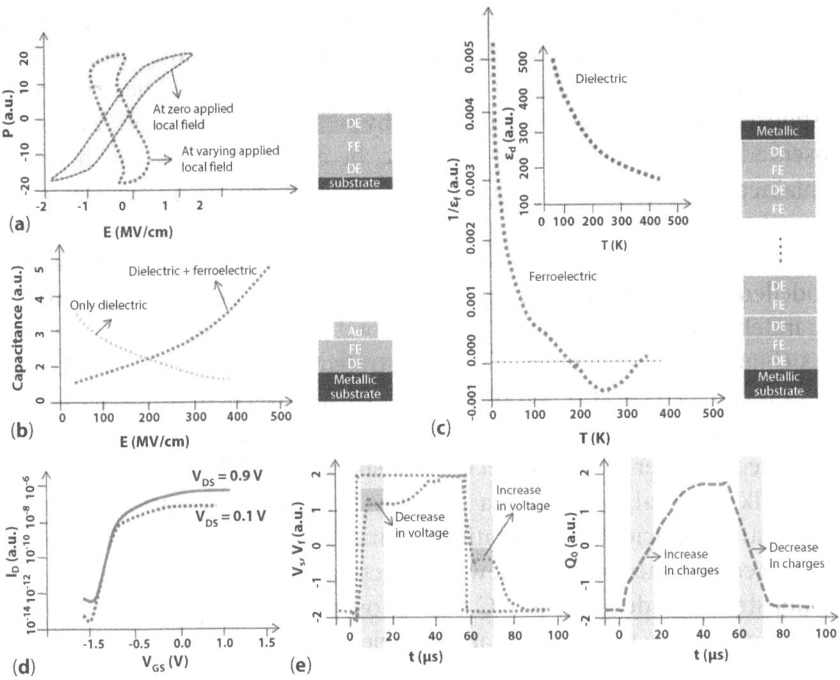

Figure 7.6 Schematic representation of NC identification in the ferroelectric used in devices, (a) the total applied voltage drops a major portion across ferroelectric capacitive layer and partially across metal ferroelectric interfacial capacitance that could be used to estimate voltage drop across ferroelectric can be estimated from the polarization as a function of applied voltage, (b) the signature feature of ferroelectric is that it enhances the capacitance as total, when ferroelectric is added in series with dielectric (c) as ferroelectric added to dielectrics in series for the superlattice formation, the ferroelectric domains gets decoupled due to formation of dense lattice, their effective permittivity can be measured by the evaluation of total capacitance in series, the inverse permittivity of ferroelectrics reflect negative values whereas for dielectric it is normally decreasing from high to low of bulk (d) the subthreshold swing slope is higher in green one as compared to red (doted) line one for a ferroelectric in FET, and (e) the voltage across ferroelectric start decreasing (shown in brown color box) in a particular time period while the total charge across it (yellow strip), is still increasing and vice versa for a segment period of time again, that reflect the NC behavior of ferroelectrics in FET.

Thus, most reports of NC evidence are based on ferroelectric-dielectric interface heterostructures that provide straightforward recognition of NC owing to depolarization field effects [77]. A hallmark of negative capacitance is that it reflects a higher capacitance in a layered structure of ferroelectric in series connection with a pure dielectric layer as compared to the capacitance of a pure dielectric layer material of the same thickness (Figure 7.6b). There are several reports on the importance of ferroelectric and dielectric series-connected capacitance measurements to enhance capacitance [18, 29, 67–76]. Although reflects static NC behavior, time dependence plays an important role in the capacitance owing to the dynamic effect of the leakage, charge trapping, and frequency dispersion in the dielectric on its response, which is generally carried out by AC methods. This NC evidence observed in the ferroelectric-dielectric heterostructure is much more explained by Maxwell–Wagner effects [77]. Furthermore, in the superlattices of the ferroelectric–dielectric, the electrostatically decoupled ferroelectric layers owing to the availability of a dense domain structure can be modelled as capacitors in series, and their effective permittivities can be separately estimated by measuring the total capacitance of a set of samples (Figure 7.6c). Zubko *et al.* demonstrated the superlattice of (Pb, Sr) $TiO_3/SrTiO_3$ (PST/STO) and observed that the inverse permittivity ($1/\varepsilon_f$) of the ferroelectric PST layers reflects a clear region of NC behavior, while the dielectric layer of STO reflect a decrease in the dielectric constant (ε_d) of a bulk material. Therefore, a carefully designed superlattice structure that ensures sufficient electrostatic coupling of ferroelectric layers in series (from a capacitance point of view) connection with the dielectric layer can demonstrate a wide range of NC regions over a long temperature range by spinning both the ferroelectric and paraelectric phases [78].

7.5.2 For Device Fabrication

Ferroelectrics in FET have attracted much attention from the scientific community owing to their memory application In order to enhance presently available computer systems and logic devices by utilizing the hysteresis features of ferroelectrics due to the available domain polarization. These domain polarizations in ferroelectrics render them sensitive to voltage and electric fields. Therefore, the presence of ferroelectricity in the gate stack plays an important role, whereas the application of external voltages improves the gate role in transistor applications [19, 20]. It improved the subthreshold swing (SS) voltage limit, which reduced the power consumption in the devices owing to its long use without power dissipation. The SS of the MOSFET is given as:

$$SS = \frac{dV_G}{dlog_{10}I_{DS}} = \left(\frac{d\psi_{MOS}}{dlog_{10}I_{DS}}\right) \times \left(\frac{dV_G}{d\psi_{MOS}}\right) = n \times m \tag{7.16}$$

where V_G, I_{DS}, and ψ_{MOS} denote the gate voltage, drain-to-current ratio, and surface potential, respectively. The first term "n" is fixed at 60 mV/decade due to Boltzmann's tyranny and second term "m," also known as the body factor, can be determined from the charge-balance condition in the device. Using the Kirchhoff's Law in capacitive model,

$$\frac{d}{dt}\left(C_{FE} \times V_{FE}\right) - \frac{d}{dt}\left(C_{MOS} \times V_{MOS}\right) = 0 \tag{7.17}$$

putting $V_{FE} = V_G - V_{MOS}$ and dividing by d (V_{MOS}) in Equation (7.17). Then Equation (7.17) becomes

$$C_{FE}\frac{dV_G}{dV_{MOS}} - C_{FE}\frac{dV_{MOS}}{dV_{MOS}} + V_{FE}\frac{dC_{FE}}{dV_{MOS}} = \frac{d\left(C_{MOS}V_{MOS}\right)}{dV_{MOS}} \tag{7.18}$$

rearranging the Equation (7.18), we have

$$m = 1 - \frac{1}{C_{FE}}\left\{V_{FE}\frac{dC_{FE}}{dV_{MOS}} - \frac{d\left(C_{MOS}V_{MOS}\right)}{dV_{MOS}}\right\} \tag{7.19}$$

where C_{MOS} is the MOS capacitance. If both C_{MOS} and C_{FE} are assumed to be constant, then this equation turns into a well-known formula of "m" which is

$$m = 1 + \frac{C_{MOS}}{C_{FE}} \tag{7.20}$$

where C_{FE} is nonlinear, time, thickness, and temperature dependent and being used communally as evidence of NC.

The experimental demonstration of steep-slope operation (less than 60 mV/decade) has focused on two methods: direct integration of the ferroelectric into the gate stack or external connection of the ferroelectric capacitor to the FET. Generally, non-hysteretic-based devices reflect an improved subthreshold swing [19, 20, 72–74]. There are reported observations on MoS_2 semiconducting channel, Al_2O_3 dielectric, and doped Si gate in stack with a (Hf, Zr)O_2 ferroelectric of an NC-field-effect transistor (FET) exhibits non-hysteretic current-voltage (I_D-V_{GS}) characteristics for different V_{DS}, with SS dropping below the Boltzmann limit (i.e., improvement in SS values), in contrast to Al_2O_3 gate (purely dielectric) associated counterpart FET [74]. This lowering (improvement) in SS is due to the negative capacitance of the ferroelectrics in the device (Figure 7.6d).

To directly observe the transient NC, the voltage drop across the ferroelectric slabs must be measured while flipping polarization or switching. Salvatore *et al.* demonstrated experimental confirmation of NC dependence on temperature in MFMIS architecture. They also monitored the voltage across the ferroelectric during the switching processes, which reflected the presence of NC behavior in the hysteresis [78]. However, the connection of the ferroelectric capacitor to an external series resistance slows down the delivery of charges on capacitor plates, and as a voltage is applied across the capacitor-resistor circuit, there is a reduction in voltage across the ferroelectric capacitor while charging of the ferroelectric capacitor continues reflecting the presence of NC features (Figure 7.6e). Many research groups have observed such a voltage drop across ferroelectric capacitors [71–74]. Resistors have been replaced by capacitors in series connection with ferroelectric capacitors in several reports, with some improved observations.

7.6 Perspectives

Among all prominent problems in the world of digital electronics, one of the most important is to reduce the supply voltage in the lower-scaling devices to reduce power dissipation owing to thermal broadening of the Fermi–Dirac distribution of electrons [78, 79]. One can now overcome this by introducing a ferroelectric layer into the gate stack of the transistor owing to the negative capacitance behavior of ferroelectrics, because of the tremendous efforts of several researchers, such as Landauer, Bratkovsky, Levanyuk, Salahuddin, and data on negative capacitance observation for the very first time in ferroelectric materials after almost a century interval of discovery of ferroelectric materials [6–10]. Therefore, in the present decade, this negative capacitance technology has gained momentum, driven by the development of hafnia-related ferroelectric materials that are comfortable to be placed on the layer of the gate stack in the present scenario of FET commercial transistors. Presently, there are several simulation techniques, such as the first principle, to explain the observation of negative capacitance performance in ferroelectric-dielectric superlattices. However, there are still some improvements, such as the enhancement of the negative capacitance value to maximize voltage amplification and broadening the temperature range in which the negative capacitance feature of the ferroelectric remains static. Further domain wall mobility and larger polarization in the multidomain are essential requirements for enhancing the negative capacitance-associated nanodevices [78, 79]. The domains in the ferroelectric at the nanoscale reflect wide domains that could be up to the ferroelectric size, depending on the dielectric superlattice interface potential energy.

The application of devices needs to be considered in order to improve their performance in a particular area because it defines the working environment of the ferroelectric and its responses. The implementation of ferroelectrics into the gate stack of the FET influences the mobility carriers available inside the channel, which affects the performance of the device, so it has to be optimized in every specific configuration of the device [67]. The electromagnetic responses of NC ferroelectric domains are still an important topic for broad device applications owing to the effect of the frequency response on the permittivity and refractive index. Hafnia-based ferroelectrics have many more technological advantages than previously reported ferroelectrics owing to their high NC, compatibility with commercial transistors, and ease of deposition on gate stacks. It has a different ferroelectric behavior than the present polymorphism and the existence of antiferroelectric and amorphous phases, which still need to be rigorously studied, which can reflect some more interesting features of hafnia. The recently discovered AlScN ferroelectric promising material can be added to GaN for devices to broaden the existing piezoelectric performance.

7.7 Conclusion

The discovery of some promising ferroelectric such as hafnia, AlScN, by contrast, there are several parameters which have to be studied in order to get some more efficient characteristic behavior of ferroelectric in FET are expected to come out with the passing of research time on it. However, there are some unique characteristics of ferroelectric FET, based on which the negative capacitance of the ferroelectric is detected. Some important characteristics of the NC of a ferroelectric material used in FET are observed by the presence of a negative slope of E(D), Q(V), or V(Q), the total capacitance of ferroelectric and dielectric layers, change in the permittivity, subthreshold swing slope, and ferroelectric layer voltage variation with switching/flipping of the polarization in the domain. Ferroelectric-dielectric superlattice statics mostly define the NC characteristics along with the industrial capacitance matching of the ferroelectric and dielectric layers in the devices. The local potential energy environment defines the ferroelectric layer capacitive performance in devices owing to the influence of polarization in the domain or mobility of domain walls in nano-ferroelectrics. It also influences the AlScN ferroelectric behavior due to changes in the metal polar (M-polar) and nitrogen polar (N-polar) positions, with slight tuning in the available environmental potential energy. NC optimization in hafnia-based nanomaterials to take advantage of them in the present scenario of a device requires deep and broad studies on their ferroelectric and dielectric performance.

References

1. Devadas, S., Keutzer, K., White, J., Estimation of power dissipation in CMOS combinational circuits using boolean function manipulation. *IEEE Trans. Comput. Aided Des. Integr. Circuits Syst.*, 11, 3, 373–83, 1992.
2. Gu, R.X. and Elmasry, M.I., Power dissipation analysis and optimization of deep submicron CMOS digital circuits. *IEEE J. Solid-State Circuits*, 31, 5, 707–13, 1996.
3. Verman, L.C., Negative circuit constants. *Proc. Inst. Radio Eng.*, 19, 4, 676–81, 1931.
4. Behr, L. and Tarpley, R.E., Design of resistors for precise high-frequency measurements. *Proc. Inst. Radio Eng.*, 20, 7, 1101–13, 1932.
5. Terman, F.E., Variable reactance circuit. US Patent 1950759, 1934.
6. Landauer, R., Can capacitance be negative? *Collect. Phenom.*, 2, 167, 167–70, 1976.
7. Bratkovsky, A.M. and Levanyuk, A.P., Very large dielectric response of thin ferroelectric films with the dead layers. *Phys. Rev. B*, 63, 13, 132103, 2001.
8. Bratkovsky, A.M. and Levanyuk, A.P., Depolarizing field and real hysteresis loops in nanometer-scale ferroelectric films. *Appl. Phys. Lett.*, 89, 253108, 2006.
9. Salahuddin, S. and Datta, S., Use of negative capacitance to provide voltage amplification for low power nanoscale devices. *Nano Lett.*, 8, 2, 405–10, 2008.
10. Padhi, S.K., Narendar, V., Nishad, A.K., A novel step architecture based negative capacitance (SNC) FET: Design and circuit level analysis. *Microelectron. J.*, 146, 106139, 2024.
11. Valasa, S., Kotha, V.R., Vadthiya, N., Pushing the boundaries: design and simulation approach of negative capacitance nanosheet FETs with ferroelectric

and dielectric spacers at the sub-3 nm technology node for analog/RF/mixed signal applications. *ACS Appl. Electron. Mater.*, 6, 5, 3206–15, 2024.

12. Wei, W., Lü, W., Han, Y., Zhang, C., Chen, D., A novel recessed-source negative capacitance gate-all-around tunneling field effect transistor for low-power applications. *Microelectron. J.*, 145, 106126, 2024.
13. Kwon, D., Chatterjee, K., Tan, A.J., Yadav, A.K., Zhou, H., Sachid, A.B., Dos Reis, R., Hu, C., Salahuddin, S., Improved subthreshold swing and short channel effect in FDSOI n-channel negative capacitance field effect transistors. *IEEE Electron Device Lett.*, 39, 2, 300–3, 2017.
14. Böscke, T.S., Teichert, S., Bräuhaus, D., Müller, J., Schröder, U., Böttger, U., Mikolajick, T., Phase transitions in ferroelectric silicon doped hafnium oxide. *Appl. Phys. Lett.*, 99, 112904, 2011.
15. Muller, J., Boscke, T.S., Schroder, U., Mueller, S., Brauhaus, D., Bottger, U., Frey, L., Mikolajick, T., Ferroelectricity in simple binary ZrO_2 and HfO_2. *Nano Lett.*, 12, 8, 4318–23, 2012.
16. Pradhan, D.K., Moore, D.C., Kim, G., He, Y., Musavigharavi, P., Kim, K.H., Sharma, N., Han, Z., Du, X., Puli, V.S., Stach, E.A., A scalable ferroelectric non-volatile memory operating at 600° C. *Nat. Electron.*, 7, 348–355, 2024.
17. Salvatore, G.A., Lattanzio, L., Bouvet, D., Stolichnov, I., Setter, N., Ionescu, A.M., Ferroelectric transistors with improved characteristics at high temperature. *Appl. Phys. Lett.*, 97, 053503, 2010.
18. Dasgupta, S., Rajashekhar, A., Majumdar, K., Agrawal, N., Razavieh, A., Trolier-Mckinstry, S., Datta, S., Sub-kT/q switching in strong inversion in $PbZr_{0.52}Ti_{0.48}O_3$ gated negative capacitance FETs. *IEEE J. Explor. Solid-State Comput. Devices Circuits*, 1, 43–8, 2015.
19. Upadhyay, A.K., Rahi, S.B., Tayal, S., Song, Y.S., Recent progress on negative capacitance tunnel FET for low-power applications: Device perspective. *Microelectron. J.*, 129, 105583, 2022.
20. Rahi, S.B., Tayal, S., Upadhyay, A.K., A review on emerging negative capacitance field effect transistor for low power electronics. *Microelectron. J.*, 116, 105242, 2021.
21. Guenifi, N., Rahi, S.B., Benmahdi, F., Chabane, H., Optimization of tunneling current in ferroelectric tunnel FET using genetic algorithm. *J. Supercomput.*, 79, 14, 15773–89, 2023.
22. Guenifi, N., Rahi, S.B., Benmahdi, F., Chaabane, H., Optimization for device figure of merit of ferroelectric tunnel FET using genetic algorithm. *ECS J. Solid State Sci. Technol.*, 12, 2, 023001, 2023.
23. Mikolajick, T., Schroeder, U., Slesazeck, S., The past, the present, and the future of ferroelectric memories. *IEEE Trans. Electron Devices*, 67, 4, 1434–43, 2020.
24. Auciello, O., A critical comparative review of PZT and SBT-based science and technology for non-volatile ferroelectric memories. *Integr. Ferroelectr.*, 15, 1-4, 211–20, 1997.

25. Quhe, R., Di, Z., Zhang, J., Sun, Y., Zhang, L., Guo, Y., Wang, S., Zhou, P., Asymmetric conducting route and potential redistribution determine the polarization-dependent conductivity in layered ferroelectrics. *Nat. Nanotechnol.*, 19, 2, 173–80, 2024.
26. Bind, U.C. and Rahi, S.B., Basic operational principle of anti-ferroelectric and ferroelectric materials, in: *Negative Capacitance Field Effect Transistors*, pp. 62–82, CRC Press, Taylor & Francis Group, 2023.
27. Song, Y.S., Tayal, S., Rahi, S.B., Upadhyay, A.K. (Eds.), *Negative capacitance field effect transistors: Physics, Design, Modeling and Applications*, CRC Press, Taylor & Francis Group, 2023.
28. Bohr, M.T., Chau, R.S., Ghani, T., Mistry, K., THE HIGH-k SOLUTION: Microprocessors entering production this year are the result of the biggest transistor redesign in 40 years. *IEEE Spectr.*, 44, 10, 23–9, 2007.
29. Fichtner, S., Wolff, N., Lofink, F., Kienle, L., Wagner, B., AlScN: A III-V semiconductor based ferroelectric. *J. Appl. Phys.*, 125, 114103, 2019.
30. Guan, Z., Hu, H., Shen, X., Xiang, P., Zhong, N., Chu, J., Duan, C., Recent progress in two-dimensional ferroelectric materials. *Adv. Electron. Mater.*, 6, 1, 1900818, 2020.
31. Osada, M. and Sasaki, T., The rise of 2D dielectrics/ferroelectrics. *APL Mater.*, 7, 120902, 2019.
32. Landau, L. and Lifshitz, E., *Electrodynamics of Continuous Media*, Pergamon, Elsevier, 1960.
33. Li, S., Wang, F., Wang, Y., Yang, J., Wang, X., Zhan, X., He, J., Wang, Z., Van der Waals ferroelectrics: Theories, Materials, and Device Applications. *Adv. Mater.*, 36, 22, 2301472, 2024.
34. Liu, J., Su, L., Zhang, X., Shtansky, D.V., Fang, X., Ferroelectric–optoelectronic hybrid system for photodetection. *Small Methods*, 8, 2, 2300319, 2024.
35. Li, L., Richter, C., Paetel, S., Kopp, T., Mannhart, J., Ashoori, R.C., Very large capacitance enhancement in a two-dimensional electron system. *Science*, 332, 6031, 825–8, 2011.
36. Mehta, R.R., Silverman, B.D., Jacobs, J.T., Depolarization fields in thin ferroelectric films. *J. Appl. Phys.*, 44, 8, 3379–85, 1973.
37. Junquera, J. and Ghosez, P., Critical thickness for ferroelectricity in perovskite ultrathin films. *Nature*, 422, 6931, 506–9, 2003.
38. Stengel, M. and Spaldin, N.A., Origin of the dielectric dead layer in nanoscale capacitors. *Nature*, 443, 7112, 679–82, 2006.
39. Neaton, J.B. and Rabe, K.M., Theory of polarization enhancement in epitaxial $BaTiO_3/SrTiO_3$ superlattices. *Appl. Phys. Lett.*, 82, 10, 1586–8, 2003.
40. Dawber, M., Stucki, N., Lichtensteiger, C., Gariglio, S., Ghosez, P., Triscone, J.M., Tailoring the properties of artificially layered ferroelectric superlattices. *Adv. Mater.*, 3, 23, 4153–9, 2007.
41. Valasek, J., Piezo-electric and allied phenomena in Rochelle salt. *Phys. Rev.*, 17, 4, 475, 1921.

42. Busch, G. and Scherrer, P., A new significant electrical substance. *Naturwissenschaften*, 23, 43, 737, 1935.
43. Kołtunowicz, T.N., Fedotova, J.A., Zhukowski, P., Saad, A., Fedotov, A., Kasiuk, J.V., Larkin, A.V., Negative capacitance in (FeCoZr)–(PZT) nanocomposite films. *J. Phys. D: Appl. Phys.*, 46, 12, 125304, 2013.
44. Jiang, C., Liang, R., Wang, J., Xu, J., Simulation-based study of negative capacitance double-gate junctionless transistors with ferroelectric gate dielectric. *Solid State Electron.*, 126, 130–5, 2016.
45. Ya-Fei, H., Wei-Li, L., Tian-Dong, Z., Yang, Y., Ren-Lu, H., Wei-Dong, F., Negative capacitance in $BaTiO_3/BiFeO_3$ bilayer capacitors. *ACS Appl. Mater. Interfaces*, 8, 34, 22354–22360, 2016.
46. Ke, Q., Lou, X., Yang, H., Kumar, A., Zeng, K., Wang, J., Negative capacitance induced by redistribution of oxygen vacancies in the fatigued $BiFeO_3$-based thin film. *Appl. Phys. Lett.*, 101, 022904, 2012.
47. Böscke, T.S., Müller, J., Bräuhaus, D., Schröder, U., Böttger, U., Ferroelectricity in hafnium oxide: CMOS compatible ferroelectric field effect transistors, in: *International electron devices meeting*, IEEE, pp. 24–5, 2011.
48. Merz, J. and Anderson, J.R., Ferroelectric storage devices. *Bell Lab. Rec.*, 33, 335–42, 1955.
49. Martin, D., Müller, J., Schenk, T., Arruda, T.M., Kumar, A., Strelcov, E., Yurchuk, E., Müller, S., Pohl, D., Schröder, U., Kalinin, S.V., Ferroelectricity in Si-doped HfO_2 revealed: a binary lead-free ferroelectric. *Adv. Mater.*, 26, 48, 8198–202, 2014.
50. Mueller, S., Mueller, J., Singh, A., Riedel, S., Sundqvist, J., Schroeder, U., Mikolajick, T., Incipient ferroelectricity in Al-doped HfO_2 thin films. *Adv. Funct. Mater.*, 22, 11, 2412–7, 2012.
51. Starschich, S. and Boettger, U., An extensive study of the influence of dopants on the ferroelectric properties of HfO_2. *J. Mater. Chem. C*, 5, 2, 333–8, 2017.
52. Hyuk Park, M., Joon Kim, H., Jin Kim, Y., Lee, W., Moon, T., Seong Hwang, C., Evolution of phases and ferroelectric properties of thin $Hf_{0.5}Zr_{0.5}O_2$ films according to the thickness and annealing temperature. *Appl. Phys. Lett.*, 102, 242905, 2013.
53. Xu, L., Nishimura, T., Shibayama, S., Yajima, T., Migita, S., Toriumi, A., Kinetic pathway of the ferroelectric phase formation in doped HfO_2 films. *J. Appl. Phys.*, 122, 124104, 2017.
54. Wei, Y., Nukala, P., Salverda, M., Matzen, S., Zhao, H.J., Momand, J., Everhardt, A.S., Agnus, G., Blake, G.R., Lecoeur, P., Kooi, B.J., A rhombohedral ferroelectric phase in epitaxially strained $Hf_{0.5}Zr_{0.5}O_2$ thin films. *Nat. Mater.*, 17, 12, 1095–100, 2018.
55. Sang, X., Grimley, E.D., Schenk, T., Schroeder, U., LeBeau, J.M., On the structural origins of ferroelectricity in HfO_2 thin films. *Appl. Phys. Lett.*, 106, 162905, 2015.

56. Park, M.H., Schenk, T., Schroeder, U., Dopants in atomic layer deposited HfO_2 thin films, in: *Ferroelectricity in Doped Hafnium Oxide: Materials, Properties and Devices*, pp. 49–74, Woodhead Publishing, 2019, https://doi.org/10.1016/B978-0-08-102430-0.00005-X.
57. Materlik, R., Künneth, C., Kersch, A., The origin of ferroelectricity in $Hf_{1-x}Zr_xO_2$: A computational investigation and a surface energy model. *J. Appl. Phys.*, 117, 134109, 2015.
58. Zhou, D., Xu, J., Li, Q., Guan, Y., Cao, F., Dong, X., Müller, J., Schenk, T., Schröder, U., Wake-up effects in Si-doped hafnium oxide ferroelectric thin films. *Appl. Phys. Lett.*, 103, 192904, 2013.
59. Pešić, M., Fengler, F.P., Larcher, L., Padovani, A., Schenk, T., Grimley, E.D., Sang, X., LeBeau, J.M., Slesazeck, S., Schroeder, U., Mikolajick, T., Physical mechanisms behind the field-cycling behavior of HfO_2-based ferroelectric capacitors. *Adv. Funct. Mater.*, 26, 25, 4601–12, 2016.
60. Kim, H.J., Park, M.H., Kim, Y.J., Lee, Y.H., Moon, T., Do Kim, K., Hyun, S.D., Hwang, C.S., A study on the wake-up effect of ferroelectric $Hf_{0.5}Zr_{0.5}O_2$ films by pulse-switching measurement. *Nanoscale*, 8, 3, 1383–9, 2016.
61. Grimley, E.D., Schenk, T., Sang, X., Pešić, M., Schroeder, U., Mikolajick, T., LeBeau, J.M., Structural changes underlying field-cycling phenomena in ferroelectric HfO_2 thin films. *Adv. Electron. Mater.*, 2, 9, 1600173, 2016.
62. Zhang, S., Holec, D., Fu, W.Y., Humphreys, C.J., Moram, M.A., Tunable optoelectronic and ferroelectric properties in Sc-based III-nitrides. *J. Appl. Phys.*, 114, 133510, 2013.
63. Tasnádi, F., Alling, B., Höglund, C., Wingqvist, G., Birch, J., Hultman, L., Abrikosov, I.A., Origin of the anomalous piezoelectric response in wurtzite $Sc_xAl_{1-x}N$ alloys. *Phys. Rev. Lett.*, 104, 13, 137601, 2010.
64. Konishi, A., Ogawa, T., Fisher, C.A., Kuwabara, A., Shimizu, T., Yasui, S., Itoh, M., Moriwake, H., Mechanism of polarization switching in wurtzite-structured zinc oxide thin films. *Appl. Phys. Lett.*, 109, 102909, 2016.
65. Takeuchi, N., First-principles calculations of the ground-state properties and stability of ScN. *Phys. Rev. B*, 65, 4, 045204, 2002.
66. Farrer, N. and Bellaiche, L., Properties of hexagonal ScN versus wurtzite GaN and InN. *Phys. Rev. B*, 66, 20, 201203, 2002.
67. Bind, U.C., Ion implanted modification of nanomaterials for semiconductor devices and other applications, in: *Handbook of Emerging Materials for Semiconductor Industry*, pp. 179–198, Springer Nature Singapore, Singapore, 2024.
68. Appleby, D.J., Ponon, N.K., Kwa, K.S., Zou, B., Petrov, P.K., Wang, T., Alford, N.M., O'Neill, A., Experimental observation of negative capacitance in ferroelectrics at room temperature. *Nano Lett.*, 14, 7, 3864–8, 2014.
69. Gao, W., Khan, A., Marti, X., Nelson, C., Serrao, C., Ravichandran, J., Ramesh, R., Salahuddin, S., Room-temperature negative capacitance in a ferroelectric–dielectric superlattice heterostructure. *Nano Lett.*, 14, 10, 5814–9, 2014.

70. Khan, A.I., Chatterjee, K., Duarte, J.P., Lu, Z., Sachid, A., Khandelwal, S., Ramesh, R., Hu, C., Salahuddin, S., Negative capacitance in short-channel FinFETs externally connected to an epitaxial ferroelectric capacitor. *IEEE Electron Device Lett.*, 37, 1, 111–4, 2015.
71. Chi, M., Li, A., Zhang, X., Li, Z., Jia, M., Wang, J., Wang, Z.L., Zhai, J., Strain tuning on Van der Waals negative capacitance transistors. *Nano Energy*, 126, 109640, 2024.
72. Dey, K., Das, B., Hazra, P.K., Kundu, T., Naskar, S., Das, S., Maity, S., Maji, P., Karmakar, B., Paramanik, R., Datta, S., Negative capacitance for stabilizing the logic state in a tunnel field-effect transistor. *ACS Appl. Nano Mater.*, 2024, https://doi.org/10.1021/acsanm.3c06311.
73. McGuire, F.A., Lin, Y.C., Price, K., Rayner, G.B., Khandelwal, S., Salahuddin, S., Franklin, A.D., Sustained sub-60 mV/decade switching via the negative capacitance effect in MoS_2 transistors. *Nano Lett.*, 17, 8, 4801–6, 2017.
74. Si, M., Su, C.J., Jiang, C., Conrad, N.J., Zhou, H., Maize, K.D., Qiu, G., Wu, C.T., Shakouri, A., Alam, M.A., Ye, P.D., Steep-slope hysteresis-free negative capacitance MoS2 transistors. *Nat. Nanotechnol.*, 13, 1, 24–8, 2018.
75. Islam Khan, A., Bhowmik, D., Yu, P., Joo Kim, S., Pan, X., Ramesh, R., Salahuddin, S., Experimental evidence of ferroelectric negative capacitance in nanoscale heterostructures. *Appl. Phys. Lett.*, 99, 113501, 2011.
76. Catalan, G., O'neill, D., Bowman, R.M., Gregg, J.M., Relaxor features in ferroelectric superlattices: A Maxwell–Wagner approach. *Appl. Phys. Lett.*, 77, 19, 3078–80, 2000.
77. Zubko, P., Wojdeł, J.C., Hadjimichael, M., Fernandez-Pena, S., Sené, A., Luk'yanchuk, I., Triscone, J.M., Íñiguez, J., Negative capacitance in multidomain ferroelectric superlattices. *Nature*, 534, 7608, 524–8, 2016.
78. Theis, T.N. and Solomon, P.M., It's time to reinvent the transistor. *Science*, 327, 5973, 1600–1, 2010.
79. Zhirnov, V.V. and Cavin, R.K., Negative capacitance to the rescue? *Nat. Nanotechnol.*, 3, 2, 77–8, 2008.
80. Singh, S. and Raj, B., A 2-D Analytical surface potential and drain current modeling of double-gate vertical t-shaped Tunnel FET. *J. Comput. Electron.*, Springer, 19, 1154–1163, Apl 2020.
81. Singh, J. and Raj, B., An Accurate and generic window function for nonlinear memristor model. *J. Comput. Electron.*, Springer, 18, 2, 640–647, June 2019.
82. Kaur, M., Gupta, N., Kumar, S., Raj, B., Singh, A.K., Comparative RF and crosstalk analysis of carbon based nano interconnects. *IET Circuits Devices Syst.*, 15, 6, 493–503, Feb 2021.
83. Kandasamy, N., Ahmad, F., Ajitha, D., Raj, B., Telagam, N., Quantum dot cellular automata based scan flip flop and boundary scan register. *IETE J. Res.*, 66, 535–548, 2020.

84. Singh, J. and Raj, B., Design and investigation of 7T2M NVSARM with enhanced stability and temperature impact on store/restore energy. *IEEE Trans. Very Large Scale Integr. Syst.*, 27, 6, 1322–1328, June 2019.
85. Bhardwaj, A.K., Gupta, S., Raj, B., Singh, A., Impact of double gate geometry on the performance of carbon nanotube field effect transistor structures for low power digital design. *Comput. Theor. Nanosci.*, ASP, 16, 1813–1820, 2019.
86. Jain, N. and Raj, B., Thermal stability analysis and performance exploration of asymmetrical dual-k underlap spacer (ADKUS) SOI FinFET for security and privacy applications. *Indian J. Pure Appl. Phys. (IJPAP)*, 57, 352–360, May 2019.
87. Jain, N. and Raj, B., Dual-k spacer region variation at the drain side of asymmetric SOI FinFET structure: performance analysis towards the analog/rf design applications. *J. Nanoelectron. Optoelectron.*, American Scientific Publishers, 14, 349–359, March 2019.
88. Singh, J., Sharma, S., Raj, B., Khosla, M., Analysis of barrier layer thickness on performance of $In_{1-x}Ga_xAs$ based gate stack cylindrical gate nanowire MOSFET. *JNO*, ASP, 13, 1473–1477, Oct 2018.
89. Jain, N. and Raj, B., Analysis and performance exploration of high-k SOI FinFETs over the conventional low-k SOI FinFET toward Analog/RF Design. *J. Semicond. (JoS)*, IOP Science, 39, 12, 124002-1-7, Dec 2018.
90. Goyal, C., Ubhi, J.S., Raj, B., A reliable leakage reduction technique for approximate full adder with reduced ground bounce noise. *Math. Probl. Eng.*, Hindawi, 2018, Article ID 3501041, 16, 15 Oct 2018.
91. Anuradha, Singh, J., Raj, B., Khosla, M., Design and performance analysis of nano-scale memristor-based nonvolatile SRAM. *Sens. Lett.*, American Scientific Publishers, 16, 798–805, Oct 2018.
92. Wadhwa, G. and Raj, B., Parametric variation analysis of charge-plasma-based dielectric modulated JLTFET for biosensor application. *IEEE Sens. J.*, 18, 15, 6070–6077, AUGUST 1 2018, (SCI).
93. Singh, J. and Raj, B., Comparative analysis of memristor models for memories design. *JoS*, IoP, 39, 7, 074006-1-12, July 2018.
94. Yadav, D., Chouhan, S.S., Vishvakarma, S.K., Raj, B., Application specific microcontroller design for IoT based WSN. *Sens. Lett.*, ASP, 16, 374–385, May 2018.
95. Singh, G., Sarin, R.K., Raj, B., Fault-tolerant design and analysis of quantum-dot cellular automata based circuits. *IEEE/IET Circuits Devices Syst.*, 12, 638–64, 2018.
96. Singh, J. and Raj, B., Modeling of mean barrier height levying various image forces of metal insulator metal structure to enhance the performance of conductive filament based memristor model. *IEEE Nanotechnol.*, 17, 2, 268–267, March 2018.

97. Jain, A., Sharma, S., Raj, B., Analysis of triple metal surrounding gate (TM-SG) III-V nanowire MOSFET for photosensing application. *Opto-Electron. J.*, Elsevier, 26, 2, 141–148, May 2018.
98. Jain, A., Sharma, S., Raj, B., Design and analysis of high sensitivity photo-sensor using cylindrical surrounding gate MOSFET for low power sensor applications. *Eng. Sci. Technol. Int. J.*, Elsevier's, 19, 4, 1864–1870, December 2016.
99. Saiphani Kumar, G., Singh, A., Raj, B., Design and analysis of gate all around CNTFET based SRAM cell design. *J. Comput. Electron.*, Springer, 17, 1, 138–145, March 2018.
100. Singh, G.P., Sohi, B.S., Raj, B., Material properties analysis of graphene base transistor (GBT) for VLSI analog circuits. *Indian J. Pure Appl. Phys. (IJPAP)*, 55, 896–902, Dec-2017.
101. Kumar, S. and Raj, B., Estimation of stability and performance metric for inward access transistor based 6T SRAM cell design using n-type/p-type DMDG-GDOV TFET. *IEEE VLSI Circuits Syst. Lett.*, 3, 2, 25–39, June2017.
102. Sharma, S., Kumar, A., Pattanaik, M., Raj, B., Forward body biased multi-mode multi-threshold CMOS technique for ground bounce noise reduction in static CMOS adders. *Int. J. Inf. Electr. Eng.*, 33, 567–572, 2013.
103. Singh, H., Kumar, P., Raj, B., Performance analysis of majority gate SET based 1-bit full adder. *Int. J. Comput. Commun. Eng. (IJCCE)*, IACSIT Press Singapore, 2, 4, 2933–2940, 2013, ISSN: 2010-3743.
104. Bhardwaj, A.K., Gupta, S., Raj, B., Investigation of parameters for schottky barrier (SB) height for schottky barrier based carbon nanotube field effect transistor device. *J. Nanoelectron. Optoelectron.*, ASP, 15, 783–791, July 2020.
105. Bansal, P. and Raj, B., Memristor: a versatile nonlinear model for dopant drift and boundary issues. *JCTN*, American Scientific Publishers, 14, 5, 2319–2325, May 2017.
106. Jain, N. and Raj, B., An analog and digital design perspective comprehensive approach on Fin-FET (Fin-Field Effect transistor) Technology - A Review. *Rev. Adv. Sci. Eng. (RASE)*, ASP, 5, 1–14, 2016.
107. Raj, B., Saxena, A.K., Dasgupta, S., Analytical modeling for the estimation of leakage current and subthreshold swing factor of nanoscale double gate FinFET device. *Microelectron. Int.*, UK, 26, 53–63, 2009.
108. Soniya, Singh, S., Wadhwa, G., Raj, B., An analytical modeling for dual source vertical tunnel field effect transistor. *Int. J. Recent Technol. Eng. (IJRTE)*, 8, 2, 603–608, July 2019.
109. Singh, S. and Raj, B., Design and analysis of hetrojunction vertical t-shaped tunnel field effect transistor. *J. Electron. Mater.*, Springer, 48, 10, 6253–6260, October 2019.
110. Goyal, C., Ubhi, J.S., Raj, B., A low leakage CNTFET based inexact full adder for low power image processing applications. *Int. J. Circuit Theory Appl.*, Wiley, 47, 9, 1446–1458, September 2019.

111. Raj, B., Saxena, A.K., Dasgupta, S., A compact drain current and threshold voltage quantum mechanical analytical modeling for FinFETs. *J. Nanoelectron. Optoelectron. (JNO)*, USA, 3, 2, 163–170, 2008.
112. Wadhwa, G. and Raj, B., An analytical modeling of charge plasma based tunnel field effect transistor with impacts of gate underlap region. *Superlattices Microstruct.*, Elsevier, 142, 106512, June 2020.
113. Singh, S. and Raj, B., Modeling and simulation analysis of sige hetrojunction double gatevertical t-shaped tunnel FET. *Superlattices Microstruct.*, Elsevier, 142, 106496, June 2020.
114. Jain, N. and Raj, B., Device and circuit co-design perspective comprehensive approach on FinFET technology - A Review. *J. Electron Devices*, 23, 1, 1890–1901, 2016.
115. Kumar, S. and Raj, B., Analysis of I_{ON} and ambipolar current for dual-material gate-drain overlapped DG-TFET. *J. Nanoelectron. Optoelectron.*, American Scientific Publishers, USA, 11, 323–333, June 2016.
116. Anjum, N., Bali, T., Raj, B., Design and simulation of handwritten multiscript character recognition. *Int. J. Adv. Res. Comput. Commun. Eng.*, 2, 7, 2544–2549, July 2013.
117. Singh, K. and Raj, B., Performance and analysis of temperature dependent multi-walled carbon nanotubes as global interconnects at different technology nodes. *J. Comput. Electron.*, Springer, 14, 2, 469–476, June 2015.
118. Kumar, S. and Raj, B., Compact channel potential analytical modeling of DG-TFET based on evanescent–mode approach. *J. Comput. Electron.*, Springer, 14, 2, 820–827, July 2015.
119. Singh, K. and Raj, B., Temperature dependent modeling and performance evaluation of multi-walled cnt and single-walled CNT as global interconnects. *J. Electron. Mater.*, Springer, 44, 12, 4825–4835, Dec 2015.
120. Sharma, V.K., Pattanaik, M., Raj, B., INDEP approach for leakage reduction in nanoscale CMOS circuits. *Int. J. Electron.*, Taylor Francis, 102, 2, 200–215, 2014.
121. Singh, K. and Raj, B., Influence of temperature on MWCNT bundle, SWCNT bundle and copper interconnects for nanoscaled technology nodes. *J. Mater. Sci. Mater. Electron.*, Springer, 26, 8, 6134–6142, 2015.
122. Anjum, N., Bali, T., Raj, B., Design and simulation of handwritten gurumukhi and devanagri numerical recognition. *Int. J. Comput. Appl.*, Published by Foundation of Computer Science, New York, USA, 73, 12, 16–21, 2013.
123. Khandelwal, S., Gupta, V., Raj, B., Gupta, R.D., Process variability aware low leakage reliable nano scale DG-FinFET SRAM cell design technique. *J. Nanoelectron. Optoelectron.*, 10, 6, 810–817, December 2015.
124. Sharma, V.K., Pattanaik, M., Raj, B., ONOFIC approach: low power high speed nanoscale VLSI circuits design. *Int. J. Electron.*, Taylor Francis, 101, 1, 61–73, 2014.

125. Khandelwal, S., Raj, B., Gupta, R.D., FinFET based 6T SRAM cell design: analysis of performance metric, process variation and temperature effect. *J. Comput. Theor. Nanosci.*, ASP, USA, 12, 2500–2506, 2015.
126. Singh, S., Yadav, S., Rahul, J., Srivastava, A., Raj, B., Impact of HfO_2 in graded channel dual insulator double gate MOSFET. *J. Comput. Theor. Nanosci.*, American Scientific Publishers, 12, 6, 950–953, April 2015.
127. Sharma, V.K., Pattanaik, M., Raj, B., PVT variations aware low leakage INDEP approach for nanoscale CMOS Circuits. *Microelectron. Reliab.*, Elsevier, 54, 90–99, 2014.
128. Raj, B., Saxena, A.K., Dasgupta, S., Quantum mechanical analytical modeling of nanoscale DG FinFET: evaluation of potential, threshold voltage and source/drain resistance. *Elsevier's J. Mater. Sci. Semicond. Process.*, Elsevier, 16, 4, 1131–1137, 2013.
129. Gopal, M., Siva Sankar Prasad, D., Raj, B., 8T SRAM cell design for dynamic and leakage power reduction. *Int. J. Comput. Appl.*, Published by Foundation of Computer Science, New York, USA, 71, 9, 43–48, June 2013.
130. Pattanaik, M., Raj, B., Sharma, S., Kumar, A., Diode based trimode multi-threshold CMOS technique for ground bounce noise reduction in static CMOS adders. *Adv. Mater. Res.*, Trans Tech Publications, Switzerland, 548, 885–889, 2012.
131. Raj, B., Saxena, A.K., Dasgupta, S., Nanoscale FinFET based SRAM cell design: analysis of performance metric, process variation, underlapped FinFET and temperature effect. *IEEE Circuits Syst. Mag.*, 11, 2, 38–50, 2011.
132. Sharma, V.K., Pattanaik, M., Raj, B., Leakage current ONOFIC approach for deep submicron VLSI circuit design. *Int. J. Electr. Comput. Electron. Commun. Eng.*, World Academy of Sciences, Engineering and Technology, 7, 4, 239–244, 2013.
133. Kaur, P., Gill, S.S., Raj, B., Comparative analysis of OFETs materials and devices for sensor applications. *J. Silicon*, Springer, 14, 4463–4471, 2022.
134. Sharma, S.K., Kumar, P., Raj, B., Raj, B., $In_{1-x}Ga_xAs$ double metal gate-stacking cylindrical nanowire MOSFET for highly sensitive photo detector. *J. Silicon*, Springer, 14, 3535–3541, 2022.
135. Raj, B., Saxena, A.K., Dasgupta, S., Analytical modeling of quasi planar nanoscale double gate FinFET with source/drain resistance and field dependent carrier mobility: A quantum mechanical study. *J. Comput. (JCP)*, Academy Publisher Finland, 4, 9, 1–8, 2009.
136. Bhushan, S., Khandelwal, S., Raj, B., Analyzing different mode FinFET based memory cell at different power supply for leakage reduction. *Seventh International Conference on Bio-Inspired Computing: Theories and Application*, (BIC-TA 2012) Advances in Intelligent Systems and Computing, vol. 202, pp. 89–100, 2012, 2013.
137. Singh, J. and Raj, B., Temperature dependent analytical modeling and simulations of nanoscale memristor. *Eng. Sci. Technol. Int. J.*, Elsevier's, 21, 862–868, Oct 2018.

138. Singh, S., Bala, S., Raj, B., Raj, B., Improved sensitivity of dielectric modulated junctionless transistor for nanoscale biosensor design. *Sens. Lett.*, ASP, 18, 328–333, Apl 2020.
139. Kumar, V., Vishvakarma, S.K., Raj, B., Design and performance analysis of ASIC for IoT applications. *Sens. Lett.* ASP, 18, 31–38, Jan 2020.
140. Jaiswal, A., Sarin, R.K., Raj, B., Sukhija, S., A novel circular slotted microstrip-fed patch antenna with three triangle shape defected ground structure for multiband applications. *Adv. Electromagn. (AEM)*, 7, 3, 56–63, Aug 2018.
141. Wadhwa, G. and Raj, B., Label free detection of biomolecules using charge-plasma-based gate underlap dielectric modulated junctionless TFET. *J. Electron. Mater. (JEMS)*, Springer, 47, 8, 4683–4693, August 2018.
142. Singh, G., Sarin, R.K., Raj, B., Design and performance analysis of a new efficient coplanar quantum-dot cellular automata adder. *Indian J. Pure Appl. Phys. (IJPAP)*, 55, 97–103, February 2017.
143. Kaur, P., Pandey, V., Raj, B., Comparative study of efficient design, control and monitoring of solar power using IoT. *Sens. Lett.*, ASP, 18, 419–426, May 2020.
144. Bhardwaj, A.K., Gupta, S., Raj, B., Development & analysis of compact model for double gate schottky barrier CNTFET. *J. Nanoelectron. Optoelectron.*, ASP, 15, 1199–1208, Aug 2020.
145. Wadhwa, G., Kamboj, P., Singh, J., Raj, B., Design and investigation of junctionless DGTFET for biological molecule recognition. *Trans. Electr. Electron. Mater.*, Springer, 22, 282–289, 2021.
146. Verma, S.K., Singh, S., Wadhwa, G., Raj, B., Detection of biomolecules using charge-plasma based gate underlap dielectric modulated dopingless TFET. *Trans. Electr. Electron. Mater. (TEEM)*, Springer, 21, 528–535, June 2020.
147. Jain, N. and Raj, B., Impact of underlap spacer region variation on electrostatic and analog/RF performance of symmetrical high-k SOI FinFET at 20 nm channel length. *J. Semicond. (JoS)*, IOP Science, 38, 12, 122002, Dec 2017.
148. Singh, S. and Raj, B., Analytical modeling and simulation analysis of t-shaped III-V heterojunction vertical T-FET. *Superlattices Microstruct.*, Elsevier, 147, 106717, Nov 2020.
149. Singh, G., Sarin, R.K., Raj, B., Design and analysis of area efficient QCA based reversible logic gates. *Microprocess. Microsyst.*, Elsevier, 52, 59–68, May 2017.

8

Tunnel Field Effect Transistors and Their Application in Biosensors

K. Manikanta[1], Umakanta Nanda[2*], Pratikhya Raut[3] and Biswajit Jena[4]

[1]Department of Electronics and Communication Engineering, Marri Laxman Reddy Institute of Technology and Management, Hyderabad, India
[2]School of Electronics Engineering, VIT-AP University, Amaravati, Andhra Pradesh, India
[3]Electronics and Communication Engineering, VR Siddhartha Engineering College, Vijayawada, Andhra Pradesh, India
[4]School of Electronics Engineering, VIT Chennai, Tamil Nadu, India

Abstract

Tunnel field-effect transistor (TFET)-based biosensors represent a cutting-edge technology with immense potential for revolutionizing the landscape of sensing and diagnostics. Their versatility, sensitivity, and compatibility with emerging nanotechnologies position TFET biosensors at the forefront of innovation, opening new avenues for applications in healthcare, environmental monitoring, and other fields. As research in this field continues to progress, TFET biosensors are poised to play a pivotal role in shaping the future of the analytical and diagnostic sciences. TFET biosensors operate based on the principle of detecting changes in the electrical characteristics of a semiconductor channel induced by biomolecular interactions at the sensor surface. The intrinsic sensitivity of TFETs to charge variations makes them ideal candidates for real-time, label-free detection of various biomolecules, such as proteins, nucleic acids, and small molecules.

***Keywords*:** Biosensor, immense, TFET, sensitivity, label-free detection

**Corresponding author*: uk_nanda@yahoo.co.in

Balwinder Raj, Shiromani Balmukund Rahi and Nandakishor Yadav (eds.) FeFET Devices, Trends, Technology and Applications, (185–204) © 2025 Scrivener Publishing LLC

8.1 Introduction

Currently, human life is becoming very frightening because of the invisible viruses around us. Due to these pandemics, human life and the economy of the world have become very atrocious. To detect such viruses, we need to contribute a device from the semiconductor world [1, 2]. Additionally, technological development has made it possible for biological warheads to replace conventional warheads, paving the way for the development of biological weapons. These biological warheads or weapons are created from bacteria or viruses that spread covertly and obliterate unwary people along the frontier [3].

In an era of fast-advancing technology, it is difficult for all nations to detect biohazards (dangerous gas compounds). By offering a methodical approach for the identification of biomolecules, biosensors have made significant advancements in this area. The excellent qualities of biosensors have led to a vast range of applications in the medical profession, including drug delivery, food unit processing, environment monitoring, security, and surveillance, as well as early-stage detection and diagnosis [4–7].

Based on the degree of inclusion of the distinct component, such as the method of connecting the biological receptor molecule or bio-recognition towards the element of the base transducer, biosensors can be divided into three generations. In the first generation, the bioreceptor molecule is physically imprisoned in the region of the base sensor after a discriminating membrane similar to a dialysis membrane [8]. In the second generation, immobilization can be achieved either by incorporation into a polymer matrix on the transducer surface or by covalent bonding at a well-tailored transducer interface.

Although these criteria may have been intended for enzyme electrode systems, similar classifications are typically applicable to biosensors. In the third generation, the molecule-like bioreceptor becomes a crucial component of the base-sensing element. The main development attempt are currently visible in the second- and third-generation families. The constituent components of the second generation, such as electrodes, biomolecules, and control circuits, remain distinct [9]. A biosensor is a device that can churn out an electrical signal from a biomolecular physiochemical response [10].

The two main phases of the sensing mechanism for targeted biomolecules are the detection of biomolecules and their dissemination. The target biomolecules were analyzed. This physiochemical reaction occurs during the detection stage, and the transduction stage transforms it into visible electrical data, which can be analyzed further.

After Clark *et al.* discovered the first enzyme-based biosensor in 1962 [11], which has drawn a lot of interest from researchers worldwide in creating precise and trustworthy biosensors. Numerous studies have reported the use of a wide range of biosensors for various applications that require quick, precise, and label-free detection [12].

8.2 What is Biosensor: Types and its Principle

A sensor module and transducer are illustrations of biological detecting components that are combined to form biosensors, which are analytical instruments. These sensors are more sophisticated in terms of selectivity and sensitivity than any other diagnostic tool available in the market. These biosensors are mostly used in food and agricultural industries to monitor environmental pollution control. The stability, cost, sensitivity, and repeatability of biosensors are key characteristics. A representation of the different types of biosensors is shown in Figure 8.1.

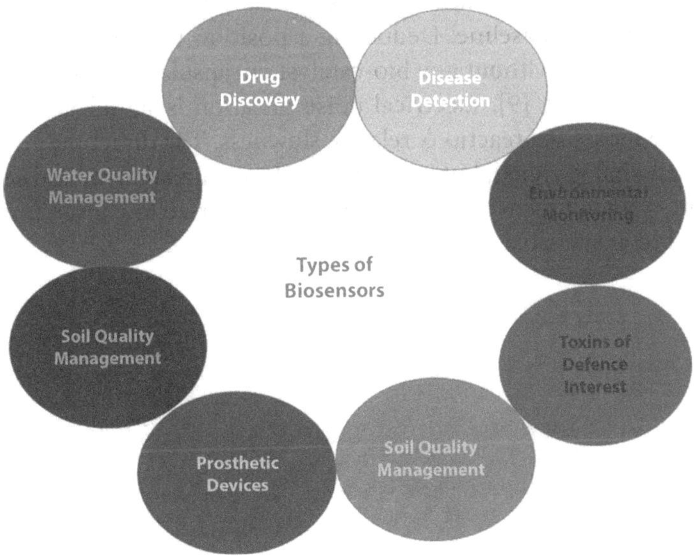

Figure 8.1 Types of biosensors.

Biosensor are acronyms for biological sensors. Hence, a biological component of this sensor could be an enzyme, nucleic acid, or antibody [13–18]. First, the bioelement of a particular substance is allowed to contact the device. The response from the device is then converted to an electrical signal using a transducer. Biosensors can be categorized into different types based on their applications. Based on their applications, biosensors are categorized into several types, including bio-computers, chemical canaries, glucometers, bio-chips, resonant mirrors, immunologic devices, and optrodes.

8.3 Components of Biosensors

The biosensor block diagram comprises three components: sensor, transducer, and related electrons. A sensor is a biological component that responds to the first segment. The detector component alters the signal resulting from the analyte's interaction in the second segment, and the results are clearly shown [19–26]. The processor, display unit, and amplifier, also known as signal-conditioning circuits, are included in the last part. Figure 8.2 shows a typical block diagram of the biosensors.

The transducer's electrical signal is typically weak and lies on top of an extremely high baseline. Deducting a position baseline signal from a related transducer without any bio-catalyst encapsulation is typically part of signal processing [9]. Electrical noise filtration is significantly simplified by the biosensor reaction's relative slowness. The direct output at this point is an analog signal, but it is converted to a digital form and sent to a microprocessor phase where it is advanced, converted to chosen units, and delivered to a data store [27–33].

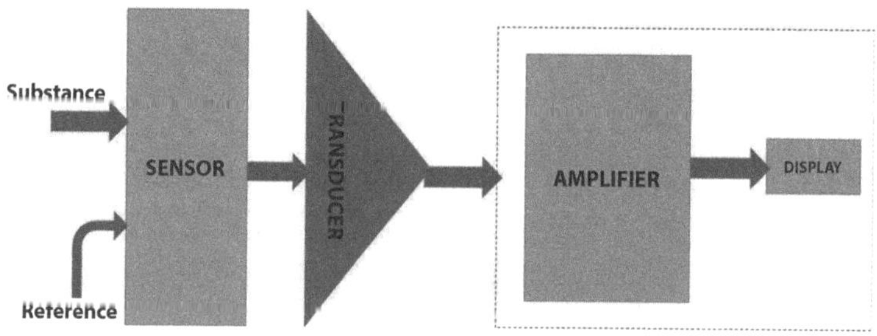

Figure 8.2 Blocks of biosensor.

8.4 Application of FET in Biosensors

Recently, FET-based biosensors have become more attractive because of their stability, reliability, cost, accuracy, label-free detection, on-chip integration of amplification circuitry and sensors [10–13], and assembly-line production. To detect the target biomolecules using FET, cavities are placed in the device, mainly in the oxide layer. The cavities are bio-receptors/bio-recognizers. By targeting the biomolecules on the device cavities, the device will generate some electrochemical reactions in a semiconductor [13, 14] device, which will affect the electrical characteristics of the device. The device's electrical properties are altered by the gating effect, which alters the sensitivity parameters for the detection of biomolecules [8]. This detection is performed before and after the receptors capture the desired biomolecules. Numerous variables can be used to quantify sensitivity, such as the gain ratio (current) (I_{ON}/I_{OFF}), switched threshold voltage (V_T), and ON current (I_{ON}) [15].

FET exhibit more advantages in building biosensors than others. However, because of some issues faced with FETs such as Channel length reduction, Short Channel Effects, and the lowest subthreshold swing that can be achieved theoretically (S.S >60 mV/dec), the performance of the device is decreasing gradually, and also due to the thermionic electron orientation, which will increase the power utilization [16–20]. Therefore, researchers have developed a new technology called TFET to regulate and boost device performance [21–23].

TFET-based biosensors improve and overcome all defects in standard FET (SFET) owing to their band-to-band tunneling and steep subthreshold voltage [2]. Another important parameter of biosensors is the response time. To achieve a quick response, the device must have a low subthreshold voltage [15]. The TFET provides a subthreshold voltage of S.S <60mV/decade compared to (SFET) [24–27]. Due to their rapid development, researchers have turned their focus to TFET-based biosensors. Different research chapters and literature surveys are available on FET-based biosensors [28]. However, the major issue in TFET is ambipolarity, which can be prevented by using less doping in the drain region than in the source region [29–31].

8.5 How TFET Works as a Biosensor and its Structure

Generally, the structural view of a TFET resembles that of a FET device. However, there is a certain change in the doping mechanism and electron movement mechanism across the source-channel and drain-channel intersections. The Standard TFET (STFET) incorporates three regions called the gate, source, drain, and two junction source and channel and drain and channel junctions such as the FET device; however, in the FET device, the source and drain are doped with the same materials, either P-type or N-type, but in the case of TFET, the doping mechanism for the source and drain is not the same. The channel FET device is p-type if it is N-P-N type and n-type if it is P-N-P type, whereas the TFET channel is doped with intrinsic material or lightly doped. Thus, TFET doping regions are also called P-I-N-type doping regions. Hence TFET resembles the P-I-N diode with the gate terminal [34–39].

In SFET, electron movement mainly occurs owing to the application of an external voltage, and this type of mechanism is called thermionic emission. However, for the TFET, the electron flow is due to tunneling at a very low external voltage. Hence, TFETs are also known as low-voltage devices. Here, the junction width is too small to allow electron tunneling across each junction. Hence, the source and drain regions are doped with a high configuration. However, tunneling is more at the source and channel sides because the source is doped higher compared with the drain region.

To determine the performance of a transistor, the current characteristics of the device must be studied under various biasing conditions. The TFET-based biosensor has three terminals, namely the gate, source, drain, and the region between the source and drain called the channel, for detecting biomolecules. A biorecognizer was placed between the source and drain. This biorecognizer is associated with the targeted biomolecules and observes the changes in electrical parameters. Thus, the biosensor will directly transfer the biological parameters to the measurable quantity [40–44]. The operation and behavior of the TFET-based biosensor were recognized in three stages, as mentioned below.

- i) Charge carrier concentration change at the surface of the channel
- ii) The change that occurs in this leads to effective gate voltage change called gating effect
- iii) Because of the gating effect, the effective tunneling length of the drain decreases, which leads to an increase in the drain current.

8.6 Recent Structures of TFET-Based Bio-Sensors

In this chapter, the latest TFET-based structure Bio-Fet device is explained based on the biorecognition element placed in the device structure. The detection of biomolecules is of two types namely i) label-based [32] and ii) label-free-based detection [33]. Designing label-based biosensors is difficult, time-consuming, and ineffective due to the change in the biorecognition element for every sample detection. Different label-based detection methods available for magnetic, fluorescence, and electrochemical techniques can alter the fundamental characteristics of biomolecules. These methods are time-consuming and yield inaccurate results [34–37]. However, label-free detection of biomolecules is possible using dielectric-modulated devices. This increased the detection capability compared to label-based detection. The food industry, environmental monitoring systems, the medical industry, and other industries use biosensors [37–45].

Different types of TFET-based dielectric modulated Bio-Fets are available, but it is not simple to explain all the devices in one chapter; instead, the latest TFET-based Bio-Fets are reviewed. Kumar *et al.* [46] designed a TFET-based Bio-Fet using a dielectric modulation technique. The device has a dual channel using a trench gate type [47–56]. In this device, the gate was placed perpendicular to the trench to create dual channels on either side of the gate. Due to this trench position of the gate, the current in the dual-channel trench gate tunnel field-effect transistor is increased because of the parallel conduction of the two channels [57–72]. A schematic of the DM-DC-TG-TFET is shown in Figure 8.3.

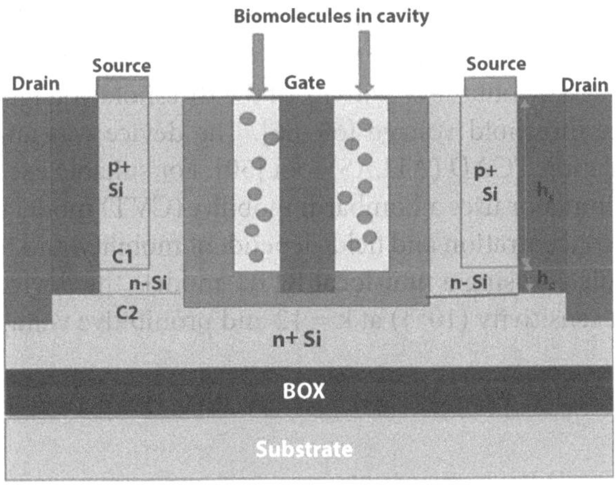

Figure 8.3 Cross-section view of DM-DCTGTFET biosensor [46].

For the detection of biomolecules, two cavities were placed on both sides of the gate oxide, as shown in the figure. Owing to these dual cavities, the detection mechanism of biomolecules is accurate and correct [73–89]. Here, the metalwork function of the gate is 4.2 eV and the height of the source is 70 nm, which is taken on either side of the gate symmetrically to increase the tunneling probability across the junction [90–108]. The channel height was set at 20 nm. Here, the device cavities are occupied by immobilized biomolecules, and these biomolecules become stable because of the dielectric constant. Due to the dielectric change occurring in the device, the tunneling width of the device changes [109–127]. This tunneling change affects the device's electrical performance mainly, the drain current of the device. The effect of the drain current leads to the device's electrical performance.

The utmost problem of TFET is ambipolarity/OFF state current is the ambipolarity/OFF state current increases, and the sensitivity of the device will affect [128–136]. To overcome this effect in the proposed device (DM-DCTGTFET), the dielectric constant of the biomolecules was increased. Due to the increase in the dielectric constant, the barrier of the source-channel junction decreases in the OFF state [137–140]. The next parameter of the biosensor was the sensitivity increment. Here, owing to the dual-channel placement, the drain current and current ratio also increase, which leads to a good sensitivity to the device. Different values of the dielectric constants (k = 3,4,5,7,10,12) of the biomolecules are shown in Figure 8.4a. Here, the drain current increases as the dielectric constant of the biomolecule increases and enhances the ON and OFF current values. The foremost focus of this TFET is to improve the current-ratio threshold voltage, sensitivity, and reduction in tunneling width. The threshold voltage variation is shown in Figure 8.4b. As the dielectric constant of the biomolecule increases, but the sensitivity of the threshold voltage is reciprocal to the device threshold voltage [60–66]. The device was analyzed using the 2D simulation TCAD (ATLAS) tool [50]. For suitable execution of the device, the simulator uses a Lombardi mobility (CVT) model to determine the effect of concentration and field-dependent mobility, and the tunneling effect is simulated using a non-local BTBT model. The device exhibited a high current sensitivity (10^{12}}) at k = 12 and prohibitive voltage sensitivity (5.2 V) [46].

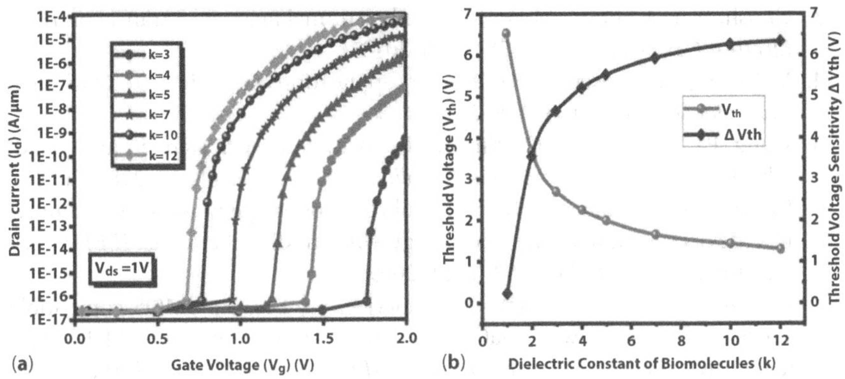

Figure 8.4 (a) Drain current versus gate voltage and (b) relationship between threshold voltage and threshold sensitivity with a dielectric constant of the biomolecule on the X-axis for DM-DCTGTFET [46].

8.7 Conclusion

This chapter concentrates on the biosensor components, principles, and applications of FETs type biosensor semiconductor devices. A simple diagram of the biosensor is depicted and explained. Here, the TFET-based biosensor principle is also explained to determine the difference between a normal biosensor and TFET-based biosensors. A recent dielectrically modulated TFET structure, named DM-DCTGTFET, is explained by its drain current and threshold voltage sensitivity plots. This device exhibited a drain current ratio of approximately 10^{13}, and the threshold sensitivity of the device increased with an increase in the dielectric constant.

References

1. Sarkar, D. and Banerjee, K., Proposal for tunnel-field-effect-transistor as ultra-sensitive and label-free biosensors. *Appl. Phys. Lett.*, 100, 14, 143108, 2012.
2. Nair, P.R. and Alam, M.A., Screening-limited response of nanobiosensors. *Nano Lett.*, 8, 5, 1281–1285, 2008.

3. Ghosh, B. and Akram, M.W., Junctionless tunnel field effect transistor. *IEEE Electron Device Lett.*, 34, 5, 584–586, 2013.
4. Mehrotra, P., Biosensors and their applications–a review. *J. Oral Biol. Craniofac. Res.*, 6, 2, 153–159, 2016.
5. Wang, P.F., Hilsenbeck, K., Nirschl, T., Oswald, M., Ch. stepper, m. weis, d. schmitt-landsiedel, and w. hansch. *Solid State Electron.*, 48, 12, 2281, 2004.
6. Khatami, Y. and Banerjee, K., Steep subthreshold slope n-and p-type tunnel-FET devices for low-power and energy-efficient digital circuits. *IEEE Trans. Electron Devices*, 56, 11, 2752–2761, 2009.
7. Gandhi, R., Chen, Z., Singh, N., Banerjee, K., Lee, S., CMOS-compatible vertical-silicon-nanowire gate-all-around p-type tunneling FETs with leq50-mv/decade subthreshold swing. *IEEE Electron Device Lett.*, 32, 11, 1504–1506, 2011.
8. Narang, R., Saxena, M., Gupta, M., Comparative analysis of dielectric-modulated FET and TFET-based biosensor. *IEEE Trans. Nanotechnol.*, 14, 3, 427–435, 2015.
9. https://www.elprocus.com.
10. Mohanty, S.P. and Kougianos, E., Biosensors: A tutorial review. *IEEE Potentials*, 25, 2, 35–40, 2006.
11. Clark Jr, L.C. and Lyons, C., Electrode systems for continuous monitoring in cardiovascular surgery. *Ann. N. Y. Acad. Sci.*, 102, 1, 29–45, 1962.
12. Hernandez-Vargas, G., Sosa-Hernández, J.E., Saldarriaga-Hernandez, S., Villalba-Rodríguez, A.M., Parra-Saldivar, R., Iqbal, H.M.N., Electrochemical biosensors: A solution to pollution detection with reference to environmental contaminants. *Biosensors*, 8, 2, 29, 2018.
13. Gao, A., Lu, N., Wang, Y., Li, T., Robust ultrasensitive tunneling-FET biosensor for point-of-care diagnostics. *Sci. Rep.*, 6, 1, 1–9, 2016.
14. Chandan, B.V., Nigam, K., Sharma, D., Junctionless based dielectric modulated electrically doped tunnel FET based biosensor for label-free detection. *Micro Nano Lett.*, 13, 4, 452–456, 2018.
15. Narang, R., Sasidhar Reddy, K.V., Saxena, M., Gupta, R.S., Gupta, M., A dielectric-modulated tunnel-FET-based biosensor for label-free detection: Analytical modeling study and sensitivity analysis. *IEEE Trans. Electron Devices*, 59, 10, 2809–2817, 2012.
16. Cao, W., Sarkar, D., Khatami, Y., Kang, J., Banerjee, K., Subthreshold-swing physics of tunnel field-effect transistors. *AIP Adv.*, 4, 6, 067141, 2014.
17. Singh, A., Pandey, C.K., Nanda, U., Performance analysis of Silicon nanotube dielectric pocket Tunnel FET for reduced ambipolar conduction. *Microelectron. J.*, 126, 105512, 2022.
18. Pandey, C.K., Kumar, D.A., Nanda, U., Improved DC Performance of Nanotube Junctionless Field-Effect Transistors with Dielectric Pocket Integration. *ECS J. Solid State Sci. Technol.*, 11, 8, 081003, 2022.
19. Das, S.K., Nanda, U., Biswal, S.M., Pandey, C.K., Giri, L.I., Performance analysis of gate-stack dual-material DG MOSFET using work-function

modulation technique for lower technology nodes. *Silicon*, 14, 6, 2965–2973, 2022.

20. Singh, A. and Pandey, C.K., Improved dc performances of gate-all-around si-nanotube tunnel fets using gate-source overlap. *Silicon*, 14, 4, 1463–1470, 2022.
21. Sarkar, D. and Banerjee, K., Fundamental limitations of conventional-FET biosensors: Quantum-mechanical-tunneling to the rescue, in: *70th Device Research Conference*, EEE, pp. 83–84, 2012.
22. Kanungo, S., Chattopadhyay, S., Gupta, P.S., Sinha, K., Rahaman, H., Study and analysis of the effects of sige source and pocket-doped channel on sensing performance of dielectrically modulated tunnel FET-based biosensors. *IEEE Trans. Electron Devices*, 63, 6, 2589–2596, 2016.
23. Kim, J.H., Kim, H.W., Song, Y.S., Kim, S., Kim, G., Analysis of current variation with work function variation in l-shaped tunnel-field effect transistor. *Micromachines*, 11, 8, 780, 2020.
24. Kanungo, S., Chattopadhyay, S., Gupta, P.S., Rahaman, H., Comparative performance analysis of the dielectrically modulated full-gate and short-gate tunnel FET-based biosensors. *IEEE Trans. Electron Devices*, 62, 3, 994–1001, 2015.
25. Kranti, A., Applicability of Transconductance-to-Current Ratio (gm/Ids) as a Sensing Metric for Tunnel FET Biosensors. *IEEE Sens. J.*, 17, 4, February 15, 2017.
26. Song, Y.S., Kim, J.H., Kim, G., Kim, H.-M., Kim, S., Park, B.-G., Improvement in self-heating characteristic by incorporating hetero-gate-dielectric in gate-all-around MOSFETs. *IEEE J. Electron Devices Soc.*, 9, 36–41, 2020.
27. Song, Y.S., Hwang, S., Min, K.K., Jang, T., Choi, Y., Yu, J., Lee, J.-H., Park, B.-G., Electrical and thermal performances of omega-shaped-gate nanowire field effect transistors for low power operation. *J. Nanosci. Nanotechnol.*, 20, 7, 4092–4096, 2020.
28. Wadhera, T., Kakkar, D., Wadhwa, G., Raj, B., Recent advances and progress in development of the field effect transistor biosensor: A review. *J. Electron. Mater.*, 48, 12, 7635–7646, 2019.
29. Vladimirescu, A., Amara, A., Anghel, C., *et al.*, An analysis on the ambipolar current in Si double-gate tunnel FETs. *Solid State Electron.*, 70, 67–72, 2012.
30. Nigam, K., Kondekar, P., Sharma, D., Approach for ambipolar behavior suppression in tunnel FET by workfunction engineering. *Micro Nano Lett.*, 11, 8, 460–464, 2016.
31. Abdi, D.B. and Jagadesh Kumar, M., Controlling ambipolar current in tunneling FETs using overlapping gate-on-drain. *IEEE J. Electron Devices Soc.*, 2, 6, 187–190, 2014.
32. Im, H., Huang, X.-J., Gu, B., Choi, Y.-K., A dielectric-modulated field-effect transistor for biosensing. *Nat. Nanotechnol.*, 2, 7, 430–434, 2007.
33. G.W. and Raj, B., Design, simulation and performance analysis of JLTFET biosensor for high sensitivity. *IEEE Trans. Nanotechnol.*, 18, 567–574, 2019.

34. Marrazza, G., Chianella, I., Mascini, M., Disposable DNA electrochemical sensor for hybridization detection. *Biosens. Bioelectron.*, 14, 1, 43–51, 1999.
35. Bras, M., Cloarec, J.-P., Bessueille, F., Souteyrand, E., Martin, J.-R., Chauvet, J.-P., Control of immobilization and hybridization on DNA chips by fluorescence spectroscopy. *J. Fluoresc.*, 10, 3, 247–247, 2000.
36. Miller, M.M., Sheehan, P.E., Edelstein, R.L., Tamanaha, C.R., Zhong, L., Bounnak, S., Whitman, L.J., Colton, R.J., A DNA array sensor utilizing magnetic microbeads and magnetoelectronic detection. Technical report, NAVAL RESEARCH LAB, WASHINGTON DC, 2001.
37. Bergveld, P., Thirty years of ISFETOLOGY: What happened in the past 30 years and what may happen in the next 30 years. *Sens. Actuators B: Chem.*, 88, 1, 1–20, 2003.
38. Singh, S., Kondekar, P.N., Jaiswal, N.K., Label-free biosensor using nanogap embedded dielectric modulated schottky tunneling source impact ionization MOS. *Microelectron. Eng.*, 149, 129–134, 2016.
39. Verma, M., Tirkey, S., Yadav, S., Sharma, D., Yadav, D.S., Performance assessment of a novel vertical dielectrically modulated TFET-based biosensor. *IEEE Trans. Electron Devices*, 64, 9, 3841–3848, 2017.
40. Goswami, R. and Bhowmick, B., Comparative analyses of circular gate TFET and heterojunction TFET for dielectric-modulated label-free biosensing. *IEEE Sens. J.*, 19, 21, 9600–9609, 2019.
41. Yotter, R.A. and Wilson, D.M., A review of photodetectors for sensing light-emitting reporters in biological systems. *IEEE Sens. J.*, 3, 3, 288–303, 2003.
42. Jang, D.-Y., Kim, Y.-P., Kim, H.-S., Ko Park, S.-H., Choi, S.-Y., Choi, Y.-K., Sublithographic vertical gold nanogap for label-free electrical detection of protein-ligand binding. *J. Vac. Sci. Technol. B: Microelectron. Nanometer Struct. Process. Meas. Phenom.*, 25, 2, 443–447, 2007.
43. Schaertel, B.J. and Firstenberg-Eden, R., Biosensors in the food industry: present and future. *J. Food Prot.*, 51, 10, 811–820, 1988.
44. Nguyen, V.-T., Kwon, Y.S., Gu, M.B., Aptamer-based environmental biosensors for small molecule contaminants. *Curr. Opin. Biotechnol.*, 45, 15–23, 2017.
45. Malhotra, B.D., Kumar, S., Pandey, C.M., Nanomaterials based biosensors for cancer biomarker detection. *J. Phys. Conf. Ser.*, 704, 012011, IOP Publishing, 2016.
46. Kumar, S., Singh, Y., Singh, B., Tiwari, P.K., Simulation study of dielectric modulated dual channel trench gate TFET-based biosensor. *IEEE Sens. J.*, 20, 21, 12565–12573, 2020.
47. Wangkheirakpam, V.D., Bhowmick, B., Pukhrambam, P.D., N+ pocket doped vertical TFET based dielectric-modulated biosensor considering non-ideal hybridization issue: A simulation study. *IEEE Trans. Nanotechnol.*, 19, 156–162, 2020.

48. Ahn, J.-H., Choi, S.-J., Han, J.-W., Park, T.J., Lee, S.Y., Choi, Y.-K., Double-gate nanowire field effect transistor for a biosensor. *Nano Lett.*, 10, 8, 2934–2938, 2010.
49. Im, M., Ahn, J.-H., Han, J.-W., Park, T.J., Lee, S.Y., Choi, Y.-K., Development of a point-of-care testing platform with a nanogap-embedded separated double-gate field effect transistor array and its readout system for detection of avian influenza. *IEEE Sens. J.*, 11, 2, 351–360, 2010.
50. Int SILVACO, ATLAS User's Manual; Device Simulation Software, Santa Clara, CA, USA, 2016.
51. Bhardwaj, A., Kumar, P., Raj, B., Anand, S., Design and Performance Optimization of doping-less Vertical Nanowire TFET using Gate Stack Technique. *J. Electron. Mater. (JEMS)* Springer, 41, 7, 4005–4013, 2022.
52. Singh, J. and Raj, B., Tunnel Current Model of Asymmetric MIM Structure Levying Various Image Forces to Analyze the Characteristics of Filamentary Memristor. *Appl. Phys. A*, Springer, 125, 3, 203.1–203.11, Feb 2019.
53. Goyal, C., Ubhi, J.S., Raj, B., Low Leakage Zero Ground Noise Nanoscale Full Adder using Source Biasing Technique. *J. Nanoelectron. Optoelectron.*, American Scientific Publishers, 14, 360–370, March 2019.
54. Singh, G., Sarin, R.K., Raj, B., A Novel Robust Exclusive-OR Function Implementation in QCA Nanotechnology with Energy Dissipation Analysis. *J. Comput. Electron.*, Springer, 15, 2, 455–465, June 2016.
55. Wadhwa, G., Kamboj, P., Raj, B., Design optimisation of junctionless TFET biosensor for high sensitivity. *Adv. Nat. Sci.: Nanosci. Nanotechnol.*, 10, 7, 045001, 2019.
56. Bansal, P. and Raj, B., MEMRISTOR MODELING AND ANALYSIS FOR LINEAR DOPANT DRIFT KINETICS. *J. Nanoeng. Nanomanuf.*, American Scientific Publishers, 6, 1–7, 2016.
57. Ashima, Vaithiyanathan, D., Raj, B., Performance Analysis of Charge Plasma induced Graded Channel Si nanotube. *J. Eng. Res. (JER)*, EMSME Special Issue, 146–154, Aug 2021.
58. Tomar, A.S., Magraiya, V.K., Raj, B., Scaling of Access and Data Transistor for High Performance DRAM Cell Design. *Quantum Matter*, 2, 412–416, Oct. 2013.
59. Jain, N. and Raj, B., Parasitic Capacitance and Resistance Model Development and Optimization of Raised Source/Drain SOI FinFET Structure for Analog Circuit Applications. *J. Nanoelectron. Optoelectron.*, ASP, USA, 13, 531–539, Apl 2018.
60. Singh, S., Vishvakarma, S.K., Raj, B., Analytical Modeling of Split-Gate Junction-Less Transistor for a Biosensor Application. *Sens. Bio-Sens.*, Elsevier, 18, 31–36, April 2018.
61. Gopal, M. and Raj, B., Low Power 8T SRAM Cell Design for High Stability Video Applications. *ITSI Trans. Electr. Electron. Eng.*, 1, 5, 91–97, 2013.
62. Raj, B., Mitra, J., Bihani, D.K., Rangharajan, V., Saxena, A.K., Dasgupta, S., Analysis of Noise Margin, Power and Process Variation for 32 nm FinFET

Based 6T SRAM Cell. *J. Comput. (JCP)*, Academy Publisher, FINLAND, 5, 6, 1–8, 2010.

63. Sharma, D., Mehra, R., Raj, B., Comparative Analysis of Photovoltaic Technologies for High Efficiency Solar Cell Design. *Superlattices Microstruct.*, Elsevier, 153, 106861, May 2021.
64. Kaur, P., Buttar, A.S., Raj, B., A comprehensive Analysis of Nanoscale Transistor based Biosensor: A Review. *Indian J. Pure Appl. Phys.*, 59, 304–318, April 2021.
65. Yadav, D., Raj, B., Raj, B., Design and Simulation of Low Power Microcontroller for IoT Applications. *Sens. Lett.*, ASP, 18, 401–409, May 2020.
66. Singh, S. and Raj, B., A 2-D Analytical Surface Potential and Drain current Modeling of Double-Gate Vertical t-shaped Tunnel FET. *J. Comput. Electron.*, Springer, 19, 1154–1163, Apl 2020.
67. Singh, J. and Raj, B., An Accurate and Generic Window function for Nonlinear Memristor Model. *J. Comput. Electron.*, Springer, 18, 2, 640–647, June 2019.
68. Kaur, M., Gupta, N., Kumar, S., Raj, B., Singh, A.K., Comparative RF and Crosstalk Analysis of Carbon Based Nano Interconnects. *IET Circuits Devices Syst.*, 15, 6, 493–503, Feb 2021.
69. Kandasamy, N., Ahmad, F., Ajitha, D., Raj, B., Telagam, N., Quantum Dot Cellular Automata based Scan Flip Flop and Boundary Scan Register. *IETE J. Res.*, 66, 535–548, 2020.
70. Singh, J. and Raj, B., Design and Investigation of 7T2M NVSARM with Enhanced Stability and Temperature Impact on Store/Restore Energy. *IEEE Trans. Very Large Scale Integr. Syst.*, 27, 6, 1322–1328, June 2019.
71. Bhardwaj, A.K., Gupta, S., Raj, B., Singh, A., Impact of Double Gate Geometry on the Performance of Carbon Nanotube Field Effect Transistor Structures for Low Power Digital Design. *Comput. Theor. Nanosci.*, ASP, 16, 1813–1820, 2019.
72. Jain, N. and Raj, B., Thermal Stability Analysis and Performance Exploration of Asymmetrical Dual-k underlap Spacer (ADKUS) SOI FinFET for Security and Privacy Applications. *Indian J. Pure Appl. Phys. (IJPAP)*, 57, 352–360, May 2019.
73. Jain, N. and Raj, B., Dual-k Spacer Region Variation at the Drain Side of Asymmetric SOI FinFET Structure: Performance Analysis towards the Analog/RF Design Applications. *J. Nanoelectron. Optoelectron.*, American Scientific Publishers, 14, 349–359, March 2019.
74. Singh, J., Sharma, S., Raj, B., Khosla, M., Analysis of barrier layer thickness on performance of In1-xGaxAs based Gate Stack Cylindrical Gate Nanowire MOSFET. *JNO*, ASP, 13, 1473–1477, Oct 2018.
75. Jain, N. and Raj, B., Analysis and Performance Exploration of High-k SOI FinFETs Over the Conventional Low-k SOI FinFET toward Analog/RF Design. *J. Semicond. (JoS)*, IOP Science, 39, 12, 124002-1-7, Dec 2018.

76. Goyal, C., Ubhi, J.S., Raj, B., A reliable leakage reduction technique for approximate full adder with reduced ground bounce noise. *Math. Probl. Eng.*, Hindawi, 2018, Article ID 3501041, 16, 15 Oct 2018.
77. Anuradha, Singh, J., Raj, B., Khosla, M., Design and Performance Analysis of Nano-scale Memristor-based Nonvolatile SRAM. *Sens. Lett.*, American Scientific Publishers, 16, 798–805, Oct 2018.
78. Wadhwa, G. and Raj, B., Parametric Variation Analysis of Charge-Plasma-based Dielectric Modulated JLTFET for Biosensor Application. *IEEE Sens. J.*, 18, 15, 6070–6077, AUGUST 1 2018, (SCI).
79. Singh, J. and Raj, B., Comparative Analysis of Memristor Models for Memories Design. *JoS*, IoP, 39, 7, 074006-1-12, July 2018.
80. Yadav, D., Chouhan, S.S., Vishvakarma, S.K., Raj, B., Application Specific Microcontroller Design for IoT based WSN. *Sens. Lett.*, ASP, 16, 374–385, May 2018.
81. Singh, G., Sarin, R.K., Raj, B., Fault-Tolerant Design and Analysis of Quantum-Dot Cellular Automata Based Circuits. *IEEE/IET Circuits Devices Syst.*, 12, 638–64, 2018.
82. Singh, J. and Raj, B., Modeling of Mean Barrier Height Levying Various Image Forces of Metal Insulator Metal Structure to Enhance the Performance of Conductive Filament Based Memristor Model. *IEEE Nanotechnol.*, 17, 2, 268–267, March 2018.
83. Jain, A., Sharma, S., Raj, B., Analysis of Triple Metal Surrounding Gate (TM-SG) III-V Nanowire MOSFET for Photosensing Application. *Opto-Electron. J.*, Elsevier, 26, 2, 141–148, May 2018.
84. Jain, A., Sharma, S., Raj, B., Design and Analysis of High Sensitivity Photosensor Using Cylindrical Surrounding Gate MOSFET For Low Power Sensor Applications. *Eng. Sci. Technol. Int. J.*, Elsevier's, 19, 4, 1864–1870, December 2016.
85. Saiphani Kumar, G., Singh, A., Raj, B., Design and Analysis of Gate All Around CNTFET based SRAM cell Design. *J. Comput. Electron.*, Springer, 17, 1, 138–145, March 2018.
86. Singh, G.P., Sohi, B.S., Raj, B., Material Properties Analysis of Graphene Base Transistor (GBT) for VLSI Analog Circuits. *Indian J. Pure Appl. Phys. (IJPAP)*, 55, 896–902, Dec-2017.
87. Kumar, S. and Raj, B., Estimation of Stability and Performance metric for Inward Access Transistor based 6T SRAM Cell Design using n-type/p-type DMDG-GDOV TFET. *IEEE VLSI Circuits Syst. Lett.*, 3, 2, 25–39, June 2017.
88. Sharma, S., Kumar, A., Pattanaik, M., Raj, B., Forward Body Biased Multimode Multi-Threshold CMOS Technique for Ground Bounce Noise Reduction in Static CMOS Adders. *Int. J. Inf. Electr. Eng.*, 3, 3, 567–572, 2013.
89. Singh, H., Kumar, P., Raj, B., Performance Analysis of Majority Gate SET Based 1-bit Full Adder. *Int. J. Comput. Commun. Eng. (IJCCE)*, IACSIT Press Singapore, 2, 4, 2013, ISSN: 2010-3743.

90. Bhardwaj, A.K., Gupta, S., Raj, B., Investigation of Parameters for Schottky Barrier (SB) Height for Schottky Barrier Based Carbon Nanotube Field Effect Transistor Device. *J. Nanoelectron. Optoelectron.*, ASP, 15, 783–791, July 2020.
91. Bansal, P. and Raj, B., MEMRISTOR: A VERSATILE NONLINEAR MODEL FOR DOPANT DRIFT AND BOUNDARY ISSUES. *JCTN*, American Scientific Publishers, 14, 5, 2319–2325, May 2017.
92. Jain, N. and Raj, B., An Analog and Digital Design Perspective Comprehensive Approach on Fin-FET (Fin-Field Effect transistor) Technology - A Review. *Rev. Adv. Sci. Eng. (RASE)*, ASP, 5, 1–14, 2016.
93. Raj, B., Saxena, A.K., Dasgupta, S., Analytical Modeling for the Estimation of Leakage Current and Subthreshold Swing Factor of Nanoscale Double Gate FinFET Device. *Microelectron. Int.*, UK, 26, 53–63, 2009.
94. Soniya, Singh, S., Wadhwa, G., Raj, B., An Analytical Modeling for Dual Source Vertical Tunnel Field Effect Transistor. *Int. J. Recent Technol. Eng. (IJRTE)*, 8, 2, 603–608, July 2019.
95. Singh, S. and Raj, B., Design and analysis of hetrojunction Vertical T-shaped Tunnel Field Effect Transistor. *J. Electron. Mater.*, Springer, 48, 10, 6253–6260, October 2019.
96. Goyal, C., Ubhi, J.S., Raj, B., A Low Leakage CNTFET based Inexact Full Adder for Low Power Image Processing Applications. *Int. J. Circuit Theory Appl.*, Wiley, 47, 9, 1446–1458, September 2019.
97. Raj, B., Saxena, A.K., Dasgupta, S., A Compact Drain Current and Threshold Voltage Quantum Mechanical Analytical Modeling for FinFETs. *J. Nanoelectron. Optoelectron. (JNO)*, USA, 3, 2, 163–170, 2008.
98. Wadhwa, G. and Raj, B., An Analytical Modeling of Charge Plasma based Tunnel Field Effect Transistor with Impacts of Gate underlap Region. *Superlattices Microstruct.*, Elsevier, 142, 106512, June 2020.
99. Singh, S. and Raj, B., Modeling and Simulation analysis of SiGe hetrojunction Double GateVertical t-shaped Tunnel FET. *Superlattices Microstruct.*, Elsevier, 142, 106496, June 2020.
100. Jain, N. and Raj, B., Device and Circuit Co-Design Perspective Comprehensive Approach on FinFET Technology - A Review. *J. Electron Devices*, 23, 1, 1890–1901, 2016.
101. Kumar, S. and Raj, B., Analysis of I_{ON} and Ambipolar Current for Dual-Material Gate-drain Overlapped DG-TFET. *J. Nanoelectron. Optoelectron.*, American Scientific Publishers, USA, 11, 323–333, June 2016.
102. Anjum, N., Bali, T., Raj, B., Design And Simulation Of Handwritten Multiscript Character Recognition. *Int. J. Adv. Res. Comput. Commun. Eng.*, 2, 7, 2544–2549, July 2013.
103. Singh, K. and Raj, B., Performance and analysis of temperature dependent Multi-walled Carbon Nanotubes as Global Interconnects at different technology nodes. *J. Comput. Electron.*, Springer, 14, 2, 469–476, June 2015.

104. Kumar, S. and Raj, B., Compact channel potential analytical Modeling of DG-TFET based on Evanescent-mode Approach. *J. Comput. Electron.*, Springer, 14, 2, 820–827, July 2015.
105. Singh, K. and Raj, B., Temperature Dependent Modeling and Performance Evaluation of Multi-Walled CNT and Single-Walled CNT as Global Interconnects. *J. Electron. Mater.*, Springer, 44, 12, 4825–4835, Dec 2015.
106. Sharma, V.K., Pattanaik, M., Raj, B., INDEP approach for leakage reduction in nanoscale CMOS circuits. *Int. J. Electron.*, Taylor Francis, 102, 2, 200–215, 2014.
107. Singh, K. and Raj, B., Influence of Temperature on MWCNT bundle, SWCNT bundle and Copper interconnects for Nanoscaled Technology nodes. *J. Mater. Sci. Mater. Electron.*, Springer, 26, 8, 6134–6142, 2015.
108. Anjum, N., Bali, T., Raj, B., Design and Simulation of Handwritten Gurumukhi and Devanagri Numerical Recognition. *Int. J. Comput. Appl.*, Published by Foundation of Computer Science, New York, USA, 73, 12, 16–21, 2013.
109. Khandelwal, S., Gupta, V., Raj, B., Gupta, R.D., Process Variability Aware Low Leakage Reliable Nano Scale DG-FinFET SRAM Cell Design Technique. *J. Nanoelectron. Optoelectron.*, 10, 6, 810–817, December 2015.
110. Sharma, V.K., Pattanaik, M., Raj, B., ONOFIC Approach: Low Power High Speed Nanoscale VLSI Circuits Design. *Int. J. Electron.*, Taylor Francis, 101, 1, 61–73, 2014.
111. Khandelwal, S., Raj, B., Gupta, R.D., FinFET based 6T SRAM Cell Design: Analysis of Performance Metric, Process Variation and Temperature Effect. *J. Comput. Theor. Nanosci.*, ASP, USA, 12, 2500–2506, 2015.
112. Singh, S., Yadav, S., Rahul, J., Srivastava, A., Raj, B., Impact of HfO_2 in Graded Channel Dual Insulator Double Gate MOSFET. *J. Comput. Theor. Nanosci.*, American Scientific Publishers, 12, 6, 950–953, April 2015.
113. Sharma, V.K., Pattanaik, M., Raj, B., PVT variations aware low leakage INDEP approach for nanoscale CMOS Circuits. *Microelectron. Reliab.*, Elsevier, 54, 90–99, 2014.
114. Raj, B., Saxena, A.K., Dasgupta, S., Quantum Mechanical Analytical Modeling of Nanoscale DG FinFET: Evaluation of Potential, Threshold Voltage and Source/Drain Resistance. *Elsevier's J. Mater. Sci. Semicond. Process.*, Elsevier, 16, 4, 1131–1137, 2013.
115. Gopal, M., Prasad, S.S.D., Raj, B., 8T SRAM Cell Design for Dynamic and Leakage Power Reduction. *Int. J. Comput. Appl.*, Published by Foundation of Computer Science, New York, USA, 71, 9, 43–48, June 2013.
116. Wadhwa, G., Kamboj, P., Singh, J., Raj, B., Design and Investigation of Junctionless DGTFET for Biological Molecule Recognition. *Trans. Electr. Electron. Mater.*, Springer, 22, 282–289, 2021.
117. Verma, S.K., Singh, S., Wadhwa, G., Raj, B., Detection of Biomolecules using Charge-Plasma based Gate Underlap Dielectric Modulated Dopingless

TFET. *Trans. Electr. Electron. Mater. (TEEM)*, Springer, 21, 528–535, June 2020.

118. Jain, N. and Raj, B., Impact of underlap spacer region variation on Electrostatic and analog/RF performance of symmetrical high-k SOI FinFET at 20 nm channel length. *J. Semicond. (JoS)*, IOP Science, 38, 12, 122002, Dec 2017.
119. Singh, S. and Raj, B., Analytical Modeling and Simulation analysis of T-shaped III-V heterojunction Vertical T-FET. *Superlattices Microstruct.*, Elsevier, 147, 106717, Nov 2020.
120. Singh, G., Sarin, R.K., Raj, B., Design and Analysis of Area Efficient QCA Based Reversible Logic Gates. *J. Microprocess. Microsyst.*, Elsevier, 52, 59–68, May 2017.
121. Singh, S. and Raj, B., Parametric variation analysis on hetero-junction Vertical t-shape TFET for Supressing ambipolar conduction. *Indian J. Pure Appl. Phys.*, 58, 478–485, June 2020.
122. Soniya, Singh, S., Wadhwa, G., Raj, B., Design and Analysis of Dual Source Vertical Tunnel Field Effect Transistor for high performance. *Trans. Electr. Electron. Mater.*, Springer, 21, 74–82, Oct 2019.
123. Kaur, M., Gupta, N., Kumar, S., Raj, B., Singh, A.K., RF Performance Analysis of Intercalated Graphene Nanoribbon Based Global Level Interconnects. *J. Comput. Electron.*, Springer, 19, 1002–1013, June 2020.
124. Wadhwa, G. and Raj, B., Design and Performance Analysis of Junctionless TFET Biosensor for high sensitivity. *IEEE Nanotechnol.*, 18, 567–574, 2019.
125. Singh, J. and Raj, B., Enhanced Nonlinear Memristor Model Encapsulating Stochastic Dopant Drift. *JNO*, ASP, 14, 958–963, 2019.
126. Pattanaik, M., Raj, B., Sharma, S., Kumar, A., Diode Based Trimode Multi-Threshold CMOS Technique for Ground Bounce Noise Reduction in Static CMOS Adders. *Adv. Mater. Res.*, Trans Tech Publications, Switzerland, 548, 885–889, 2012.
127. Raj, B., Saxena, A.K., Dasgupta, S., Nanoscale FinFET Based SRAM Cell Design: Analysis of Performance metric, Process variation, Underlapped FinFET and Temperature effect. *IEEE Circuits Syst. Mag.*, 11, 2, 38–50, 2011.
128. Sharma, V.K., Pattanaik, M., Raj, B., Leakage Current ONOFIC Approach for Deep Submicron VLSI Circuit Design. *Int. J. Electr. Comput. Electron. Commun. Eng.*, World Academy of Sciences, Engineering and Technology, 7, 4, 239–244, 2013.
129. Kaur, P., Gill, S.S., Raj, B., Comparative Analysis of OFETs Materials and Devices for Sensor Applications. *J. Silicon*, Springer, 14, 4463–4471, 2022.
130. Sharma, S.K., Kumar, P., Raj, B., Raj, B., $In_{1-x}Ga_xAs$ Double Metal Gate-Stacking Cylindrical Nanowire MOSFET for highly sensitive Photo detector. *J. Silicon*, Springer, 14, 3535–3541, 2022.
131. Raj, B., Saxena, A.K., Dasgupta, S., Analytical Modeling of Quasi Planar Nanoscale Double Gate FinFET with Source/Drain Resistance and field

dependent carrier mobility: A quantum mechanical study. *J. Comput. (JCP)*,Academy Publisher, FINLAND, 4, 9, 1–8, 2009.

132. Bhushan, S., Khandelwal, S., Raj, B., Analyzing Different Mode FinFET based Memory Cell at Different Power Supply for Leakage Reduction. *Seventh International Conference on Bio-Inspired Computing: Theories and Application, (BIC-TA 2012)* Advances in Intelligent Systems and Computing, vol. 202, pp. 89–100, 2012, 2013.
133. Singh, J. and Raj, B., Temperature Dependent Analytical Modeling and Simulations of Nanoscale Memristor. *Eng. Sci. Technol. Int. J.*, Elsevier's, 21, 862–868, Oct 2018.
134. Singh, S., Bala, S., Raj, B., Raj, B., Improved Sensitivity of Dielectric Modulated Junctionless Transistor for Nanoscale Biosensor Design. *Sens. Lett.*, ASP, 18, 328–333, Apl 2020.
135. Kumar, V., Vishvakarma, S.K., Raj, B., Design and Performance Analysis of ASIC for IoT Applications. *Sens. Lett.* ASP, 18, 31–38, Jan 2020.
136. Jaiswal, A., Sarin, R.K., Raj, B., Sukhija, S., A Novel Circular Slotted Microstrip-fed Patch Antenna with Three Triangle Shape Defected Ground Structure for Multiband Applications. *Adv. Electromagn. (AEM)*, 7, 3, 56–63, Aug 2018.
137. Wadhwa, G. and Raj, B., Label Free Detection of Biomolecules using Charge-Plasma-Based Gate Underlap Dielectric Modulated Junctionless TFET. *J. Electron. Mater. (JEMS)*, Springer, 47, 8, 4683–4693, August 2018.
138. Singh, G., Sarin, R.K., Raj, B., Design and Performance Analysis of a New Efficient Coplanar Quantum-Dot Cellular Automata Adder. *Indian J. Pure Appl. Phys. (IJPAP)*, 55, 97–103, February 2017.
139. Kaur, P., Pandey, V., Raj, B., Comparative Study of Efficient Design, Control and Monitoring of Solar Power using IoT. *Sens. Lett.*, ASP, 18, 419–426, May 2020.
140. Bhardwaj, A.K., Gupta, S., Raj, B., Development & Analysis of compact model for Double Gate Schottky Barrier CNTFET. *J. Nanoelectron. Optoelectron.*, ASP, 15, 1199–1208, Aug 2020.

9

Transparent Conducting Oxides: Introduction, Types, Deposition Techniques and Applications

Isha Arora* **and Rishi Kant**

School of Natural Sciences, GNA University, Phagwara, Punjab, India

Abstract

This chapter elaborates on a distinct class of semiconductor materials known as transparent conducting oxides (TCO). These materials have novel characteristics because they exhibit an unusual combination of two contemporary properties: high optical transparency and high electrical conductivity. The unique optoelectronic properties of these materials depend on different parameters, which are discussed in detail in the chapter. For the optical properties, contributing factors, such as optical gap, hybridization of the orbitals, defects, and doping, are discussed in detail. On a similar platform, the effective masses and electron distribution laws were discussed to justify the electrical properties of the TCOs. Furthermore, different types of TCO materials are discussed in this section. The materials are elaborated with their structures, which are responsible for imparting unique properties to a specific material. Transparent conducting oxides form the backbone of modern technology because of their unique combination of transparency and conductivity. The chapter also describes different routes for depositing transparent conducting oxides in the form of nanostructures, such as thin films and nanoparticles. Deposition techniques play a significant role in imparting specific properties to deposited nanostructures. Two types of deposition techniques, i.e., physical deposition techniques and chemical deposition techniques, along with their specific requirements, are explored in this chapter. Among the chemical deposition techniques, this chapter lays special emphasis on the sol–gel-based spin-coating technique because it is inexpensive and easy to handle. Furthermore, the importance of transparent conducting oxides from an application perspective is

**Corresponding author*: isha.arora@gnauniversity.edu.in

Balwinder Raj, Shiromani Balmukund Rahi and Nandakishor Yadav (eds.) FeFET Devices, Trends, Technology and Applications, (205–238) © 2025 Scrivener Publishing LLC

discussed. Application of transparent conducting oxides in photovoltaics, display devices, sensors and multifunctional coatings are also elaborated in detail.

Keywords: TCOs, sol gel, spinel structure, solar cells, nanostructured films, UV-Vis, zinc stannate, FET devices

9.1 Introduction

Transparent conducting oxides (TCOs) are semiconductor materials with large band gaps (Figure 9.1). They are generally binary or ternary compounds containing one or two metallic elements, with relatively high carrier concentrations and large optical transparencies. Some examples include doped ZnO, In_2O_3, SnO_2, and CdO. TCOs at their best can be described as materials with transparency above 80% and resistivity of at least ~10^{-3} Ω-cm [1]. The electrical conduction in TCOs can be estimated from the available number and mobility of charge carriers [2]. The rare combination of complementary properties is responsible for making TCO materials an important class of materials that form the backbone of optoelectronic devices, such as solar cells, display devices, light-emitting diodes (LEDs), gas sensors, and field effect transistors (FETs) [3–5]. CdO, the first ever TCO reported by the German physicist Karl Bedakar in 1907, has a small band gap (2.3 eV) with

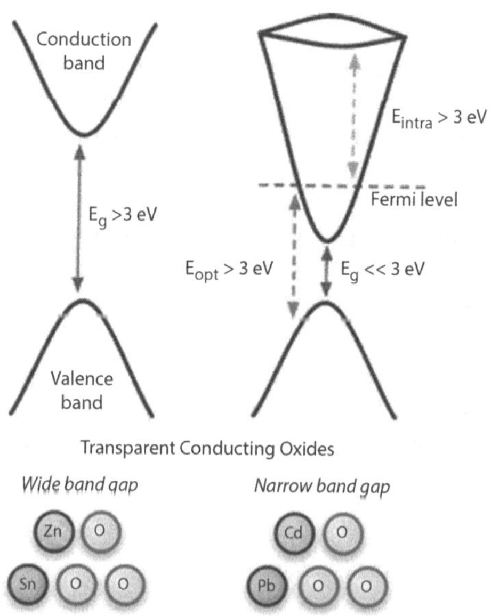

Figure 9.1 Energy band diagrams of transparent conducting oxides.

low transparency. Later, in 1947, an American physicist, Harold McMaster, investigated SnO_2 coatings deposited via CVD and patented its use as a TCO for de-icing windows at high altitudes [6]. Zinc oxide (ZnO), which is popularly known as an intrinsic n-type semiconductor, was recognized as an efficient TCO by the studies made by Minami *et al.* [7]. However, CdO-based TCO materials, which are toxic in nature, have not been extensively investigated; however, ZnO-based TCOs and indium-doped tin oxide have gained immense popularity and remain the best choice on the basis of their industrial performance. Apart from various important applications, ZnO and Indium-Gallium-Zinc-Oxide nanowire-based structures are being investigated in nanoscale field-effect transistors (FETs). The surface morphology of these oxides favors the formation of such contacts with conformal geometry in FETs, where the simple classical theory of metal-semiconductor planar contact can be applied [8, 9].

Extensive studies in the last decade have explored the functionalities of this special class of semiconductor materials. Industrial applications of these materials include field-effect transistors (FETs) and transparent electrodes. These applications involve the production of heterojunctions, for which the tuning of structural defects is a fundamental requirement. The above-mentioned parameters are crucial for a better performance of the device. The need for higher transparency in the visible spectrum and electrical conductivity are the two primary properties of any TCO. Usually, there is a trade-off between the two, as they are mutually exclusive properties.

9.2 Physical Characteristics of TCOs

9.2.1 Optical Properties

TCOs are special members of category of semiconductor materials whose transmission in the visible portion of the spectrum is usually above 80%, which is not expected from typical semiconductors. Several factors contribute to this anomalous property of the TCO's as discussed below:

(a) *Optical Gap*: The optical gap is the most fundamental property, and there is a need to tailor the optical gap for TCOs to exploit their full potential in emerging optoelectronic devices. The bandgap measures the energy requirement of an electron to jump from the valence band (VB) to the conduction band (CB). For instance, metals possess a very small band gap (0.3 eV), whereas insulators have a much higher band gap (4–12 eV). The difference in electronegativities

[10, 11] of participating atoms, hybridization of orbitals, and doping are prominent factors affecting the band gap of a material. In addition, spin–orbit coupling is responsible for reducing the band gaps, as it results in the upliftment of the valence band and hence reducing the same. The role of the spin–orbit interaction is significant in the case of heavier elements and can be neglected in the case of lighter elements. Band alignment is a requirement that must be carefully performed when designing semiconductor heterojunctions. Two parameters, ionization potential (IP) and electron affinity (EA), are the prime factors used when discussing the band gaps and their alignment.

(b) *Electronegativity:* Most electrons jump from the VB to the CB, which is adversely affected by the electronegativity of the anions and cations involved which is responsible for the band gap of semiconductors. The Pauli scale considers the dissociation energies (DE) of the chemical bonds between atoms involved, and not on the individual atoms. Despite the efficient calculation of the band gap of binary oxides [12], the degree of covalency could not be estimated with the help of the Pauli scale. A solution to this was found by introducing the concept of optical electronegativity based on the promotion of electrons in the solid state. Optical electronegativity is a better option because it considers the solid-state phase of binary semiconductor oxides [13].

(c) *Conglomeration of Orbitals*: Certain materials such as chalcopyrites semiconductors such as $AgAlS_2$, $CuAlS_2$, and $ZnSnP_2$ are reported to have narrow band gaps compared to semiconductor oxides such as ZnO and TiO_2 [14]. This abnormality can be attributed to the p-d overlapping effect, electronegativity of cations, and structural defects. The overlapping effect of orbitals is more pronounced in chalcopyrites originating from groups I–III–VI elements and has very little impact for the elements from groups II–IV–V, where the main dominating factor becomes the electronegativity of the cations [15–19]. It is shown that d-electrons contribute significantly to the reduction of the optical bandgap through p–d overlapping. Orbitals merging gives rise to two states, a lower-energy cation d-state and a higher-energy cation p-state, which are present just above the VB. According to perturbation theory, states

with the same symmetry tend to repel each other, so the reduction in the optical gap can be attributed to the repulsion between the symmetric p and d states [13].

(d) *Doping and Defect Formation Energy*: Point defects play a vital role in semiconductor materials owing to their diversity [15]. The fundamental property associated with these defects is formation energy. Here, the defect formation energies together with the configurational entropy, help in determining the equilibrium defect concentration and density of the charge carriers (Figure 9.2). The thermodynamic transition levels corresponding to the acceptor, donor, and deep traps were also estimated by the energies of multiple charge states [20–26]. The mid-gap defects that arise from structural imperfections can also be responsible for light absorption in the case of transparent conducting oxides. Semiconductors can have two types of defects: native or intrinsic defects and extrinsic defects introduced as a result of doping. Apart from Schottky and interstitial defects in the crystal lattice, substitution defects arise as a result of doping, where ions or atoms substitute ions or atoms that are usually found in the lattice [27–35]. The role of doping is crucial for manipulating the electrical and optical behaviors of any semiconductor, especially those of TCOs. Therefore, doping is normally aimed at improving both the conductivity and increasing the optical gap of the TCOs.

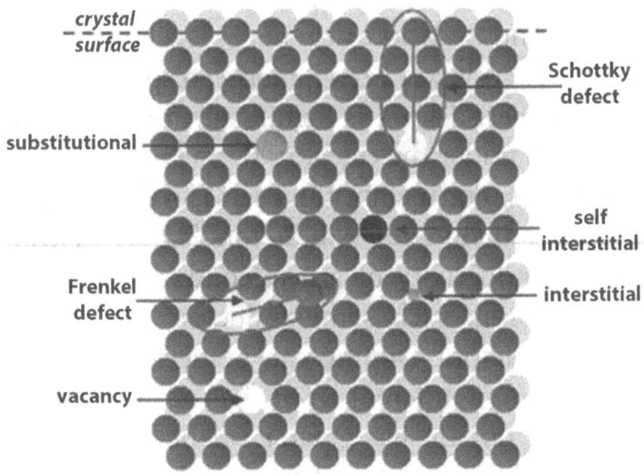

Figure 9.2 Elaboration of different kinds of defects in crystal lattice.

9.2.2 Electrical Properties

The conductivity (σ) of TCO material is evaluated by the number of charge carriers (n), their mobility (μ) and the electron charge (e) as given below:

$$\sigma = en\mu \tag{9.1}$$

The electrical resistance of uniformly thick thin films is calculated as the product of thickness and sheet resistance (Rs = ρ / d); alternatively, the sheet resistance is independent of the film thickness. Increasing the carrier concentration through substitutional doping, vacancy formation, or interstitial implantation is the initial step towards improving the conductivity [36–41]. The factors influencing the electrical characteristics of the TCO materials are discussed below:

(a) *Electron distribution law and Fermi level:* The electric properties, including conductivity, number of charge carriers, and mobility, are the primary factors affecting the performance of any TCO. Thermal excitations are temperature-dependent and the number of electrons in the excited state at any temperature is not governed by the Boltzmann distribution law as the electrons are indistinguishable; hence, a better explanation is given by the Fermi–Dirac equation as:

$$F(E) = \frac{1}{1 + e^{\frac{E-E_F}{KT}}} \tag{9.2}$$

where F(*E*) and E_F denote the energy level fraction (allowed) with the E energy and Fermi energy, respectively. The position of this Fermi level is significant for making different materials fall under the category of metals, semiconductors, and insulators. For metals, the Fermi level is within a band, which makes them good conductors [42–49]. However, the electrical conductivity of a semiconductor can be modulated by doping; such semiconductors are called extrinsic semiconductors. Doping can be performed with atoms having a valency different from that of the host atom; it is n-type if the dopant has greater valency and p-type if the dopant atom has less valency compared to that of the parent atom. Fermi levels therefore shift in n-type semiconductors towards the conduction band and in p-type semiconductors downward towards the valence band. The conductivity of TCO in intrinsic semiconductors is primarily ascribed to the existence of native defects, such as oxygen vacancies, while in extrinsic semiconductors, conductivity is linked to the presence of both native defects and impurity atoms [50–57].

(b) *Effective masses:* Effective masses of the charge carriers (holes and electrons) are crucial because they are related to the relative mobility of charge carriers, affecting the properties of any semiconductor or device fabricated from the same. Typical evaluations of effective masses include band curvature analysis and wave vector studies. The density of states (DOS) effective mass can also be approximated using the theoretical electronic DOS. The electrical properties of TCOs are important for optoelectronic devices. The electrical parameters that are crucial are electrical resistivity, carrier concentration, and mobility. An increase in carrier concentration results in the enhancement of scattering centers, which leads to loss of transparency [58–65]. The best way to enhance the electrical performance of any TCO is to increase the mobility of the charge carriers which is a direct function of the effective mass. As the effective masses are less for the negatively charged electrons compared to the positive charge carrier, holes, it is usually the electrons that are the lead charge carriers in most TCOs. Depending on the type of charge carrier, TCOs can be divided into the following two classes: n-type and p-type TCO materials.

N-type TCO materials are currently used in a majority of applications and are mainly binary oxides of zinc, tin, indium, and cadmium. Various dopants have been added to base materials (ternary and quaternary oxides) to achieve the desired properties on a commercial scale [16]. These materials produce TCOs with high conductivity and high optical transparency with a free carrier density of $\sim 10^{20}$ cm^{-3} or higher with the drawback of being unstable at high temperatures. The doping of metal cations significantly decreases the electrical resistivity, specifically when using dopants with an $(n - 1)d^{10}(n)s^0$ electronic configuration (where $n > 5$). The successful doping of In_2O_3 with Sn to form tin-doped indium oxide (ITO) results in a remarkable reduction in resistivity from $\sim 10^{-3}$ to 10^{-5} Ω^{-cm} [17]. Complementary metal-oxide semiconductor (CMOS) devices rely heavily on P-type TCO materials. Since the discovery of NiO, the first known p-type TCO with a resistivity of ~10^{-1} Ω^{-cm} and transparency of ~40%, in 1997, a number of studies have been conducted to identify p-type TCOs with better properties. $CuAlO_2$, reported in 1997, exhibited a higher transparency of ~60% with a resistivity as high as three to four times greater than that of standard ITO [19]. Since then, various deposition techniques have been used to reduce the resistivity of CuAlO2 for commercial applications [20]. A computational approach using density functional theory (DFT) has also been adopted to identify materials with the potential to be p-type TCOs for emerging devices [66–74].

9.3 Types of Transparent Conductors

Transparent conducting films (TCFs) have emerged as the backbone of modern technology because they combine electrical conduction and optical transparency. TCFs are important components in devices such as organic light-emitting diodes (OLEDs), liquid crystal displays (LCDs), photovoltaic devices, coatings, and touch screens. Furthermore, transparent conductors are of different types, as described below:

(i) Metal Oxides

These materials possess a wide band gap in the range of 3–4 eV. Binary metal oxides such as ZnO, SnO_2, In_2O_3, and TiO_2 are popular metal oxides in the TCOs category. The requirement of band alignment and compatibility in high-end devices is the primary need to improve the efficiency [75–83]. An improvement in the performance of optoelectronic devices has triggered interest in new ternary oxides such as Al:ZnO, F:SnO_2, Sn:In_2O_3,

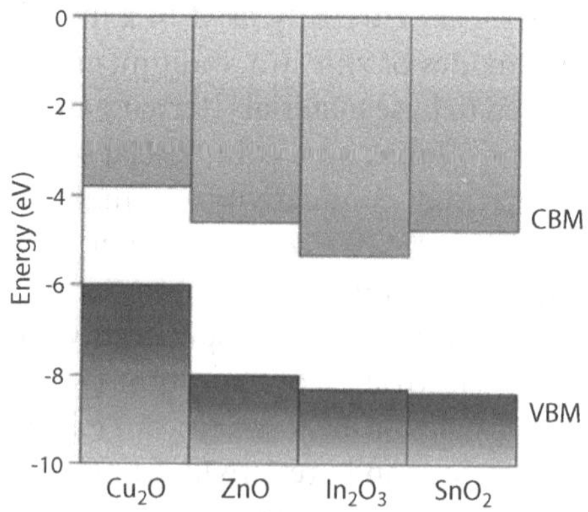

Figure 9.3 Estimated band gaps of popular metal based TCO host structures.

and Sn:CdO, which have emerged as promising TCO with characteristics that meet ondustrial requirements. Major areas of application include their use as transparent electrodes in photovoltaics, smart windows, coatings, OLEDs, and gas sensors. Figure 9.3 illustrates the different metal oxides and their bandgaps.

(ii) Conductive Polymers

Conductive polymers, especially organic polymers (Figure 9.4), can conduct electricity with conductivities that lie between those of metals and semiconductors [21]. Conducting polymers do not exhibit moldable properties, and their electrical properties can be optimized using suitable synthesis techniques. The large-scale applications of these materials include their use in organic solar cells, organic light-emitting diodes, electrochromism, and chemical sensors [22].

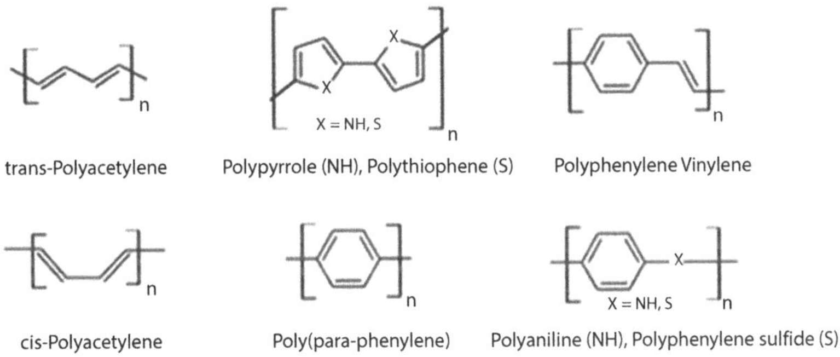

Figure 9.4 Structure and examples of conductive polymers.

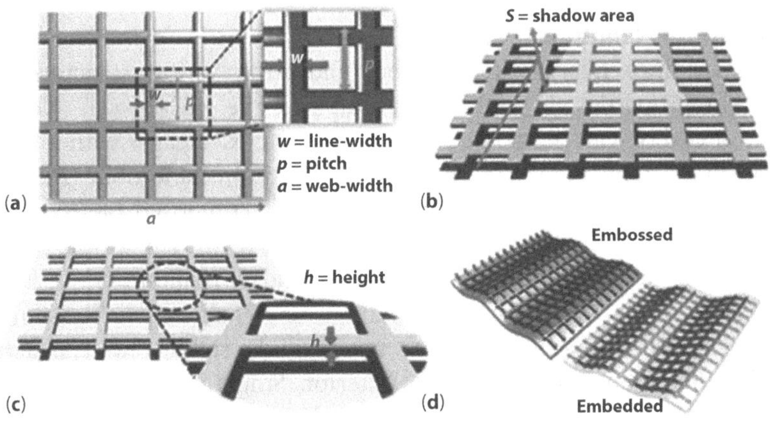

Figure 9.5 Structure of metal grid and its types.

This renders the synthesis and processing of these resins difficult and expensive. In fact, their melt viscosities are often so high that injection molding and similar processing methods are neither feasible nor practical [84–91]. A wide range of electrically conducting and electroluminescent polymers can be employed in applications, such as chemical sensors, electromagnetic shielding, antistatic coatings, corrosion inhibitors, smart or electrochromic windows, and compact electronic devices, such as polymer-based transistors, LEDs, and lasers. Selected compositions with unmatched properties may also find applications in the electronics industry.

(iii) Metal Grids and Metallic Networks

Transparent conducting oxides are highly transparent owing to their wide bandgap and low electrical resistance. However, most TCO materials are brittle, limiting their use in flexible electronics. Metals, on the other hand, possess excellent conductivity and can be easily welded together at the nanoscale to overcome resistance at the junction; thus, metal nanowires can be easily used in applications where flexibility is an essential requirement [23]. The typical structure of a metal nanogrid is that with periodic metal lines with the flexibility to manipulate the charge carriers to achieve properties that cannot be met with ordinary flat TCO electrodes. Metal lines are larger than metal films; they offer conductivity close to that of the bulk material [92–97]. Scattering of light and coupling nanogrids is more suitable for their application in devices such as solar cells. Two critical parameters in designing a metal grid for maximum transparency and uniform current across the active layer are the linewidth of the metal mesh and the period of the mesh. Figure 9.5 shows the structure of the metal grid and its two types: embossed and embedded.

(iv) Carbon Nanotubes (CNTs)

Carbon with sp^2 hybridization can be molded into various structures with different properties. In addition to graphite, the well-known allotropic form of carbon can be molded into open and closed cages with a honeycomb-like arrangement [24]. Multi-walled carbon nanotubes (MWNTs) are elongated cylindrical-shaped nano objects composed of sp^2 carbon. Their diameter falls in the range of 3–30 nm with their aspect (length-to-diameter ratio) ratio varying between 10 and 10 million [98–103]. They vary from single-walled carbon nanotubes on behalf of their Russian doll structure. Other distinguishing factors include rigidity, wall structure, small outer diameter, and hollow interior. Single-walled carbon nanotubes (SWNTs) are one-dimensional carbon-based nanomaterials with walls up to one atom thick. Currently, MWNTs and SWNTs are being

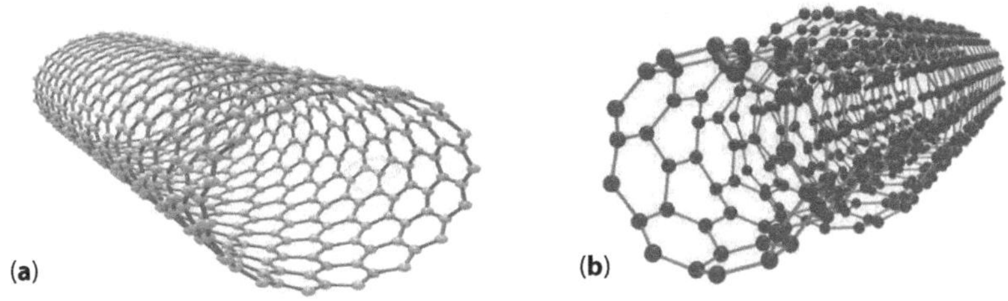

Figure 9.6 Structure of (a) single walled and (b) multi-walled CNTs.

prepared by catalytic growth, arc discharge, and laser ablation techniques. Characterization of the laboratory-prepared nanotubes was carried out using Raman, electronic, and optical spectroscopies. Important information is obtained through mechanical, electrical, and thermal measurements. The high Young's modulus and tensile strength of the nanotubes have led to their application in composite materials with improved mechanical properties. Nanotubes are suitable as electron field emitters because of their ultra-small size, perfect structure, low resistivity, and stability in harsh chemical environments, with applications in display devices [104–109]. Multiwall nanotubes have been employed in fuel cells because they can be used as catalysts during the oxygen reduction reaction. Further applications of nanotubes include their use as primary components of electronic devices, such as single-electron transistors, field-effect transistors (FETs), logic circuits, and rectifying diodes. Figure 9.6 shows the components of the single-walled CNTs and multi-walled CNTs.

(v) Graphene

It is a magical material in which a combination of carbon atoms is organized in a sheet that is one atom-thick. The material was very thin and could be used to conduct electricity. In addition to having good conductivity, the material also has mechanical flexibility, optical transparency, and stretchability [25]. With immense potential properties, the material is a boon to the world of applications in ultra-fast electronics, hyper-sensitive sensors, and extra durable materials. Monolayer extraction of graphene from bulk materials has revolutionized the industry of two-dimensional (2D) materials, such as hexagonal boron nitride (h-BN) and transition metal dichalcogenides (e.g., MoS_2). The entire spectrum of 2D materials includes a range of materials, from good conductors, such as graphene, to semiconducting materials, such as MoS_2 to completely insulating h-BN [110–115]. The structures of these 2D materials can be thoroughly studied to exploit their useful properties. The structure of 2D crystals has a

(a)

Figure 9.7 (a) Structure of graphene.

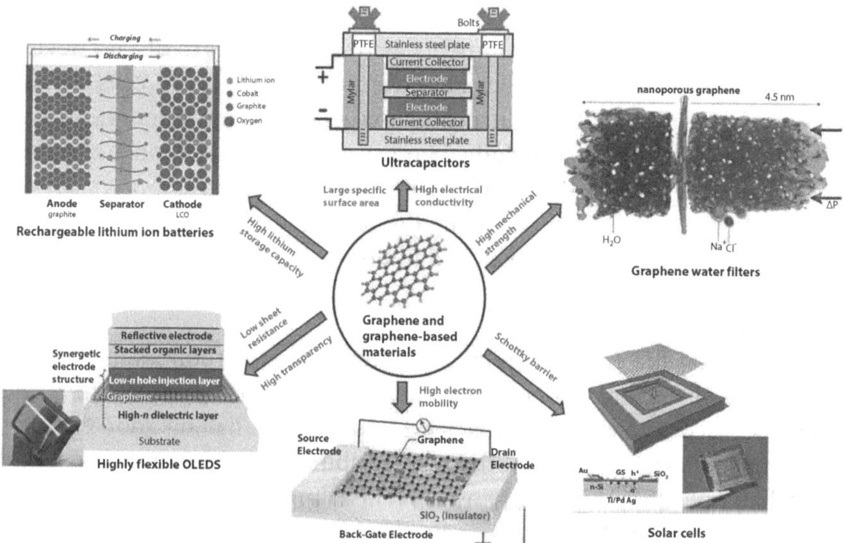

(b)

Figure 9.7 (b) Graphene based materials and their applications.

distinctive combination of properties with increased strength and stiffness but decreased rigidity. Graphene, which is just one atom thick, can be stored in other graphitic structures, such as fullerene, CNTs, and thin graphene. It is often called a zero-gap semiconductor with a very high charge density and ballistic transport. Furthermore, the passage of zero-mass electrons through the honeycomb structure without scattering leads to the investigation of nano effects in graphene. Figure 9.7(a) shows the structure of graphene, which has a hexagonal geometry with six carbon atoms. Figure 9.7(b) shows the importance of graphene from an application perspective [116–122].

9.4 Deposition Techniques

Functional thin films can be deposited using several methods. The deposition of TCO materials in the form of nanostructures such as thin films and nanoparticles is crucial from an application point of view. In general, physical and chemical deposition techniques are employed now a day [26, 27]. Physical deposition techniques require a special vacuum arrangement and hence are much more costly than chemical deposition techniques. The two techniques differ in the approach used to deposit films. While chemical methods use a fluid to react chemically with the substrate, physical methods use of vacuum-based deposition techniques or mechanical methods to deposit the films. Examples of chemical deposition techniques include wet chemical synthesis such as sol–gel dip coating, spray pyrolysis, or chemical bath deposition. Deposition techniques such as molecular beam epitaxy, sputtering, pulsed laser, cathodic arc, and electro-hydrodynamic deposition are in the category of physical deposition techniques. A brief overview of all deposition methods, along with their merits and de-merits is presented below.

(i) Physical Deposition Techniques

Physical deposition techniques are also called vacuum-based syntheses where deposition occurs under high vacuum. These techniques are much cleaner than solution-based techniques, and hence, there is a negligible chance of any foreign or undesired material being deposited in the process. In addition, these techniques are costly, as are a number of accessories like vacuum pumps, gauges, and gas purging facilities. Examples of physical deposition techniques are discussed in detail below [123–127].

(a) **Evaporation techniques:** The mechanism of this technique involves changing the solid phase of the material to the vapor phase and converting the vapors again to a solid phase on the desired substrate. There are of two types of evaporation techniques. In the case of thermal evaporation, the material to be deposited is thermally vaporized with the help of a high-current filament in the high-vacuum coating chamber. Vacuum generally favors the transportation of vapor particles to reach the substrate directly. The energetic impact of the electron beam is responsible for increasing the temperature of the target material to create a hot zone on the material. All evaporators were attached to the flux monitoring plates to measure the ion current in the beam.

(b) **Sputtering:** In this process, energetic ions bombard target material atoms in the solid target material and are driven by the exchange of momentum between atoms and ions due to collision. The plasma of gases such as argon (non-reactive) is generated by a potential difference inside the vacuum chamber, which is then collected on the surface to be coated. The technique is basically used to deposit device-quality thin films [128–135].

(c) **Pulsed Laser Deposition (PLD):** PLD works on the principle of ablation and employs a laser beam of high-power pulses to strike the target material, leading to deposition of the material on the substrate. The deposition rate and film composition were controlled by the presence of neutral gases such as Ar. PLD has been utilized to deposit high-quality crystalline films and materials with complex stoichiometry.

(d) **Molecular beam epitaxy (MBE):** MBE is an ultrahigh-vacuum-based technique for producing high mono-structural layers of the material. This technique was applied to the ordered crystalline growth of materials on a crystalline substrate. The process is a type of interface between a thin film and substrate. This technique features the epitaxial growth of materials via the chemical interaction of one of several molecules or atomic beams of variable intensities and compositions that occur on the surface of the substrate. The biggest advantage of this technique is that the deposited film has a lattice structure and orientation identical to that of the substrate on which the film is grown. If the film is deposited on a substrate of the same composition, the process is called homoepitaxy; otherwise, it is called heteroepitaxy.

(e) **Chemical vapor deposition (CVD):** As the name suggests, the CVD process uses a precursor in gaseous form that is made to reach the surface of the substrate placed in a chamber maintained at a high temperature. Upon reaching the substrate, the gaseous precursor reacts with the substrate surface and the process continues until the required thickness is obtained. However, in plasma-enhanced CVD (PECVD), the plasma of ions, radicals, neutral atoms, and molecules is formed in the reaction chamber that interacts with the substrate at a comparatively low temperature (300–350 °C) as compared to CVD (600–900 °C).

(f) **Atomic layer deposition (ALD):** This technique is applicable to thin-film deposition in a wide range of applications. The ALD technique uses more than one gaseous precursor that reacts with the substrate consecutively. The precursors are separated by inert gas, and the process comprises successive, self-terminated surface reactions leading to the controlled growth of the desired material. As the parameters are very much under control, the self-limiting growth mechanism leads to the formation of films with uniform thickness, even on complicated structures. This technique is highly recommended in the field of microelectronics owing to its superior atomic-level control and uniformity. In addition, the covalent bonding of ALD coatings offers better adhesion. ALD is very similar to the CVD process, except that the ALD process comprises a two-step process that keeps the precursor materials separate during the reaction.

(ii) Chemical Deposition Techniques

These techniques are based on chemical reactions in the liquid phase; therefore, they are the simplest and cheapest, yet efficient, in synthesizing a range of materials. However, the shortcomings of solution-based syntheses process however include the production of undesired particles within the synthesized material, in addition to being not as clean and incompatible with modern solid-state device technology in comparison to vacuum-based deposition techniques.

(a) **Sol–gel method:** The sol–gel process is based on the polycondensation of metal oxides. In the process, “sol” is obtained in the form of colloidal suspension and “gel” is obtained by aging the “sol” to get its thick version. Sol–gel techniques work on low-temperature processing and it has two routes to prepare metal oxides. The first route involves the preparation of inorganic precursors via inorganic salts in aqueous solutions, and the second involves preparation of metal alkoxides via metal alkoxides in non-aqueous solvents.

(b) **Chemical bath deposition:** Chemical bath deposition is another liquid-based technique for depositing thin films or nanomaterials. The process is applied specifically for large-area batch processing where continuous deposition is required. The technique begins with the initial step of nucleation, followed by the particle growth mechanism. In this procedure, the material to be coated was immersed in a solution containing the precursor. The primary factors affecting the properties of the resultant coating are bath temperature, pH, concentration, and immersion time. The biggest advantage of this technique lies in its simplicity of setup, requiring only the solution containers and the substrate holder, in addition to providing a stable and uniform film. The drawbacks include wastage of the solution after each deposition.

(c) **Electrodeposition:** Electrodeposition is another liquid-based deposition process that operates through a polar liquid by applying an electric potential between the two electrodes. The passage of current leads to the fragmentation of the liquid into various charged species that are deposited on the electrodes, depending upon on potential between the electrodes. This technique is commonly used to fabricate electroplated materials, i.e., by depositing a relatively thin layer of metal with the desired property, which is lacking in the material without coating. Another important application of this technique is to provide the desired thickness on the undersized parts. The technique offers many advantages, such as control of film thickness, uniformity, and control of the deposition rate, along with being inexpensive. Even substrates with complicated shapes are best coated with the help of this technique.

9.5 Sol–Gel Deposition

Sol–gel technique comes under the chemical deposition technique for the depositing the nanomaterials. The technique involves the formation of a colloidal suspension known as the sol and allows the sol to thicken to evolve a network, leading to the formation of a gel. The precursors used to form a colloidal sol are generally metallic ions and ligands, including metal alkoxides and alkoxysilanes, because of their affinity towards water.

In situations where alkoxides are not completely soluble in water, alcohols are utilized as solvents to ease the miscibility of such materials. This process can be briefly described in terms of four basic steps:

- Hydrolysis or alcoholysis
- Condensation of water and alcohol followed by the particle formation as a result of
- Polymerization of monomers
- Growth of the particles
- Agglomeration of particles results in the formation of a thick gel.

The structural morphology of the sol–gel-derived network as well as the optoelectronic properties are largely dependent on the initial conditions such as the reaction temperature, reactant molarity, reaction time, presence of the catalyst or its concentration, pH value of the sol, and aging time for the formation of the gel. Control of the initial synthesis conditions and the formation of end products are also possible in this technique. Depending on the drying process, the synthesized gel can be converted into a range of products with different structures and properties. Rapid drying of the gel results in the formation of an aerogel, while slow drying leads to a xerogel, and further calcination forms a dense ceramic material. Various surfactants can be utilized to form gelled spheres, as these are responsible for modifying the surface tension of dispersed particles, preventing them from agglomeration. Again, the gelled spheres formed nanopowders when treated at high temperatures. The choice of surfactant and its precise use determines the level of agglomeration of the dispersed particles, and its concentration controls the size of the nanoparticles to produce powdered material. Sol–gel can be converted into nanowires or nanorods with the help of porous matrix and templates. Sol–gel process offers versatilities of the end products that can be obtained at the termination of the process. Another important aspect of sol–gel-based techniques is that they can be used to coat materials via dip coating, spin coating, or spray pyrolysis. Figure 9.8 shows step by step synthesis process of the sol–gel method to give different varieties of end products.

Process Description: The chemical reaction that takes place in the sol–gel process involves major initial conditions, such as pH, cation concentration, temperature, and time, which play a significant role in controlling the properties of the deposited material. The chemical reaction can be

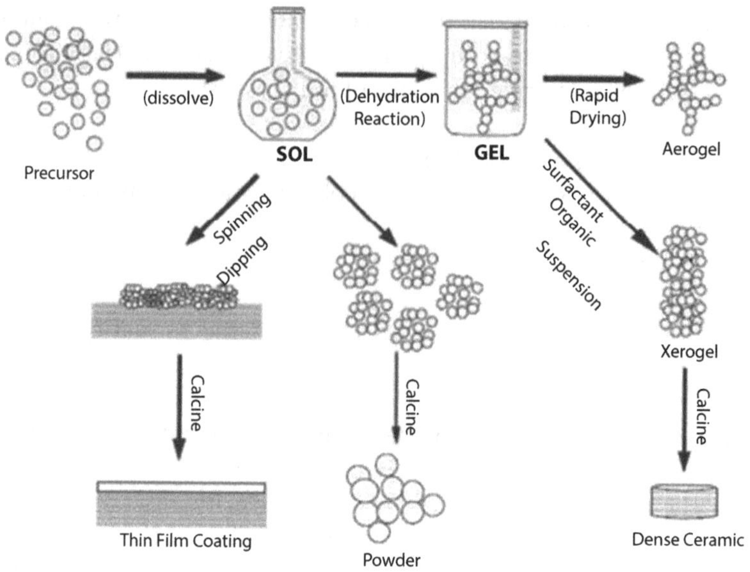

Figure 9.8 Schematic diagram of the sol–gel synthesis process.

described as follows. The reaction illustrates the formation of silica gel via the sol–gel process. TEOS is used as a source of Si and is dissolved in water. During hydrolysis, the alkoxide group (OR) of TEOS is replaced by a hydroxyl group (OH). The hydrolysis process is then followed by a condensation reaction forming the silanol group (Si-OH), which produces siloxane bonds (Si-O-Si), accompanied by water and alcohol formation.

The synthesized sol was then aged for a certain time to obtain a thick gel by the formation of long chains of siloxane bonds, which ended by the hydroxyl group. Figure 9.9 described the process in terms of different chemical reactions prevailing in the sol–gel process.

The sol–gel technique offers multiple advantages, and the major advantages are listed below;

- As the process is based on chemical reactions in the liquid phase, it is a simple technique.
- It is also cost-effective and requires simple accessories for chemical reactions and deposition procedures.
- The doping level was easier to achieve in this process.
- Using this technique, higher-purity materials can be obtained.
- Owing to the chemistry involved in this process, a large range of materials can be deposited using this procedure.

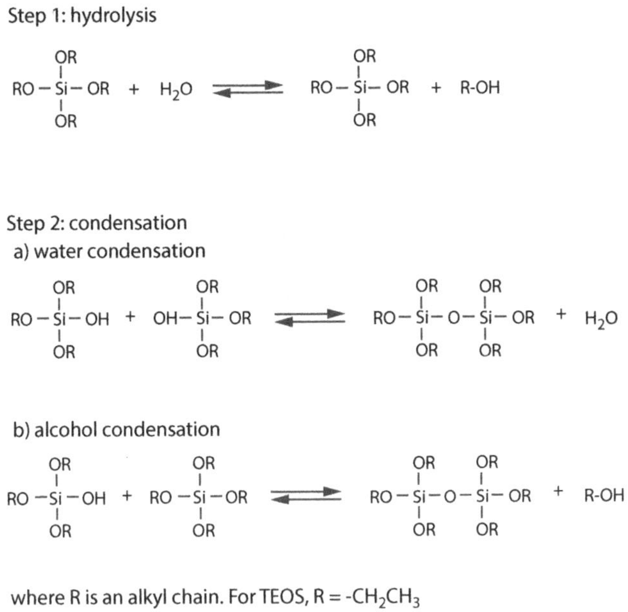

Figure 9.9 Reaction scheme of the sol–gel process.

In addition, this process has some disadvantages, as mentioned below:

- Much efforts and time are required in optimization of the sol and its durability.
- The alkoxide precursors used in this process are generally expensive.

9.6 Applications of TCOs

Owing to their interesting optical and electrical properties, transparent conducting oxides have a wide range of applications in photovoltaics (PV), display devices, organic light-emitting diodes (LEDs), and gas sensors.

(a) Photovoltaics: Photovoltaic (PV) devices are the ones that can directly convert sunlight into electricity through an electronic process that is inherent in certain type of materials known as semiconductors. Solar energy is responsible for setting the free electrons which are then directed to travel across an electric circuit, powering electrical devices. Solar cells are commonly used to supply small amounts of energy to power everyday

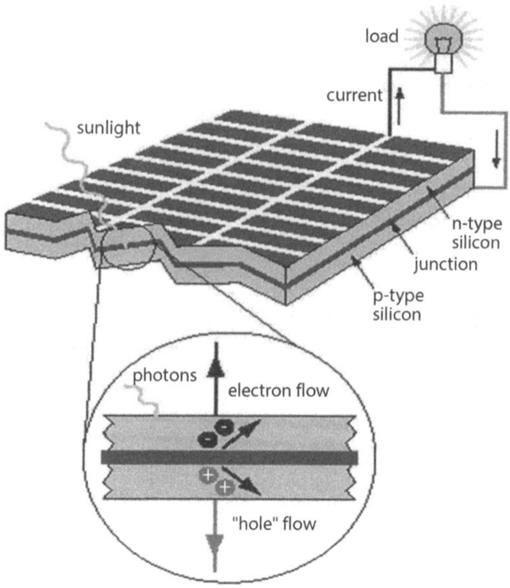

Figure 9.10 Working mechanism of a solar cell.

appliances and large amounts of energy to critical power grids. Solar cells are commonly referred to as photovoltaic (PV) cells by the scientific community. Solar cells are commonly referred to as photovoltaic (PV) cells in the scientific community. A variety of materials are being utilized to develop this unique energy-producing technology. The physics of solar cells rely heavily on their reliability and efficiency. PV cells are rapidly becoming a significant technology in the renewable energy industry as the costs gradually decline. PV cells are made of a variety materials, including silicon, cadmium telluride, gallium arsenide, copper indium diselenide, light-absorbing dyes, and organic and polymer substances [28–31]. Figure 9.10 depicts the working mechanism of the solar cell.

(b) Display devices: The display devices are output devices that display information in visual form using various types of display systems. In digital instruments, the output device indicates the value of the measured quantity using a digital display device. LEDs are most commonly used in digital displays to display alphanumeric characters. LEDs have advantages such as low voltage, long life, high reliability, low cost, and fast switching characteristics. Similarly, other devices requiring TCOs include liquid

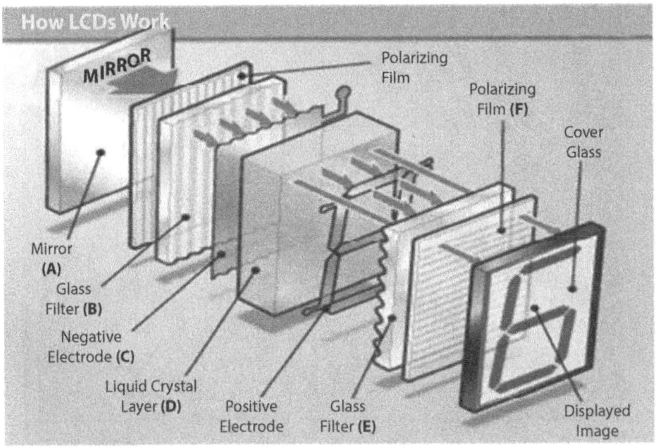

Figure 9.11 Components of a display device.

crystal displays (LCDs), cold cathode displays (Nixie tubes), incandescent displays, fluorescent displays, and liquid-vapor-based displays [32–36].

As shown in Figure 9.11, LCD technology is used to display an image in a notebook or other electronic devices such as minicomputers.

(c) OLEDs: OLEDs are devices constructed from thin films of materials that generate light when electricity is supplied. The advantages of using an OLED instead of conventional LEDs and LCDs include better image quality and less power consumption. These devices operate based on electrophosphorescence. To make the display colorful, multiple types of organic films are placed on the same OLED, and the brightness of the display can be increased by increasing the applied current. Figure 9.12 shows a conventional OLED and the arrangement of the different component layers.

(d) Coatings: A coating is a very thin layer of material deposited onto the surface of another material called a substrate. The basic function of any coating is to change the properties of the host material, i.e., the substrate. Coatings can be decorative, functional, or may serve both purposes and a number of applications in optics, electronics, and medical fields [37–41]. Antireflective (AR) and high-reflective (HR) coatings significantly change the behavior of the optics towards light. AR coatings minimize reflection and increase the transmission of light, whereas HR works completely opposite to it. TCO coatings make use of inorganic transparent conductive films, which are intensively used in the fields of electronics and optoelectronics.

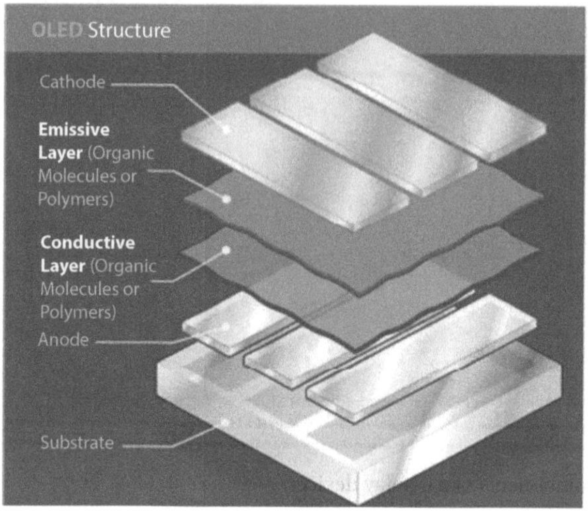

Figure 9.12 Schematic diagram showing the structure of an OLED.

ITO is a TCO coating with a high transmittance and low resistivity. DLC coatings have the benefits of hardness, lubrication, and resistance to chemicals and scratches. Figure 9.13 shows a typical TCO coating on a glass substrate.

(e) Gas Sensors: Nanomaterials have actively participated in gas-sensing devices in the field of environmental gas sensors. As technology and civilization progress, environmental pollution has become a matter of concern. The detection of harmful/polluting gases in the atmosphere and the

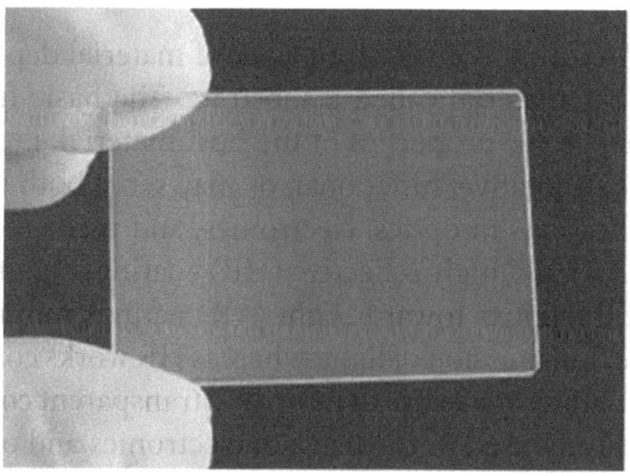

Figure 9.13 An image of TCO coatings.

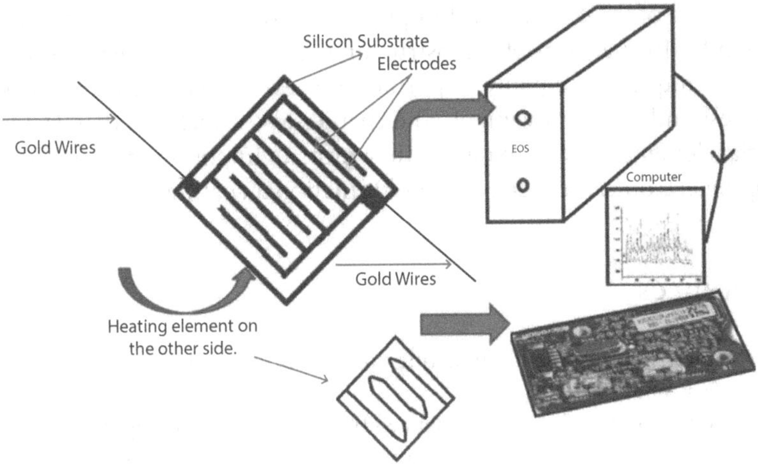

Figure 9.14 Metal oxide-based gas sensors.

resulting health hazards has aided the development of efficient gas sensors. Materials such as nanostructured metal oxide semiconductors, polymers, metals, CNTs, and graphene have been investigated for their efficiency as gas sensors for detecting traces of poisonous gases in the environment. Gases, such as hydrocarbons, nitrogen oxides, carbon monoxide, carbon dioxide, sulfur dioxide, ammonia, and hydrogen sulfide, cause serious health hazards. Gas sensors are critically important because their presence at low concentrations cannot be smelled by the human nose. The working principle of any TCO-based gas sensor is the change in conductivity of the thin film upon exposure to the target gas [42–45]. A number of metal oxides like In_2O_3, WO_3, MoO_3, SnO2, ZnO2, and TiO2 in their metastable polymorphic forms have been found to be efficient gas sensors. Figure 9.14 shows the components of any metal oxide-based gas sensor.

9.7 Conclusion

The chapter provides an elaborated introduction to an important class of materials: transparent conducting oxides (TCOs). These materials have a unique combination of two contradictory properties: conductivity and transparency. Different types of TCOs are also discussed in this chapter, and an insight into their structures is provided. TCOs are an important and essential part of transparent electronics and other state-of-the-art optoelectronic devices. The chapter also summarizes and provides a brief elaboration of the different techniques for depositing transparent conducting oxides in the form of nanostructures. Physical and chemical deposition techniques are equally important; however, chemical techniques are cheapest and most convenient. The chapter also concludes that TCOs are crucial elements in technology, as they form an important part in the fields of optoelectronics, sensors, and multifunctional coatings.

References

1. Stadler, A., Transparent conducting oxides—An up -to-date overview. *Materials*, 5, 661–683, 2012.
2. Noor, N. and Parkin, I.P., Enhanced transparent-conducting fluorine-doped tin oxide films formed by aerosol-assisted chemical vapour deposition. *J. Mater. Chem. C*, 1, 984–996, 2013.
3. Tiwana, P., Docampo, P., Johnston, M.B., Snaith, H.J., Herz, L.M., Electron mobility and injection dynamics in mesoporous ZnO, SnO2, and TiO2 films used in dye sensitized solar cells. *ACS Nano*, 5, 5158–5166, 2011.
4. Yu, X., Yu, X., Zhang, J., Zhang, D., Ni, J., Cai, H., Zhang, D., Zhao, Y., Efficient inverted polymer solar cells based on surface modified FTO transparent electrodes. *Sol. Energy Mater. Sol. Cells*, 136, 142–147, 2015.
5. Han, K., Kong, Q., Kang, Y., Sun, C., Wang, C., Zhang, J., Gong, X., Indium-gallium-zinc-oxide (IGZO) nanowire transistors. *IEEE Trans. Electron Devices*, 68, 12, 6610–6616, 2021.
6. Zhang, Z., Xu, M., Liu, L., Ruan, X., Yan, J., Zhao, W., Yun, J., Wang, Y., Qin, S., Zhang, T., Novel SnO2@ZnO hierarchical nanostructures for highly sensitive and selective NO2 gas sensinng. *Sens. Actuators B-Chem.*, 257, 714–727, 2018.
7. Beri, A., Kant, R., Banipal, T.S., Interactional behaviour of analgesic drugs in aqueous solution of caffeine at different temperatures using multi-technique approach. *Can. J. Chem.*, 102, 2, 61–67, 2023.
8. King, P.D.C. and Veal, T.D., Conductivity in transparent oxide semiconductors. *J. Phys.: Condens. Matter*, 23, 334214, 2011.
9. Dahiya, A.S., Opoku, C., Oshman, C., Poulin-Vittrant, G., Cayrel, F., Tran Huu Hue, L.P., Camara, N., Zinc oxide sheet field-effect transistors. *Appl. Phys. Lett.*, 107, 3, 033105, 2015.
10. Jin, S., Yang, Y., Medvedeva, J.E., Wang, L., Li, S., Cortes, N., Ireland, J.R., Metz, A.W., Ni, J., Hersam, M.C., Freeman, A.J., Marks, T.J., Tuning the properties of transparent oxide conductors. dopant ion size and electronic structure effects on CdO-based transparent conducting oxides. Ga- and In-doped CdO thin films grown by MOCVD. *Chem. Mater.*, 20, 220–230, 2008.
11. Ridley, B.K., Large-band gap semiconductors. *Turk. J. Phys.*, 23, 577–582, 1999.

12. Dagenais, K., Chamberlin, M., Constantin, C., Modelling energy band gap as a function of optical electronegativity for binary oxides. *J. Young Investig.*, 25, 73–78, 2013.
13. Mishra, S. and Ganguli, B., Effect of p–d hybridization, structural distortion and cation electronegativity on electronic properties of ZnSnX2 (X=P, As, Sb) chalcopyrite semiconductors. *J. Solid State Chem.*, 200, 279–286, 2013.
14. McCluskey, M.D. and Haller, E.E., *Dopants and defects in semiconductors*, 1st Edn., CRC Press, London, 2012.
15. Aggarwal, N., Kant, R., Kumar, G., James, C., Maji., S., Synthesis, characterization and biological evaluation studies of Cu (II) and Zn (II) complexes with gly-o-andn or gly-p-andn as primary ligand and N, N'donors as secondary ligand. *J. Phys. Conf. Ser.*, 1531, 1, 012111, 2020, IOP Publishing.
16. Castañeda, L., Present status of the development and application of transparent conductors oxide thin solid films. *Mater. Sci. Appl.*, 2, 1233–1242, 2011.
17. Savarimuthu, E., Lalithambika, K.C., Raj, A.M.E., Nehru, L.C., Ramamurthy, S., Thayumanavan, A., Sanjeeviraja, C., Jayachandran, M., Synthesis and materials properties of transparent conducting In2O3 films prepared by sol-gel spin coating technique. *J. Phys. Chem. Solids*, 68, 1380–1389, 2007.
18. Sato, H., Minami, T., Takata, S., Yamada, T., Transparent conducting ptype NiO thin films prepared by magnetron sputtering. *Thin Solid Films*, 236, 27–31, 1993.
19. Sheng, S., Fang, G., Li, C., Xu, S., Zhao, X., P-type transparent conducting oxides. *Phys. Status Solidi A*, 203, 1891–1900, 2006.
20. Kawazoe, H., Yasukawa, M., Hyodo, H., Kurita, M., Yanagi, H., Hosono, H., P-type electrical conduction in transparent thin films of CuAlO2. *Nature*, 389, 939–942, 1997.
21. Otero, T.F., *Conducting polymers, bio-inspired intelligent materials and devices*, Royal Society of Chemistry, Cambridge, 2016.
22. Lange, U., Roznyatovskaya, N.V., Mirsky, V.M., Conducting polymers in chemical sensors and arrays. *Anal. Chim. Acta*, 614, 1–26, 2008.
23. Guo, C.F. and Ren, Z., Flexible transparent conductors based on metal nanowire networks. *Mater. Today*, 18, 143–154, 2015.
24. Harik, V., Meachanics of carbon nanotubes: shell models, chiral formulations and nanoscale buckling, in: *Nanotechnology of carbon nanotubes*, 1st Edn., Academic Press, London, 2018.
25. Geim, A.K. and Novoselov, K.S., The rise of graphene. *Nat. Mater.*, 6, 183–191, 2007.
26. Arthur, J.R., Physical and chemical methods for thin-film deposition and epitaxial growth, in: *Specimen Handling, Preparation, and Treatments in Surface Characterization*. Methods of Surface Characterization, A.W. Czanderna, C.J. Powell, T.E. Madey, D.M. Hercules, J.T. Yates (Eds.), pp. 239–293, Springer, Boston, 2002.

27. Abegunde, O.O., Akinlabi, E.T., Oladijo, O.P., Akinlabi, S., Ude, A.U., Overview of thin film deposition techniques. *Aims Mater. Sci.*, 6, 174–199, 2019.
28. Goe, M. and Gaustad, G., Strengthening the case for recycling photovoltaics: An energy payback analysis. *Appl. Energy*, 120, 41–48, 2014.
29. Barbato, M., Artegiani, E., Bertoncello, M., Meneghini, M., Trivellin, N., Mantoan, E., Romeo, A., Mura, G., Ortolani, L., Zanoni, E., Meneghesso, G., CdTe solar cells: technology, operation and reliability. *J. Phys. D: Appl. Phys.*, 54, 333002, 2021.
30. Papez, N., Dallaev, R., Talu, S., Kastyl, J., Overview of the Current State of Gallium Arsenide-Based Solar Cells. *Materials*, 14, 3075, 2021.
31. Gunes, S., Neugebauer, H., Sariciftci, N.S., Conjugated Polymer-Based Organic Solar cells. *Chem. Rev.*, 107, 1324–1338, 2007.
32. Varanytsia, A., Weng, L., Lin, T.C., Yang, J., Chien, L.C., High-Performance and Low-Cost Aluminum Zinc Oxide and Gallium Zinc Oxide Electrodes for Liquid Crystal Displays. *J. Disp. Technol.*, 12, 1033–1039, 2016.
33. Chae, G.S., A Modified Transparent Conducting Oxide for Flat Panel Displays Only. *Jpn. J. Appl. Phys.*, 40, 1282, 2001.
34. Liu, H., Avrutin, V., Izyumskaya, N., Özgür, U., Morkoç, H., Transparent conducting oxides for electrode applications in light emitting and absorbing devices. *Superlattices Microstruct.*, 48, 458–484, 2010.
35. Santra, S., Das, N.S., Besra, N., Banerjee, D., Chattopadhyay, K.K., Graphene-Anchored p-Type CuBO2 Nanocrystals for a Transparent Cold Cathode. *Langmuir*, 33, 9961–9971, 2017.
36. Guillén, C. and Herrero, J., TCO/metal/TCO structures for energy and flexible electronics. *Thin Solid Films*, 520, 1–17, 2011.
37. Ramzan, M., Imran, M., Ullah, S., Khan, M.A., Naz, G., Ghouri, M.I., Iqbal, H.MN., Fabrication and characterization of multifunctional thin multi-layer films for transparent conducting oxides. *Prog. Org. Coat.*, 149, 105976, 2020.
38. Jana, S. and Majumdar, A., Glass and glass-ceramic coatings, versatile materials for industrial and engineering applications. *Bull. Mater. Sci.*, 24, 69–77, 2001.
39. Ganesh, R., Unlocking Multifunctional Advantages with Nanocomposites Coatings for Solar Cells: A Comprehensive Review. *Curr. Org. Chem.*, 27, 1946–1959, 2023.
40. Riapanitra, A., Asakura, Y., Yin, S., Improved thermochromic and photocatalytic activities of F–VO2/Nb–TiO2 multifunctional coating films. *Tungsten*, 1, 306–317, 2019.
41. Sarkın, A.S., Ekren, N., Saglam, S., A review of anti-reflection and self-cleaning coatings on photovoltaic panels. *Sol. Energy*, 199, 63–73, 2020.
42. Sumit, V., A Short Review on Properties and Applications of Zinc Oxide Based Thin Films and Devices: ZnO as a promising material for applications in electronics, optoelectronics, biomedical and sensors. *Johnson Matthey Technol. Rev.*, 64, 202–218, 2020.

43. Carottaa, M.C., Cervi, A., Natalea, V., Gherardi, S., Giberti, A., Guidi, V., Puzzovioa, D., Vendemiati, B., Martinelli, G., Sacerdoti, M., Calestani, D., Zappettini, A., Zhac, M., Zanotti, L., ZnO gas sensors: A comparison between nanoparticles and nanotetrapods-based thick films. *Sens. Actuators B Chem.*, 137, 164–169, 2009.
44. Zakrzewska, K., Mixed oxides as gas sensors. *Thin Solid Films*, 391, 229–238, 2001.
45. Rambu, A.P., Iftimie, N., Nica, V., Effect of In incorporation on the structural, electrical, and gas sensing properties of ZnO films. *J. Mater. Sci.*, 47, 6979–6985, 2012.
46. Bhardwaj, A., Kumar, P., Raj, B., Anand, S., Design and Performance Optimization of doping-less Vertical Nanowire TFET using Gate Stack Technique. *J. Electron. Mater. (JEMS)* Springer, 41, 7, 4005–4013, 2022.
47. Singh, J. and Raj, B., Tunnel Current Model of Asymmetric MIM Structure Levying Various Image Forces to Analyze the Characteristics of Filamentary Memristor. *Appl. Phys. A*, Springer, 125, 3, 203.1–203.11, Feb 2019.
48. Goyal, C., Ubhi, J.S., Raj, B., Low Leakage Zero Ground Noise Nanoscale Full Adder using Source Biasing Technique. *J. Nanoelectron. Optoelectron.*, American Scientific Publishers, 14, 360–370, March 2019.
49. Singh, G., Sarin, R.K., Raj, B., A Novel Robust Exclusive-OR Function Implementation in QCA Nanotechnology with Energy Dissipation Analysis. *J. Comput. Electron.*, Springer, 15, 2, 455–465, June 2016.
50. Wadhwa, G., Kamboj, P., Raj, B., Design optimisation of junctionless TFET biosensor for high sensitivity. *Adv. Nat. Sci.: Nanosci. Nanotechnol.*, 10, 045001 (7), 2019.
51. Bansal, P. and Raj, B., MEMRISTOR MODELING AND ANALYSIS FOR LINEAR DOPANT DRIFT KINETICS. *J. Nanoeng. Nanomanuf.*, American Scientific Publishers, 6, 1–7, 2016.
52. Ashima, Vaithiyanathan, D., Raj, B., Performance Analysis of Charge Plasma induced Graded Channel Si nanotube. *J. Eng. Res. (JER)*, EMSME Special Issue, 9, 146–154, Aug 2021.
53. Tomar, A.S., Magraiya, V.K., Raj, B., Scaling of Access and Data Transistor for High Performance DRAM Cell Design. *Quantum Matter*, 2, 412–416, Oct. 2013.
54. Jain, N. and Raj, B., Parasitic Capacitance and Resistance Model Development and Optimization of Raised Source/Drain SOI FinFET Structure for Analog Circuit Applications. *J. Nanoelectron. Optoelectron.*, ASP, USA, 13, 531–539, Apl 2018.
55. Singh, S., Vishvakarma, S.K., Raj, B., Analytical Modeling of Split-Gate Junction-Less Transistor for a Biosensor Application. *Sens. Bio-sensing*, Elsevier, 18, 31–36, April 2018.
56. Gopal, M. and Raj, B., Low Power 8T SRAM Cell Design for High Stability Video Applications. *ITSI Trans. Electr. Electron. Eng.*, 1, 5, 91–97, 2013.

57. Raj, B., Mitra, J., Bihani, D.K., Rangharajan, V., Saxena, A.K., Dasgupta, S., Analysis of Noise Margin, Power and Process Variation for 32 nm FinFET Based 6T SRAM Cell. *J. Comput. (JCP)*, Academy Publisher, FINLAND, 5, 6, 1–8, 2010.
58. Sharma, D., Mehra, R., Raj, B., Comparative Analysis of Photovoltaic Technologies for High Efficiency Solar Cell Design. *Superlattices Microstruct.* Elsevier, 153, 106861, May 2021.
59. Kaur, P., Buttar, A.S., Raj, B., A comprehensive Analysis of Nanoscale Transistor based Biosensor: A Review. *Indian J. Pure Appl. Phys.*, 59, 304–318, April 2021.
60. Yadav, D., Raj, B., Raj, B., Design and Simulation of Low Power Microcontroller for IoT Applications. *Sens. Lett.*, ASP, 18, 401–409, May 2020.
61. Singh, S. and Raj, B., A 2-D Analytical Surface Potential and Drain current Modeling of Double-Gate Vertical t-shaped Tunnel FET. *J. Comput. Electron.*, Springer, 19, 1154–1163, Apl 2020.
62. Singh, J. and Raj, B., An Accurate and Generic Window function for Non-linear Memristor Model. *J. Comput. Electron.*, Springer, 18, 2, 640–647, June 2019.
63. Kaur, M., Gupta, N., Kumar, S., Raj, B., Singh, A.K., Comparative RF and Crosstalk Analysis of Carbon Based Nano Interconnects. *IET Circuits Devices Syst.*, 15, 6, 493–503, Feb 2021.
64. Kandasamy, N., Ahmad, F., Ajitha, D., Raj, B., Telagam, N., Quantum Dot Cellular Automata based Scan Flip Flop and Boundary Scan Register. *IETE J. Res.*, 66, 535–548, 2020.
65. Singh, J. and Raj, B., Design and Investigation of 7T2M NVSARM with Enhanced Stability and Temperature Impact on Store/Restore Energy. *IEEE Trans. Very Large Scale Integr. Syst.*, 27, 6, 1322–1328, June 2019.
66. Bhardwaj, A.K., Gupta, S., Raj, B., Singh, A., Impact of Double Gate Geometry on the Performance of Carbon Nanotube Field Effect Transistor Structures for Low Power Digital Design. *Comput. Theor. Nanosci.*, ASP, 16, 1813–1820, 2019.
67. Jain, N. and Raj, B., Thermal Stability Analysis and Performance Exploration of Asymmetrical Dual-k underlap Spacer (ADKUS) SOI FinFET for Security and Privacy Applications. *Indian J. Pure Appl. Phys. (IJPAP)*, 57, 352–360, May 2019.
68. Jain, N. and Raj, B., Dual-k Spacer Region Variation at the Drain Side of Asymmetric SOI FinFET Structure: Performance Analysis towards the Analog/RF Design Applications. *J. Nanoelectron. Optoelectron.*, American Scientific Publishers, 14, 349–359, March 2019.
69. Singh, J., Sharma, S., Raj, B., Khosla, M., Analysis of barrier layer thickness on performance of In1-xGaxAs based Gate Stack Cylindrical Gate Nanowire MOSFET. *JNO*, ASP, 13, 1473–1477, Oct 2018.

70. Jain, N. and Raj, B., Analysis and Performance Exploration of High-k SOI FinFETs Over the Conventional Low-k SOI FinFET toward Analog/RF Design. *J. Semicond. (JoS)*, IOP Science, 39, 12, 124002-1-7, Dec 2018.
71. Goyal, C., Ubhi, J.S., Raj, B., A reliable leakage reduction technique for approximate full adder with reduced ground bounce noise. *Math. Probl. Eng.*, Hindawi, 2018, Article ID 3501041, 16, 15 Oct 2018.
72. Anuradha, Singh, J., Raj, B., Khosla, M., Design and Performance Analysis of Nano-scale Memristor-based Nonvolatile SRAM. *Sens. Lett.*, American Scientific Publishers, 16, 798–805, Oct 2018.
73. Wadhwa, G. and Raj, B., Parametric Variation Analysis of Charge-Plasma-based Dielectric Modulated JLTFET for Biosensor Application. *IEEE Sens. J.*, 18, 15, 6070–6077, AUGUST 1 2018, (SCI).
74. Singh, J. and Raj, B., Comparative Analysis of Memristor Models for Memories Design. *JoS*, IoP, 39, 7, 074006-1-12, July 2018.
75. Yadav, D., Chouhan, S.S., Vishvakarma, S.K., Raj, B., Application Specific Microcontroller Design for IoT based WSN. *Sens. Lett.*, ASP, 16, 374–385, May 2018.
76. Singh, G., Sarin, R.K., Raj, B., Fault-Tolerant Design and Analysis of Quantum-Dot Cellular Automata Based Circuits. *IEEE/IET Circuits Devices Syst.*, 12, 638–64, 2018.
77. Singh, J. and Raj, B., Modeling of Mean Barrier Height Levying Various Image Forces of Metal Insulator Metal Structure to Enhance the Performance of Conductive Filament Based Memristor Model. *IEEE Nanotechnol.*, 17, 2, 268–267, March 2018.
78. Jain, A., Sharma, S., Raj, B., Analysis of Triple Metal Surrounding Gate (TM-SG) III-V Nanowire MOSFET for Photosensing Application. *Opto-electron.*, Elsevier, 26, 2, 141–148, May 2018.
79. Jain, A., Sharma, S., Raj, B., Design and Analysis of High Sensitivity Photosensor Using Cylindrical Surrounding Gate MOSFET For Low Power Sensor Applications. *Eng. Sci. Technol. Int. J.*, Elsevier's, 19, 4, 1864–1870, December 2016.
80. Saiphani Kumar, G., Singh, A., Raj, B., Design and Analysis of Gate All Around CNTFET based SRAM cell Design. *J. Comput. Electron.*, Springer, 17, 1, 138–145, March 2018.
81. Singh, G.P., Sohi, B.S., Raj, B., Material Properties Analysis of Graphene Base Transistor (GBT) for VLSI Analog Circuits. *Indian J. Pure Appl. Phys. (IJPAP)*, 55, 896–902, Dec-2017.
82. Kumar, S. and Raj, B., Estimation of Stability and Performance metric for Inward Access Transistor based 6T SRAM Cell Design using n-type/p-type DMDG-GDOV TFET. *IEEE VLSI Circuits Syst. Lett.*, 3, 2, 25–39, June2017.
83. Sharma, S., Kumar, A., Pattanaik, M., Raj, B., Forward Body Biased Multimode Multi-Threshold CMOS Technique for Ground Bounce Noise Reduction in Static CMOS Adders. *Int. J. Inf. Electr. Eng.*, 3, 3, 567–572, 2013.

84. Singh, H., Kumar, P., Raj, B., Performance Analysis of Majority Gate SET Based 1-bit Full Adder. *Int. J. Comput. Commun. Eng. (IJCCE)* IACSIT Press Singapore, 2, 4, 2933–2940, 2013, ISSN: 2010-3743.
85. Bhardwaj, A.K., Gupta, S., Raj, B., Investigation of Parameters for Schottky Barrier (SB) Height for Schottky Barrier Based Carbon Nanotube Field Effect Transistor Device. *J. Nanoelectron. Optoelectron.*, ASP, 15, 783–791, July 2020.
86. Bansal, P. and Raj, B., MEMRISTOR: A VERSATILE NONLINEAR MODEL FOR DOPANT DRIFT AND BOUNDARY ISSUES. *JCTN*, American Scientific Publishers, 14, 5, 2319–2325, May 2017.
87. Jain, N. and Raj, B., An Analog and Digital Design Perspective Comprehensive Approach on Fin-FET (Fin-Field Effect transistor) Technology - A Review. *Rev. Adv. Sci. Eng. (RASE)*, ASP, 5, 1–14, 2016.
88. Raj, B., Saxena, A.K., Dasgupta, S., Analytical Modeling for the Estimation of Leakage Current and Subthreshold Swing Factor of Nanoscale Double Gate FinFET Device. *Microelectron. Int.*, UK, 26, 53–63, 2009.
89. Soniya, Singh, S., Wadhwa, G., Raj, B., An Analytical Modeling for Dual Source Vertical Tunnel Field Effect Transistor. *Int. J. Recent Technol. Eng. (IJRTE)*, 8, 2, 603–608, July 2019.
90. Singh, S. and Raj, B., Design and analysis of hetrojunction Vertical T-shaped Tunnel Field Effect Transistor. *J. Electron. Mater.*, Springer, 48, 10, 6253–6260, October 2019.
91. Goyal, C., Ubhi, J.S., Raj, B., A Low Leakage CNTFET based Inexact Full Adder for Low Power Image Processing Applications. *Int. J. Circuit Theory Appl.*, Wiley, 47, 9, 1446–1458, September 2019.
92. Raj, B., Saxena, A.K., Dasgupta, S., A Compact Drain Current and Threshold Voltage Quantum Mechanical Analytical Modeling for FinFETs. *J. Nanoelectron. Optoelectron. (JNO)*, USA, 3, 2, 163–170, 2008.
93. Wadhwa, G. and Raj, B., An Analytical Modeling of Charge Plasma based Tunnel Field Effect Transistor with Impacts of Gate underlap Region. *Superlattices Microstruct.*, Elsevier, 142, 106512, June 2020.
94. Singh, S. and Raj, B., Modeling and Simulation analysis of SiGe hetrojunction Double GateVertical t-shaped Tunnel FET. *Superlattices Microstruct.*, Elsevier, 142, 106496, June 2020.
95. Jain, N. and Raj, B., Device and Circuit Co-Design Perspective Comprehensive Approach on FinFET Technology - A Review. *J. Electron Devices*, 23, 1, 1890–1901, 2016.
96. Kumar, S. and Raj, B., Analysis of I_{ON} and Ambipolar Current for Dual-Material Gate-drain Overlapped DG-TFET. *J. Nanoelectron. Optoelectron.*, American Scientific Publishers, USA, 11, 323–333, June 2016.
97. Anjum, N., Bali, T., Raj, B., Design And Simulation Of Handwritten Multiscript Character Recognition. *Int. J. Adv. Res. Comput. Commun. Eng.*, 2, 7, 2544–2549, July 2013.

98. Singh, K. and Raj, B., Performance and analysis of temperature dependent Multi-walled Carbon Nanotubes as Global Interconnects at different technology nodes. *J. Comput. Electron.*, Springer, 14, 2, 469–476, June 2015.
99. Kumar, S. and Raj, B., Compact channel potential analytical Modeling of DG-TFET based on Evanescent-mode Approach. *J. Comput. Electron.*, Springer, 14, 2, 820–827, July 2015.
100. Singh, K. and Raj, B., Temperature Dependent Modeling and Performance Evaluation of Multi-Walled CNT and Single-Walled CNT as Global Interconnects. *J. Electron. Mater.*, Springer, 44, 12, 4825–4835, Dec 2015.
101. Sharma, V.K., Pattanaik, M., Raj, B., INDEP approach for leakage reduction in nanoscale CMOS circuits. *Int. J. Electron.*, Taylor & Francis, 102, 2, 200–215, 2014.
102. Singh, K. and Raj, B., Influence of Temperature on MWCNT bundle, SWCNT bundle and Copper interconnects for Nanoscaled Technology nodes. *J. Mater. Sci. Mater. Electron.*, Springer, 26, 8, 6134–6142, 2015.
103. Anjum, N., Bali, T., Raj, B., Design and Simulation of Handwritten Gurumukhi and Devanagri Numerical Recognition. *Int. J. Comput. Appl.*, Published by Foundation of Computer Science, New York, USA, 73, 12, 16–21, 2013.
104. Khandelwal, S., Gupta, V., Raj, B., Gupta, R.D., Process Variability Aware Low Leakage Reliable Nano Scale DG-FinFET SRAM Cell Design Technique. *J. Nanoelectron. Optoelectron.*, 10, 6, 810–817, December 2015.
105. Singh, S., Bala, S., Raj, B., Raj, B., Improved Sensitivity of Dielectric Modulated Junctionless Transistor for Nanoscale Biosensor Design. *Sens. Lett.*, ASP, 18, 328–333, Apl 2020.
106. Kumar, V., Vishvakarma, S.K., Raj, B., Design and Performance Analysis of ASIC for IoT Applications. *Sens. Lett. ASP*, 18, 31–38, Jan 2020.
107. Jaiswal, A., Sarin, R.K., Raj, B., Sukhija, S., A Novel Circular Slotted Microstrip-fed Patch Antenna with Three Triangle Shape Defected Ground Structure for Multiband Applications. *Adv. Electromagn. (AEM)*, 7, 3, 56–63, Aug 2018.
108. Wadhwa, G. and Raj, B., Label Free Detection of Biomolecules using Charge-Plasma-Based Gate Underlap Dielectric Modulated Junctionless TFET. *J. Electron. Mater. (JEMS)*, Springer, 47, 8, 4683–4693, August 2018.
109. Singh, G., Sarin, R.K., Raj, B., Design and Performance Analysis of a New Efficient Coplanar Quantum-Dot Cellular Automata Adder. *Indian J. Pure Appl. Phys. (IJPAP)*, 55, 97–103, February 2017.
110. Kaur, P., Pandey, V., Raj, B., Comparative Study of Efficient Design, Control and Monitoring of Solar Power using IoT. *Sens. Lett.*, ASP 18, 419–426, May 2020.
111. Bhardwaj, A.K., Gupta, S., Raj, B., Development & Analysis of compact model for Double Gate Schottky Barrier CNTFET. *J. Nanoelectron. Optoelectron.*, ASP, 15, 1199–1208, Aug 2020.

112. Wadhwa, G., Kamboj, P., Singh, J., Raj, B., Design and Investigation of Junctionless DGTFET for Biological Molecule Recognition. *Trans. Electr. Electron. Mater.*, Springer, 22, 282–289, 2021.
113. Verma, S.K., Singh, S., Wadhwa, G., Raj, B., Detection of Biomolecules using Charge-Plasma based Gate Underlap Dielectric Modulated Dopingless TFET. *Trans. Electr. Electron. Mater. (TEEM)*, Springer, 21, 528–535, June 2020.
114. Jain, N. and Raj, B., Impact of underlap spacer region variation on Electrostatic and analog/RF performance of symmetrical high-k SOI FinFET at 20 nm channel length. *J. Semicond. (JoS)*, IOP Science, 38, 12, 122002, Dec 2017.
115. Singh, S. and Raj, B., Analytical Modeling and Simulation analysis of T-shaped III-V heterojunction Vertical T-FET. *Superlattices Microstruct.*, Elsevier, 147, 106717, Nov 2020.
116. Singh, G., Sarin, R.K., Raj, B., Design and Analysis of Area Efficient QCA Based Reversible Logic Gates. *J. Microprocess. Microsyst.*, Elsevier, 52, 59–68, May 2017.
117. Singh, S. and Raj, B., Parametric variation analysis on hetero-junction Vertical t-shape TFET for Supressing ambipolar conduction. *Indian J. Pure Appl. Phys.*, 58, 478–485, June 2020.
118. Soniya, Singh, S., Wadhwa, G., Raj, B., Design and Analysis of Dual Source Vertical Tunnel Field Effect Transistor for high performance. *Trans. Electr. Electron. Mater.*, Springer, 21, 74–82, Oct 2019.
119. Kaur, M., Gupta, N., Kumar, S., Raj, B., Singh, A.K., RF Performance Analysis of Intercalated Graphene Nanoribbon Based Global Level Interconnects. *J. Comput. Electron.*, Springer, 19, 1002–1013, June 2020.
120. Wadhwa, G. and Raj, B., Design and Performance Analysis of Junctionless TFET Biosensor for high sensitivity. *IEEE Nanotechnol.*, 18, 567–574, 2019.
121. Singh, J. and Raj, B., Enhanced Nonlinear Memristor Model Encapsulating Stochastic Dopant Drift. *JNO*, ASP, 14, 958–963, 2019.
122. Sharma, V.K., Pattanaik, M., Raj, B., ONOFIC Approach: Low Power High Speed Nanoscale VLSI Circuits Design. *Int. J. Electron.*, Taylor & Francis, 101, 1, 61–73, 2014.
123. Khandelwal, S., Raj, B., Gupta, R.D., FinFET based 6T SRAM Cell Design: Analysis of Performance Metric, Process Variation and Temperature Effect. *J. Comput. Theor. Nanosci.*, ASP, USA, 12, 2500–2506, 2015.
124. Singh, S., Yadav, S., Rahul, J., Srivastava, A., Raj, B., Impact of HfO_2 in Graded Channel Dual Insulator Double Gate MOSFET. *J. Comput. Theor. Nanosci.*, American Scientific Publishers, 12, 6, 950–953, April 2015.
125. Sharma, V.K., Pattanaik, M., Raj, B., PVT variations aware low leakage INDEP approach for nanoscale CMOS Circuits. *Microelectron. Reliab.*, Elsevier, 54, 90–99, 2014.
126. Raj, B., Saxena, A.K., Dasgupta, S., Quantum Mechanical Analytical Modeling of Nanoscale DG FinFET: Evaluation of Potential, Threshold Voltage and

Source/Drain Resistance. *Elsevier's J. Mater. Sci. Semicond. Process.*, 16, 4, 1131–1137, 2013.

127. Gopal, M., Prasad, S.S.D., Raj, B., 8T SRAM Cell Design for Dynamic and Leakage Power Reduction. *Int. J. Comput. Appl.*, Published by Foundation of Computer Science, New York, USA, 71, 9, 43–48, June 2013.
128. Pattanaik, M., Raj, B., Sharma, S., Kumar, A., Diode Based Trimode Multi-Threshold CMOS Technique for Ground Bounce Noise Reduction in Static CMOS Adders. *Adv. Mater. Res.*, Trans Tech Publications, Switzerland, 548, 885–889, 2012.
129. Raj, B., Saxena, A.K., Dasgupta, S., Nanoscale FinFET Based SRAM Cell Design: Analysis of Performance metric, Process variation, Underlapped FinFET and Temperature effect. *IEEE Circuits Syst. Mag.*, 11, 2, 38–50, 2011.
130. Sharma, V.K., Pattanaik, M., Raj, B., Leakage Current ONOFIC Approach for Deep Submicron VLSI Circuit Design. *Int. J. Electr. Comput. Electron. Commun. Eng.*, World Academy of Sciences, Engineering and Technology, 7, 4, 239–244, 2013.
131. Kaur, P., Gill, S.S., Raj, B., Comparative Analysis of OFETs Materials and Devices for Sensor Applications. *J. Silicon*, Springer, 14, 4463–4471, 2022.
132. Sharma, S.K., Kumar, P., Raj, B., Raj, B., $In_{1-x}Ga_xAs$ Double Metal Gate-Stacking Cylindrical Nanowire MOSFET for highly sensitive Photo detector. *J. Silicon*, Springer, 14, 3535–3541, 2022.
133. Raj, B., Saxena, A.K., Dasgupta, S., Analytical Modeling of Quasi Planar Nanoscale Double Gate FinFET with Source/Drain Resistance and field dependent carrier mobility: A quantum mechanical study. *J. Comput. (JCP)*, Academy Publisher, FINLAND, 4, 9, 1–8, 2009.
134. Bhushan, S., Khandelwal, S., Raj, B., Analyzing Different Mode FinFET based Memory Cell at Different Power Supply for Leakage Reduction. *Seventh International Conference on Bio-Inspired Computing: Theories and Application*, (BIC-TA 2012) Advances in Intelligent Systems and Computing, vol. 202, pp. 89–100, 2012, 2013.
135. Singh, J. and Raj, B., Temperature Dependent Analytical Modeling and Simulations of Nanoscale Memristor. *Eng. Sci. Technol. Int. J.*, Elsevier's, 21, 862–868, Oct 2018.

10

Ferroelectric and FeFET Devices as Biosensors: Principle, Mechanisms and Applications in Health, Environmental, and Agricultural Monitoring

Umesh Chandra Bind[1]*, Keshav Kumar[2], Vimala Bind[3], Ajay Kumar[4] and Jyoti Nishad[5]†

[1]*Department of Applied Science and Humanities, Institute of Engineering and Technology, Dr APJA Kalam Technical University, Lucknow, UP, India*
[2]*New Chemistry Unit, International Centre for Material Science, and School of Advanced Materials, Jawaharlal Nehru Centre for Advanced Scientific Research, Jakkur, Bengaluru, India*
[3]*Department of Zoology, Navyug Kanya Mahavidyalaya, Rajendra Nagar, Lucknow, UP, India*
[4]*Department of Zoology, Tilak Dhari College, Jaunpur, UP, India*
[5]*Food Technology, ICAR-Directorate of Cashew Research, Puttur, Karnataka, India*

Abstract

Owing to their versatile ability to interact with various types of external disturbances, namely light, force, heat, and electricity, ferroelectrics plays a special role in several devices, such as traducers, ferroelectric random-access memory, FeFET, ferroelectric tunneling junctions, ferroelectric diodes, and nanobiosensors. The characteristic features of ferroelectric enhances the performance of FeFET biosensors as compared to its conventional counterparts/ones owing to improvements in the subthreshold swing slope, better control over the channel, and lower power consumption. Additionally, ferroelectric biosensors outperform traditional biosensors because of the induced spontaneous polarization resulting from interactions with biomolecular energy and/or physical stimuli. Owing to this feature of ferroelectrics, the output characteristics of ferroelectric biosensors and FeFET

**Corresponding author*: umesh.bind@gmail.com
†*Corresponding author*: jyoti.nishad@icar.gov.in

Balwinder Raj, Shiromani Balmukund Rahi and Nandakishor Yadav (eds.) FeFET Devices, Trends, Technology and Applications, (239–274) © 2025 Scrivener Publishing LLC

biosensors have been improved. This happened due to the excess availability of carriers at the interface surface resulted from the additional generation of spontaneous polarization, which induces electron-hole pairs. Consequently, this leads to enhancement in sensitivity and selectivity and quicker response times. Extensive studies are required to further understand the functionality of ferroelectrics. Herein, we summarize the ferroelectric-based biosensor design, working principle, complex interaction mechanisms, and monitoring efficiency towards a specific targeted biomolecule (or analyte)in the presence of interfering contents. In addition, a detailed discussion of its applications in human health, environmental, and agricultural monitoring, supported by recent reports, is provided. It also provides insights into the future of ferroelectricity in biosensor devices.

***Keywords*:** Agricultural monitoring, biomolecules, environmental monitoring, FeFET-biosensors, ferroelectrics, nanocavity, response time, transducer

10.1 Introduction

The main aim of research, particularly in science and technology is to build a better world for living organisms. However, people continue to encounter difficulties in accessing nutritious and well-balanced food, clean drinking water, and healthy environments. These challenges contribute to the development of life-threatening diseases despite their ability to perceive and respond to various physical cues in their surroundings, thereby engaging in appropriate physiological activities[1–7]. Over the past few decades, significant efforts have been made to understand biological processes and successfully fabricate a variety of devices with high tactile sensitivities owing to the rapid development of bioelectronics [8–26]. These approaches are achieved by integrating multiple sensors on heterogeneous substrates made up of various types of material combinations with sophisticated interconnections. Therefore, it is essential to identify suitable materials to deliver physical cues to biological systems. Thus, an advanced material with energy transduction capability is desirable to fabricate an appropriate biosensor that senses surrounding activities and monitors biological manipulations in living organisms. Cost-effective and advanced biosensor devices that are capable of monitoring human health, environmental, food, and agricultural pollutant/contaminants and are portable in nature may reduce the path of the aforementioned goals.

The ferroelectric characteristic of spontaneous polarization was first observed by J. Valasek in 1920, during an experimental study of the influence of electric field on Rochelle salt [27]. Further studies have observed several characteristic properties such as ferroelectricity, piezoelectricity, pyroelectricity, non-linear optics, and inverse-piezoelectricity reflected in ferroelectric materials (including some bionanomaterials such as biomolecules) that occur due to the conversion of one form of energy into another with the assistance of atom or molecule movements [10, 12, 13, 20, 21, 28, 29]. These properties are desirable for biocompatible and long-life stable nanomaterials for their advanced utilization in biosensors. Therefore, biocompatible ferroelectrics can be utilized for several applications such as sensing, actuation, energy harvesting, and data storage, owing to their ability to convert one physical stimulus into another [8, 13–23, 30]. In addition, there are several reports on ferroelectric applications as diagnostic biomarkers and biomedical and smart multifunctional devices such as e-skin, e-nose, and p-jet for continuous monitoring of human health, crops, and plants [9, 11, 14, 16, 18–20, 24–26, 31]. For example, Rega *et al.* (2019) have demonstrated an innovative system named as 'pyro-electrohydrodynamic jet' (p-jet) that is able to accumulate collagen molecules with the utilization of piezoelectric substrate of lithium niobate and observe improvement in the performance of this novel device almost 10-fold better than the performance of traditional enzyme-linked immunosorbent assay (ELISA) kits [31].

Herein, biosensor design and its component details, functioning along with characteristics, will be briefly mentioned first. A detailed discussion on the interaction mechanism of ferroelectrics with physical stimuli, the working principle of biosensors, and the biosensing mechanism of ferroelectrics will be addressed. Then, we will emphasize discussing ferroelectrics as biosensors in different forms, including nanorods, thin films, and in the gate or channel of Field Effect Transistors (FETs) with the latest examples (due to a lack of reports on ferroelectric FET for specific application areas, we have selected some other examples) of sensing and monitoring in health, environmental, and agricultural components. Finally, we also discuss the future perspective of the proposed advanced model of FETs for biosensing and monitoring.

10.2 Biosensors

Biosensors are one of the most important parts of human life and are used to protect, detect, and monitor human health from unwanted/harmful species available in the human body, environment, or agricultural products. So, first of all, it is necessary to understand and define 'biosensors.' According to the International Union of Pure and Applied Chemistry (IUPAC), "a biosensor is an independently integrated receptor transducer device, which is capable of providing selective quantitative

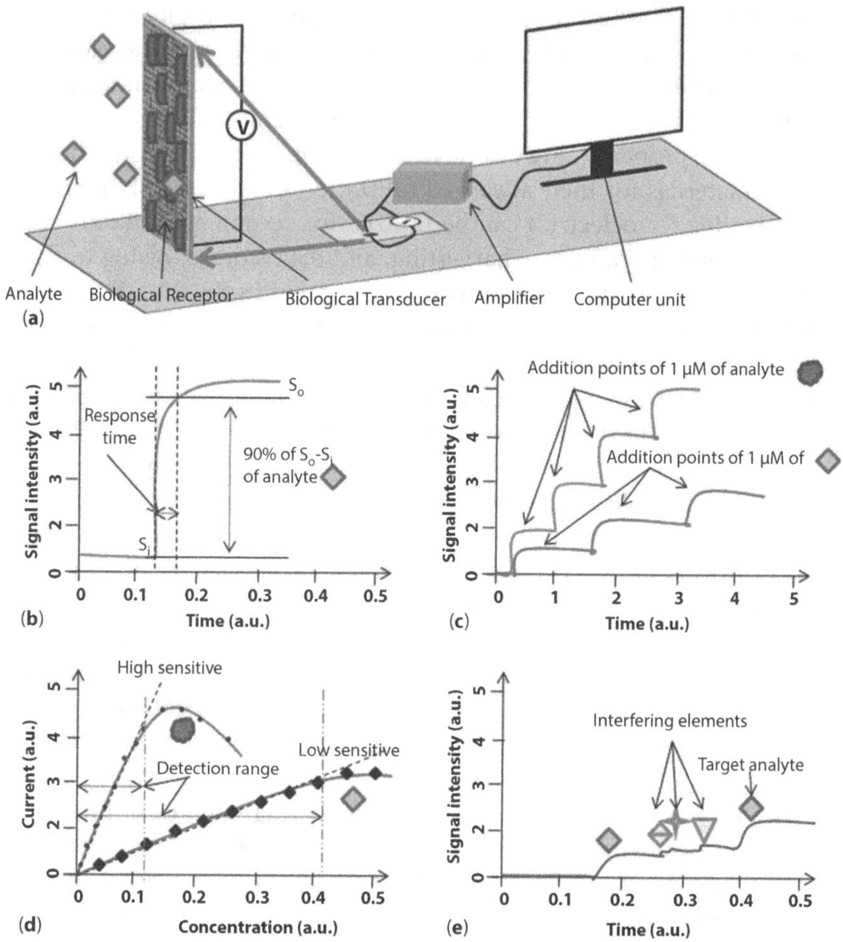

Figure 10.1 Schematic representation of (a) biosensor design and its components, (b) response time of a biomolecule, (c) two different type of analyte response and sensitivity on the surface of identical device, (d) plot of an output response (current signal) versus input concentration of analytes, it reflects the difference in sensitivity and variation in their detection range, (e) interfering ◇, ▽, ✦, and target (◆) analyte response, it identified that the target analyte output response is much higher in comparison of interfering analyte output response, reflected the detection of target analyte (e.g., biomolecule).

or semi-quantitative analytical information using a biological recognition element" [32]. In another word, "a sensor is an analytical device that is used in quantitative or semi quantitative analysis of the existing physical (such as gas, temperature), chemical species or biological (like organisms, molecules) species through its information translation to a quantifiable signal." These species are generally, known as 'analyte' who is capable to reproduce with one another and produce fertile offspring or specific forms of elements (or compounds) that could be in its isotropic composition, electronic, oxidation, or various complex states. Thus, for identification, a standard biosensor is composed of a biological receptor (i.e., sensing element) with a transducer along with a common process unit (containing a display or printer) in conjugation (Figure 10.1a).

Biological receptors: This component is responsible for detecting or sensing the presence of an analyte or substance and is also used for quantification; thus, it is commonly called a sensor or detector. It is biological or biocompatible material component that behaves as a biological receptor that senses or recognizes a target analyte [32, 33]. As analytes interact with biological receptors, they produce signals such as light, heat, charge, mass variation, and pH owing to changes in their microenvironment. The receptor materials should be immobilized, specific, and stable under the observation conditions. Additionally, the receptor should be capable of interacting with the selective analyte present in the sample. The receptor determines the sensitivity of the biosensing device by generating physiological signals that are monitored with a transducer [32–35]. These biological receptors can be categorized into two parts on the basis of their applications: i) catalytic receptor (reusable)—utilized in the continuous monitoring of concentration, for example, enzymes, tissues, microorganisms, etc., and ii) non-catalytic receptor (non-reusable)—utilized in the analysis of analytes such as steroids, drugs, and toxins, and concentration, for example, antibodies, antigens, and nucleic acids [32–37].

The most common receptors are antibodies, aptamers, cells and enzymes [32, 33, 38]. Antibodies(immunological proteins) are developed by the immune system and are highly specific to the antigen, which identifies and neutralizes the effects of foreign antigens inside the system [39]. The long process of isolation using SELEX(systematic evolution of ligands by experimental enrichment), *in vitro* selection, and production makes antibodies expensive [41]. Antibodies selectively bind molecules via aptamers, which are single- or double-stranded nucleic acids with unique sequences [40, 41]. Antibodies are thermostable and have long shelf lives. Cells have the potential to be utilized as biorecognition elements at low cost and can be prepared via a simple route. As it is a living system, it needs a controlled environment for its healthy behavior and to utilize it in a biosensor for the environment, plants, and agricultural products [1–6]. These enzymes act as biological catalysts in several biochemical reactions and can selectively identify specific target analytes or molecules [11, 18, 42]. The production of enzymes is difficult due to their sensitivity to variations in pH and temperature, so a trained one can only handle the production processor [41]. Enzymes can be reused for detecting analytes because their binding sites are safe and unaffected by analyte interactions and cannot be reused in biochemical reaction-based biosensing applications due to damage to binding sites [43].

Biological transducers: This is the second main component of biosensing devices, which converts one form of energy into another [44]. In a biosensor, the transducer converts the physiological signal received from the biological receptor after its interaction with the targeted analyte into a signal that can be measured and quantified, such as optical, electrochemical, electrical, or ferroelectric. The main role of the transducer is to detect and measure changes occurring in the receptor microenvironment after its interaction with the targeted analyte.

10.3 Characteristics of Biosensors

Analyte monitoring or quantitative analysis results depend on the characteristic features of biosensors, such as sensitivity, response time, limit of detection (LOD), selectivity, detection range, and reusability.

Response Time: The response time of a sensor is principally the time within which it changes its output with respect to the input signal (Figure 10.1b). In other words, one can define the response time as the time taken in a signal to propagate from the input to the optimized output that depends upon the user; however, researchers commonly use 90% of the output for the final response [45]. This depends on the nature of the species, the procedure, and its concentration.

Sensitivity: This is defined as the ratio of the variation in the output signal generated from the transducer to the variation in analyte concentration. For example, in the case of a particular analyte, if the variation in the output signal is very large for a very small concentration change in the input, then it is a highly sensitive device for that analyte (Figures 10.1c, d). Generally, one can say that the minimum input of the analyte can create an observable change in the output signal. Mathematically, it is defined as the slope of the output signal characteristic or derivative of the output with respect to species. This depends on the nature of the species and sensor quality.

Limit of detection: This is the lowest concentration at which the analyte can be reliably detected. This reflects the precision of the biosensor response obtained by the procedure; when the concentration of the analyte is zero and the uncertainty in the range of the measurement includes zero, we do not confirm that the species has been detected [46]. Therefore, the magnitude of the detection limit critically depends on the conditions under which the precision is estimated. Biosensors must be able to reflect the lowest detection limit, especially the one used for biomedical applications.

Detection Range: The detection range of a sensor is the concentration range of the analyte within which the sensor works properly. It is defined as "the accuracy of the signal received, in the response to a set of measurements with variation of analyte concentrations" (Figure 10.1d).

Selectivity: The ability of a sensor to discriminate or specify the targeted species among all interfering species or analytes and reflect a specific response (Figure 10.1e). In other words, it can be defined as the ability to bind or interact with a selective analyte, and respond only to the presence of other analytes. It is a critical characteristic feature of a biosensor that can be utilized as a monitoring device in the fields of health, food, agriculture, and environment because of the common availability of several hazardous substances. Therefore, a biosensor with a high selectivity response is essential for obtaining accurate information about the existence of a particular analyte among all interfering analytes.

Reusability: The reusability of a sensor is defined as the quality of the sensor that can be used multiple times in the analysis of a particular analyte after proper cleaning with suitable chemicals that do not affect the physicochemical properties of the sensors. This reduces the commercial cost and availability in specific places.

Reproducibility: It is defined as "the ability of the biosensor to produce matching output signals or data in duplicate experimental run." This depends on the precise and accurate performance of the transducer [44].

10.4 Interaction Mechanism of Ferroelectric with Physical Stimuli

To specify the biosensing applications of ferroelectric materials (within devices), it is necessary and essential to understand the fundamental

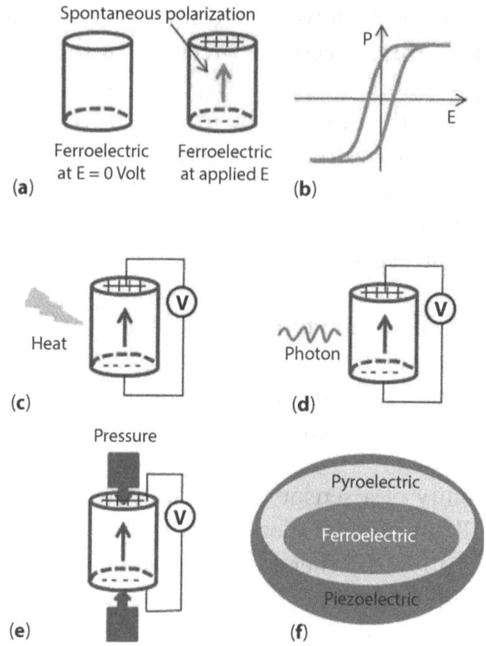

Figure 10.2 Schematic illustration of (a) ferroelectric cylindrical nanocrystal spontaneous polarization without an electric field and with electric field, (b) ferroelectric polarization-electric field loop (P–E loop), (c) heat physical stimuli interaction with ferroelectric, (d) photon physical stimuli interaction with ferroelectric, (e) pressure physical stimuli interaction with ferroelectric, and (f) relationship representation of ferroelectric, pyroelectric, and piezoelectric interactions with each other.

interaction mechanism and concept of ferroelectrics at the atomic level, that is, the means movement of atoms at the unit cell level with external energy such as heat, light, temperature, field, and forces, which leads to disturbances in the non-centrosymmetric structures of ferroelectrics. In addition, we discuss how ferroelectrics are a subset of pyroelectrics and how pyroelectrics are a subset of piezoelectric materials. It can be stated that all the characteristic features of ferroelectrics exist within pyroelectrics; however, pyroelectrics exhibit some additional features along with all the characteristic features of ferroelectrics. Similarly, all the characteristic features of pyroelectrics exist in a piezoelectric material, but piezoelectric materials reflect some additional characteristic features along with all the characteristic features of pyroelectrics. Mathematically, ferroelectrics are subset of pyroelectrics and piezoelectrics.

In other words, all ferroelectrics are pyroelectric and piezoelectric, whereas all pyroelectrics and piezoelectrics are non-ferroelectric (Figure 10.2a).

The characteristic feature of ferroelectrics is their response to the presence of an external electric field, such as the procession of spontaneous polarization, which can be reversed by the application of an external electric field, on the basis of which it reflects hysteresis loop formation (Figures 10.2b,c). This spontaneous polarization creates positive and negative bound charges on the corresponding polarized surfaces [47]. These charges are collected by external screening of surface charges, including injected charges or internal charges developed by defects or surface relaxation, and are commonly generated from absorbed species, such as protons, hydroxyl ions, and hydrocarbons [48–50].

In most ferroelectrics, the spontaneous polarization exists only for a certain temperature range within which it varies, followed by the variation in the resultant concentration of charges at the polarizing surfaces. Such ferroelectrics are known as pyroelectrics, and their behavior is called pyroelectricity (Figure 10.2d). Such behavior is also observed in some biological systems, and it is essential to detect biological activity in physiological sensing organs [51]. As the temperature increases in the pyroelectrics to a certain range, spontaneous polarization decreases to zero; this particular phase is termed paraelectrics, and a further increase in the temperature leads to the generation of spontaneous polarization. This variation in spontaneous polarization generates a surface charge that develops a potential difference across the polar axis, which is used to determine pyroelectric current (i). This output current can be express as [52].

$$i = \frac{pAdT}{dt} \tag{10.1}$$

Where, p is the pyroelectric coefficient that can be defined as $= \frac{dP_s}{dT}$, and A is the surface area of the electrode.

In some materials, spontaneous polarization is induced after the application of directional mechanical stress across the faces, leading to the development of electrical charges of equal magnitude and opposite sign on the stressed applied surfaces. This induced electric field along the direction of the applied stress and its strength is known as piezoelectric; its properties are termed piezoelectricity (Figure 10.2e). Materials reflecting the piezoelectric reversed phenomena are known as inverse-piezoelectric materials. Similar piezoelectric features are also found in several biological tissues, such as bones, teeth, muscles, nerves, and tendons [53]. For example, an applied mechanical stimulus mechanism has been reported to be used in the development of bone size growth and remodeling [54]. The mechanical and electrical responses as a function of each other in such type of materials (piezoelectric) crystals can be depicted as [55]

$$S_{ij} = d_{kij} E_k \tag{10.2}$$

where, S, d, and E represent the second-rank strain tensor, piezoelectric coefficient, and electric field, respectively, and i, j, and k denote the three coordinate axes of the system (x, y, and z coordinates in the Cartesian coordinate system). Further, as directional mechanical stress is applied across the surfaces of piezoelectric material, it will generate potential difference across the opposite surface, leads to pass of charges across the wire if it is connected through the surfaces of applied strain, this can be measured by output current (i) analysis and given by

$$i = d_{33} EA\left(\frac{d\varepsilon}{dt}\right) \tag{10.3}$$

Where, E, A, Ɛ, and t are Young modulus, cross sectional area, applied strain, and time, respectively.

In addition, it has been observed that some ferroelectrics with some band gap interact with electromagnetic radiation of approximately the same energy and produce bound surface charges due to the generation of spontaneous polarization within it and *vice versa* (Figure 10.2f). This coupling of electromagnetic waves with ferroelectrics presents a unique nonlinear optical effect and electro-optic, acoustic-optic, and photovoltaic features [56–58]. When ferroelectrics are illuminated with an electromagnetic wave (photon energy)of higher energy than the band gap energy then it generates electron-hole pair is generated inside the ferroelectrics. This leads to spontaneous separation of the electron holes due to the electric field induced by intrinsic polarization, resulting from potential development that reflects photocurrent. This mechanism is known as the ferroelectric–photovoltaic effect, which is different from the conventional photovoltaic effect in semiconductor theory [59]. The ferroelectric photovoltaic performance of a material depends on its domain walls and polarization strength [55, 56]. Generally, this type of ferroelectric–photovoltaic characteristic is observed with UV light interaction in ferroelectrics owing to the wide band gap and ability to easily produce electron-hole pairs. Lower-bandgap ferroelectrics are being generated by researchers to utilize them for solar energy [60]. Such ferroelectrics could be beneficial for biological, agricultural, environmental, and food monitoring, as well as detection applications owing to their suitable energy (order of 1.5 eV) for small systems such as biological species, molecules, and ions.

It is speculated that different ferroelectric characteristics are observed in different biological systems [28]. Previous studies have shown the influence of ferroelectricity on biomolecules and cellular processes [29]. For instance, Liu *et al.* demonstrated the switching mechanism in a biological elastin system using microscopic observation for the first time [28]. In addition, self-assembled peptide nanoparticles and inorganic biosubstances, e.g., hydroxyapatite have recently been reported for their ferroelectric features [61, 62].

10.5 Working Principle of Biosensors

Typically, it constitutes a biological receptor coupled with transducers and a processing unit for signals. It works on the basis of a transducer signal that is generated after interaction with biological molecules or species, in general, known as an analyte. The compilation of these components within a unit is related to the rapid conversion of the signal to the final output response within a minimum time interval. Therefore, each component work should be very fast (quick with time) and smart (complete transfer of information) to improve the device efficiency. Biosensors can quantify analytes by converting their biological responses into quantifiable signals. First, biomolecules present in the sample interact or bind with available biological receptors at the surface of the transducer, resulting in physiological changes. It results in recognizable alteration in the optical, mechanical or electrical properties of the transducers that further modulated into detectable electrical signal and process for analysis. This is more easily understood through a schematic representation (Figure 10.1a). Generally, transducer-generated signals are in the form of current or voltage, depending on the transducer characteristic features as well as biological receptor properties. Furthermore, analysis of these outputs in processing units using different modulation techniques leads to the output result in the quantity of biological molecules [63].

10.6 Biosensing Mechanism of Ferroelectrics

Controlled devices are well known for the internal monitoring of humans, animals, or living organisms. The nanosized ferroelectric exhibits variation in its spontaneous polarization characteristics with slight alteration of

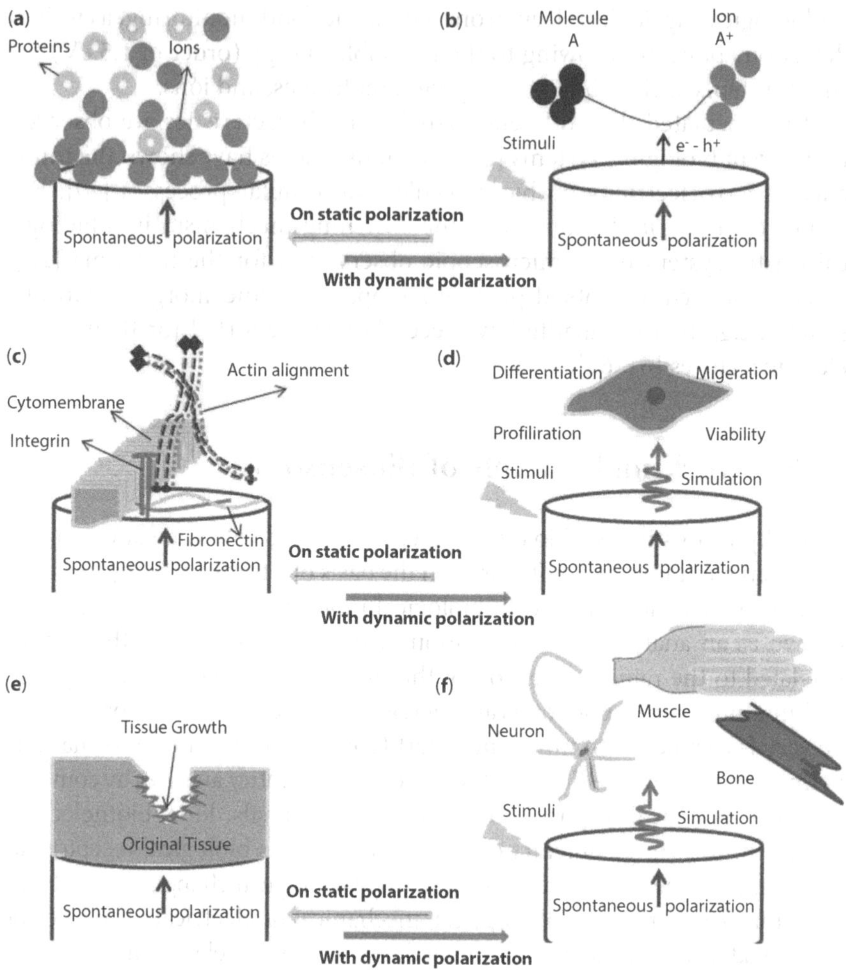

Figure 10.3 Schematic illustrations of ferroelectric utilization in (a) biomolecular adsorption, (b) polarization-stimulated biochemical reaction (i.e., oxidation reaction), (c) cellular nature, (d) cell modulation stimulated by spontaneous polarization variation, (e) tissue regeneration based on ferroelectrics, and (f) tissue regeneration and function triggered by the spontaneous polarization of ferroelectrics.

its environmental conditions owing to changes in the total energy of the nanoparticles after interaction or interchange of energy with environmental content.

Thus, this tunable variation in the nanosized ferroelectric energy alters the strength of polarization, causing a change in the resultant surface charge concentration.

It is necessary to understand the ferroelectric characteristic behavior, when it interacts with biological substances such as biomolecules, cells, and tissues. This section discusses the alteration in the system (biological substances associated with ferroelectrics) behavior after its interactions with physical stimuli.

As ferroelectrics are immersed in biological media, such as human plasma, biological species (e.g., ions and proteins) are absorbed electrostatically on the interface, resulting in the formation of a layer with an increased concentration of ions/molecules near the surface (Figure 10.3a). The layer thickness can be externally controlled owing to the variation in spontaneous polarization with the application of an electric field or voltage across the system. In the case of biological reactivity at the interface, the adsorbent analyte starts reacting with biomolecules with higher reactivity due to the excess presence of electrons and holes (Figure 10.3b). For example, an accelerated *in vitro* biomineralization was observed after the interaction of negatively charged surfaces with positively charged species (e.g., calcium ions) in simulated body fluid [64, 65]. Furthermore, the enhanced adsorption of biomolecules on the charged surfaces with preferred selectivity on the basis of its charge affinity could activate some important and additional signaling pathways in reference to cell proliferation, differentiation, and function (Figure 10.4c). For example, Tang *et al.* fabricated a P(VDF-TrFE)-based ferroelectric thin film that reflected a broad range of surface potentials for the preferred adsorption of fibronectin with integrin. They discovered that it was effective and suggested a complete, tightly bound interaction between integrin α5β1 and fibronectin at an appropriate surface potential. This interaction induces effective activation of the integrin-mediated FAK/FRK signaling pathway, thereby promoting the osteogenic differentiation of MC3T3-E1 cells. [66]. In addition, Biazquez-Castro *et al.* explained the interaction mechanism between cells and ferroelectrics [67]. This adsorption capability can be externally enhanced and controlled according to the rearrangement, migration, viability, or proliferation of cells (Figure 10.3d). Ferroelectrics exhibit spontaneous polarization, which generates an electric potential near the surface. This property can be advantageous for repairing damaged tissues by restoring their physiological and microenvironmental potential (Figure 10.3e). Wang *et al.* successfully observed this mechanism of tissue repair in defective bone with the application of a $BaTiO_3$/P(VDF-TrFE) composite film that produced a surface potential of −76.8 mV at the bone defect site, proving the strength of endogenous biopotential [68, 69]. These cells or bone responses could also be controlled externally owing to the induced potential on the ferroelectric surfaces. The repair of bone defects is also applicable to neurons and muscles (Figure 10.3f). These reflect the capacity of ferroelectrics for sensing and treatment applications, but the optimization of the surface potential with bone defects can analyze the degree of bone defects. Thus, this mechanism can be used for biosensing and monitoring.

10.7 Ferroelectrics for Biosensing

There are several compositions of ferroelectrics available for decades, such as inorganic materials barium titanate (BTO), organic materials such as polyvinylidene fluoride (PVDF), alkaline niobates (e.g., KNbO3 and LiNbO3), and perovskite-type materials that exhibit oxide-based ferroelectric performance, which are being utilized for various biomedical applications including biological sensors [58]. However, owing to the increasing demand for wearable devices and biosensors, a very new type of advanced material with lightweight, biocompatibility, long stable life, mechanical strength with flexible nature, biodegradability, and easy synthesis have been developed [10–12, 22–24]. Such materials can be developed by hybridization of organic compounds, liquid crystals, polymers, and inorganic materials for ferroelectric-based applications [8–11, 14, 16, 20, 21, 25, 26]. These organic framework ferroelectrics should have a higher ferroelectric coefficient or be comparable to inorganic (presently available) ferroelectrics in order to improve device performance. For instance, Herdia *et al.* reported a 30 pmV^{-1} excess yield value of the piezoelectric coefficient of self-assembled nanotubes compared to the inorganic niobate $LiNbO_3$ [70]. Moreover, organic ferroelectrics usually exhibit weak performance owing to their relatively low critical temperature and unwanted solubility, which leads to a reduction in the dysfunction under physiological circumstances [8, 71]. Furthermore, there are still several difficulties in achieving multiple properties, such as piezoelectricity with flexibility and degradability, in a single material system with one or more available fabrication techniques and material compatibility. To achieve this, several efforts are being made with various types of organic structures, such as porous organic frameworks and columnar liquid crystal-type soft materials [8, 71].

Recently, a super-plastic BTO membrane of a 2D ferroelectric single crystal fabricated by a damage-free lifting procedure was shown to reflect its flexibility. This is because of the advancement in ferroelectric nanodomains that can undergo bending up to 180° during *in situ* testing and can be easily integrated with polymeric stuffs [72]. This type of ferroelectric system, such as thin film-based structures, ferroelectrics with 1D, and ferroelectrics with 2D morphological features, are being used in various applications such as flexible biosensors, electronic skin, and biological generators [9, 14, 20–22, 30]. Furthermore, 3D ferroelectrics have been achieved through advancements in additive fabrication techniques. Advanced printing techniques have embodied designed anisotropy and directional piezoelectric response in 3D ferroelectric PZT that reflects the light-sensitive behavior of PZT owing to the presence of photon-sensitive highly concentrated nanoparticulate [73, 74].

The *in vivo* cell response can be monitored using the extracellular microenvironment, as it has specific structural characteristics. Thus, by selecting an appropriate ferroelectric, cell behavior can be monitored via a specific configuration of extracellular morphology and by stipulating physical stimuli (external signals). Kim *et al.* fabricated electrospun PVDF-based fibers with gradient morphological features by using modified collectors to produce a variety of micropatterns to bundle structures and used them to study cell culture behavior [75]. They observed that these morphological features with piezoelectric cues not only influenced cell growth but also activated signaling routes of neuron-specific cytoskeletons. Such advanced nanomorphological features may be functionalized with the groups required for utilization in advanced biosensor application. To obtain such functionalized novel materials with ferroelectrics, the primary and initial condition is that they should be biocompatible and biodegradable for their utilization as advanced healthcare devices or tissue engineering scaffolds. For example, Sonay et al. developed triphenylalanine peptide nanoassemblies with glycine crystals as a ferroelectric-based biomolecular system for potential application in healthcare, which is easily digested by enzymes or gradually dissolved under the present physiological conditions [76]. Hosseini *et al.* used a simple solvent casting method to fabricate free-standing and biodegradable glycine-chitosan piezoelectric films and successfully implemented it for pressure monitoring [77].

10.8 Ferroelectrics in Health Monitoring

Biosensing devices can be utilized for continuous monitoring of vital signs such as heartbeats, pulse, body temperature, blood glucose, and cell nature of the human body or other living organisms as a function of time. This real-time analysis/monitoring of vital signs can prevent the user from unexpected emergency situations without missing time.

Pressure biosensors: These devices can monitor mechanical forces as stimuli such as stress, strain, tension, torque, and pressure, and convert them into electrical signals based on the ferroelectric characteristics of the material used in its fabrication. Physical stimuli provide health information about the respiration rate, heartbeat rate, blood pressure, and muscle movement by regulating the physiological activity of the human body. The presence of ferroelectric materials i.e., piezoelectric materials with biocompatible materials, triggers the measuring efficiency even at high-frequency dynamic signals of the devices. For example, there are various piezoelectric flexible sensors such as $PbTiO_3$, ZnO, PZT, and P(VDF-TrFE) [20, 73–75], of which P(VDF-TrFE) has been reported to be an excellent flexible piezoelectric sensor owing to its high piezoelectric coefficient and remarkable stability, fabricated by an easily available electrospinning technique. It exhibits an extraordinarily small pressure detecting capability up to 0.1 Pa that suggests its ultimate application in the monitoring of human body motion or robotic electronics [73]. Sun *et al.* developed a novel, biocompatible, highly transparent, and flexible triboelectric-piezoelectric-pyroelectric hybrid nanogenerator (nanodevice) with the sharing of common electrodes of integrated polydimethylsiloxane (PDMS), polyvinylidene fluoride (PVDF), and silver nanowires, which reflects the possibilities of its massive production as well as dimension miniaturization of the device for its utilization in multiple circumstances. For instance, it can be attached to different parts of the human body to monitor various physiological conditions (heartbeat rate, breathing, and swallowing) by self-powering from the human body [76]. The self-powering in this device is due to the combined effect of the triboelectric and piezoelectric characteristic response of PVDF, which collects mechanical energy from the human body, and the pyroelectric effect in PVDF, which collects energy from temperature fluctuations between the human body and the ambient environment. They reported that a hybrid generator of size 3 cm × 3.5 cm produce open-circuit voltages up to 55 V for mechanical energy and 86 V for thermal energy [76]. They realized novel features for the very first time for a transparent flexible hybrid generator (device) after its integration with a thermochromic thin liquid crystal film (LCD). It changes color with temperature and can be used to monitor human health. Niu *et al.* proposed a thin film-based capacitive tactile sensor with high sensitivity by developing interlocked asymmetric-nanocones made up of poly(vinylidenefluoride-*co*-trifluoroethylene) (P(VDF-TrFE)) on an aluminum oxide template, which is used to produce cost effective and controlled-pattern nanocone arrays on a large scale [77]. Their finite element analysis revealed that upon the application of vertical force, the resulting stress/strain within the film could strengthen and localize the contact apexes, leading to amplification of the variation in the film permittivity and thickness owing to the piezoelectric characteristics of the P(VDF-TrFE) material.

This sensor has an ultra-low detection limit (≈3 Pa), response time (48 ms), maximum sensitivity of 6.583 kPa^{-1} in the low-pressure region (0–100 Pa), excellent stability, recovery time (36 ms), and exceptional reproducibility even after 10,000 cycles. Nguyen *et al.* reported a pressure monitoring device in a cell at a nanolevel with a piezoelectric PZT nanoribbon owing to its electromechanical characteristic features [78]. The PZT nanoribbons were suspended in PC12 cells, and after the application of electric voltage across the membrane, its output signal intensity started to vary owing to the piezoelectric nature of PZT. This provides a novel device for monitoring small changes in the cells. Thus, the stress inside the cell can be monitored based on an external signal.

Temperature biosensors: The body temperature of a living organism is one of the most omportant parameters for a healthy life. For instance, the healthy human body-core temperature is generally in the range of 36.2–37.2 °C. Abnormal changes in this temperature range indicate poor health conditions. An increase in body temperature is a signature of fever, and a decrease in body temperature signifies anemia. There are several reports on different types of pyroelectric materials, such as PZT, LiTaO3, LiNbO3, PVDF, and P(VDF-TrFE) as pyroelectric sensors owing to their characteristic feature of variation in spontaneous polarization as a function of temperature [73–75]. Some of these are flexible biosensors used in the monitoring of health and flexible electronics, e.g., P(VDF-TrFE). Li *et al.* utilized nanofibers of a PVDF/hydroxylamine hydrochloride (HHE) composite for the fabrication of soft sensors for monitoring of micro-pressure around cardiovascular walls [79]. The composite produces ultra-sensitivity owing to the uniaxial orientation of PVDF chains with high β-phase contents, which increases the piezoelectric effect. Thus, the piezoelectric signals generated by the PVDF/HHE device synchronously distinguished the variation in the elasticity of cardiovascular walls in real time and achieved early detection of thrombosis and atherosclerosis by regular monitoring.

Physiological biochemical sensors: In order to monitor health issues, one has to always monitor the vital biochemical signs, such as blood glucose, sweat, interstitial fluids, tears and saliva, variation in the body [80–86]. These are generally evaluated by monitoring their compositions and concentrations during the chemical reactions between the sensor and targeted detection, which leads to tuning of the electrical parameters of the biosensors. However, with the advancement of flexible biosensors, it is now possible to monitor these biochemical *in situ* analyses. Gai *et al.* demonstrated a highly flexible multiplexed integrated wearable sensing system for monitoring sweat. It comprises four different sensors for simultaneous and selective monitoring of biomarkers in sweat, including sodium ions, potassium ions, glucose and lactose [86]. Sophia *et al.* developed device for the monitoring of glucose molecules based on BTO thin film [87]. The glucose on the BTO thin-film surface acts as a gate potential, and the applied potential generates carriers on the piezoelectric film output. They reported that the device can achieve a detection limit of 10 μM and a detection range of up to 800 μM, which is much lower than that of existing devices.

Perk *et al.* created fingerprint-like patterns and interlocked microstructures within ferroelectric thin films known as e-skin. This advancement enhances the detection sensitivity to piezoelectric, pyroelectric, and piezoresistive static and dynamic mechanothermal signals. The e-skin was composed of a ferroelectric polymer composite comprising PVDF and reduced graphene oxide (rGO) [9]. They demonstrated that it could simultaneously analyze temperature and pressure variations and separately resolve them via different signaling pathways. It generates different modes, such as tiny and temporal pressure, sustained pressure, and temperature owing to the availability of piezoresistive, piezoelectric, and pyroelectric PVDF materials. Thus, this device can be used as a wearable sensor, robotic skin, and medical diagnostic monitoring device. Pullano *et al.* reported a thermoelectric approach based on the infrared-induced pyroelectric effect to collect thermodynamic and thermophysical data from biological samples for the identification of cellular suspensions at low concentrations [88]. The device induced a pyroelectric response using a Polyvinylidene Fluoride (PVDF) thin film, which was illuminated with infrared radiation. They used four hepatocytes, T lymphocytes, prostate cancer cells, and retinal pigment epithelium cells from different tissues as representative *in vitro* models to search for different experimental conditions for the analysis of cell activities. This analysis can monitor the thermodynamic nature, cell concentration, average single-cell thermal conductivity, and cell viability [88]. Rega *et al.* have demonstrated an innovative system (biosensor) to which they named as 'pyro electrohydrodynamic jet' (p-jet) which accumulates collagen molecules that improves the performance of this novel device almost 10-fold than performance of traditional ELISA kits [31]. In fact, they utilized the piezoelectric effect characteristics of substrate (slide) crystals, such as lithium niobate, to disperse biomolecules within a specific area of a deposition slide using an electrode and nozzle-free configuration This device has an LOD of less than 3 pg/mL. Dagdeviren *et al.* fabricated a conformal piezoelectric system that enables *in vivo* determination of soft tissue viscoelasticity in the epidermis near surface regions and attained conformal contact with the texture of target skin, underlying complex topography, and with other organ surfaces under dynamic and quasi-static conditions [89]. The device obtains information on biological tissues and organ systems in animal models by utilizing piezoelectric actuator sensors by pairing experimental and theoretical characterization responses for clinical and experimental characterization of soft tissue biomechanics [89].

10.9 Ferroelectrics for Environmental Monitoring

As human life gets associated with technological development, varieties of man-made chemicals and by-products prepared in industries and laboratories through different processes have been, and still are, released in the environment. Some of these substances are found to be contaminants, such as pesticides, heavy metals, or polychlorinated biphenyls (PCBs), which adversely affect on the quality of the present environment [1, 3, 5, 6]. Therefore, various types of biosensors have already been developed and are being used for their detection and monitoring. Various types of biosensors are being utilized for monitoring, such as electrochemical, enzymatic, and ferroelectric-based biosensors [90, 91]. For instance, Yilmaz *et al.* reported a molecular imprinting based-bacterial biosensor for the rapid detection of bacteria in water bodies. They have considered *Escherichia (E. coli)* as a model bacteria because of its abundant presence in the water, which causes severe illness and affects the agricultural runoffs, for monitoring purposes with developed sensors. They observed that it is selective, specific, and label-free rapid detection within an hour via microcontact imprinting of whole cells on both optical and mass-sensitive devices (piezoelectric devices) as compared to traditional biosensor detection, which is based on genomic analysis, antibody-based assays, culture methods, fluorescence, and microscopy in joint functioning that takes a whole week [92]. *N*-methacryloyl-L-histidine methyl ester, which is a polymerizable form of histidine immobilized on gold surfaces, has been utilized for the fabrication of biosensors for the recognition of elements (e.g., natural antibodies) with the unique characteristic features of surface plasma resonance and spontaneous polarization. Antibody detection has been reported to be improved by the presence of piezoelectric materials in fabricated biosensor devices [93]. Recently, Cezaro *et al.* developed a tyrosinase cantilever nanobiosensor for the environmental monitoring of 17β-estradiol and estrone availability in a typical water body system with a detection limit of less than 0.00051 ppb and a detection range of 0.0001–100 ppb [94]. This efficiency of the nanobiosensor in cantilever functioning surpassed that of the other enzymatic biosensors. This means that cantilever type piezoelectric biosensors can achieve better results owing to the pressure sensitivity of the ferroelectric material at the atomistic level.

10.10 Ferroelectrics for Agricultural Monitoring

Modern agricultural industries utilize various sensor technologies to gather accurate information on crop, soil, and weather conditions for quality control and monitoring purposes to enhance industrial growth [1, 2, 4–7]. These technologies are primarily employed to gather real-time data on the chemical and physical characteristics and the quality of plants, crops, and their products. They also monitored humidity levels and the release of gas effluents from agricultural products over time. Such rigorous monitoring can be affordable only with electronic nose (e-nose) devices that can be fabricated using biocompatible ferroelectric materials or ferroelectric biomaterials [95, 96]. Such information can also be collected from piezoelectric biosensor-based biomaterials that include polysaccharides and polypeptides, which are polymeric in nature and have a high piezoelectric coefficient of 2–30 pC/N. These properties of biomass-assisted biomaterials have revealed various applications in biosensors, and biomedical devices [97]. For example, Ganeshkumar *et al.* demonstrated a 1D $KNbO_3$ nanofiber, developed by a sol–gel-assisted far-field electrospinning process, as a good quality humidity sensor that reflects logarithm-linear dependence on conductance with relative humidity [98]. They observed that the conductance dramatically increases from 10^{-10} ℧ to 10^{-6} ℧ with variation in humidity from 15% to 95% at room temperature along with its ultrafast sensitivity response (≤2 s), recovery time (≤10 s), with good linearity and reproducibility [98]. $KNbO_3$ has been found to be very useful for humidity sensors, along with its different heterojunction/nanocomposite structures with TiO_2 [99–101]. A systematic evaluation of cellular responses to physical stimuli can provide a better understanding of physiology and biology.

10.11 FeFET Biosensors for Monitoring

A bio-active field-effect transistor is composed of three electrodes: source, drain, and gate. Thin films of semiconductors, semimetals, and conducting polymeric substances act as semiconducting channels in active devices [102, 103]. These FET exist as sensing devices in several configurations, such as metal oxide semiconductor FETs (MOSFETs) [104, 105, 108], thin-film transistors (TFTs) [106, 107], electrolyte-gated FETs (EG-FETs) [102], electrochemical transistors (ECTs) [109], ion-sensitive FETs (ISFETs) [110], and chemically sensitive FETs (ChemFETs) [111] with some differences in their working principle or channel materials (Figure 10.4). Metal-oxide-semiconductor FETs are transistors that consist of a gate electrode insulated from a silicon (Si) semiconducting channel by a silicon dioxide (SiO_2) layer. TFTs are composed of varieties non-Si materials that are semiconducting in nature and may be flexible, stretchable, and biocompatible. Such transistors are commonly used in biosensing devices. In some configurations, the gate electrode is isolated from the remaining

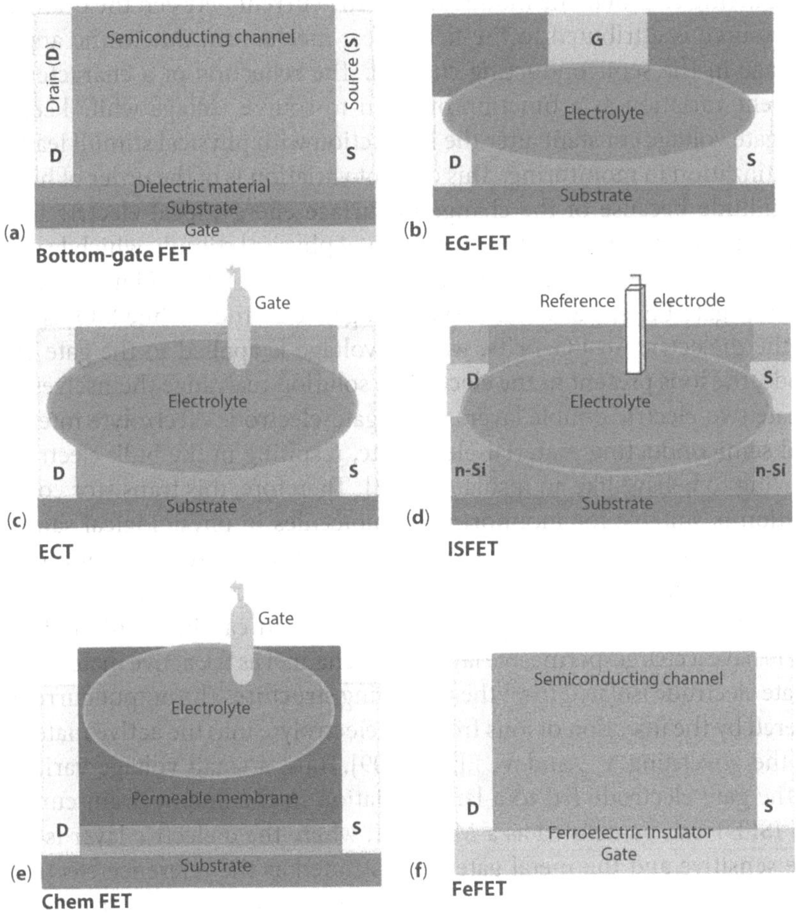

Figure 10.4 Schematic representation of some common transistor configurations being used as biosensor; (a) bottom-gate FET, (b) EG-FET, (c) ECT, (d) ISFET, (e) Chem FET, and (f) FeFET.

structure by means of an electrode that can be either a liquid solution, solid polymer, or ion gel. This electrode is a conducting medium that contains atoms and molecules with constant charges or nanomaterials with externally influenced charges, which makes it convenient to be interfaced with a physiological sample solution to the monitor concentration and nature of substances [112]. In most cases, the best-performing transistor-based nanobiosensor devices are obtained by depositing a biocompatible semiconductor material on the substrate to form a good channel.

Working mechanism: When a voltage is applied to the gate electrode, the charge carriers start to flow from the drain to the source through the semiconducting channel. The modulation of the current between the drain and the source is attributed to the field-effect mechanism due to the applied voltage in the semiconducting channel. The reflection of a characteristic current variation as a function of drain-to-source voltage while keeping the gate voltage constant after the interaction with physical stimuli leads to its utilization in monitoring. This current variation is of the order of higher magnitude because of the change in surface energy, local electric fields, and redox reactions with the interaction of physical stimuli, which leads to the generation of an electron hole inside the ferroelectric (Figure 10.5). In bottom-gate FETs, a Si wafer serves as a gate electrode, while SiO_2 is used as the dielectric. In EG-FETs, when a voltage is applied to the gate electrode, the ions present in the electrolyte solution rearrange themselves and create two electric double layers at the gate-electrode-electrolyte interface and semiconducting material-electrolyte, resulting in the bulk electrolyte starting to behave like an insulator [102]. Therefore, this transistor configuration is suitable for monitoring biomolecules in physiological samples because these samples can be easily mixed in the electrolyte solution, and any change in the electrolyte solution, such as charge/ions generation, can reflect the signal at the output of the device and can be monitored. The ECTs have a charge-permeable layer in the channel as the active material and a gate electrode isolated from the remaining structure. The output current is altered by the injection of ions from the electrolyte into the active materials by the governing V_{GS} and V_{DS} [106, 109].Thus, a small voltage variation in the gate electrode led to a large variation in the output drain current. An ISFET can be defined as a MOSFET, where the dielectric layer is analyte sensitive and the metal gate is substituted as the reference electrode. Therefore, the dielectric layer materials should be more sensitive to ions and charges to enhance the device drain current.

In addition, surplus charges can be generated by the application of physical stimuli to the novel monitoring performance of the device as a biosensor.

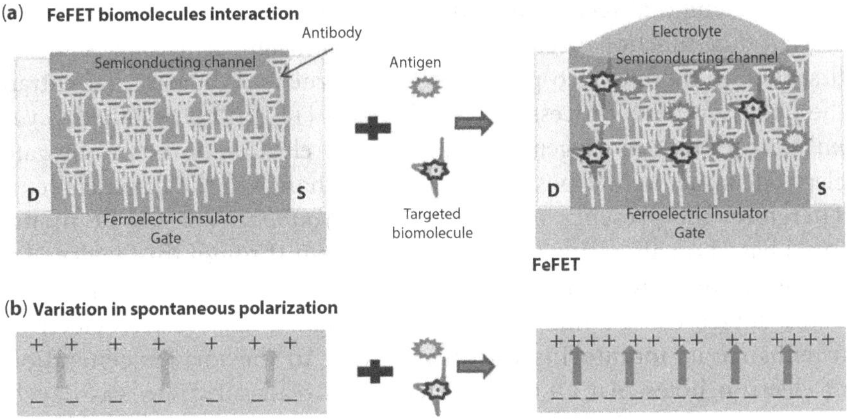

Figure 10.5 (a) Schematic illustration of the ferroelectric FETs application as biosensor; the biolecules bind/interact with antibody which transfer signal to channel, and (b) schematic representation of spontaneous polarization status with interaction/binding with antibody and biomolecules, that could be altered with variation in strength of spontaneous polarization of ferroelectric.

ChemFETs are considered ISFET with the covering of one additional layer over the dielectric layer with a membrane that is able to modulate specific chemical stimuli owing to its selective nature to a specific ion or gas analyte [111].

The selection of FET configuration is based on i) the application, i.e., the requirement to use stretchable, biocompatible, or flexible material in the device, and ii) the analyte of interest, i.e., the nature or behavior of the monitoring species. For instance, we discuss some of the reported transistors as novel biosensor devices.

Lee *et al.* has developed an artificial intelligence enabled integrated tactile learning skin (e-skin) comprising of array of P(VDF-TrFE) ferroelectric-gate based FETs equipped with dome-shaped tactile top-gate of PEDOT:PSS, P3HT as semiconductor onto Au structured source and drain, for the use as smart electronic skin device to monitor human health [113]. For demonstration, they applied pressure on a dome-shaped top gate to measure the ferroelectric residual polarization in the gate materials of the device. This indicates that there will be a variation in the remnant polarization of the gate for different applied pressures, which means that the variation in the applied pressure can be monitored by the variation in the amplitude of the drain current, thus imitating a variety of tactile and synaptic functions. This device showed excellent performance even after 10,000 cycles of long-term potentiation and depression with low variability (3.18%) [113]. The array of 4×4 devices enabled the monitoring of different handwriting patterns using

2D spatiotemporal recognition with a high (99.66%) degree of accuracy. Le Gall *et al.* functionalized gate electrodes with alginate hydrogels for the first time to monitor two pesticides (glyphosate and diuron) that entrap the photosynthesis process of *cyanobacteria* [114]. Photosynthetic *cyanobacteria* produce oxygen, which is directly electroreduced on the gate electrode under light exposure, leading to a change in the drain current. Therefore, the presence of pesticides in *cyanobacteria* can be monitored based on alterations in the drain current through this device. An important point to be noted here is that the gate electrode of this transistor device can be functionalized with light-sensitive ferroelectrics to enhance its monitoring capability owing to the surplus generation of electron holes. An OECT on a cotton thread substrate was developed with poly(3,4-ethylenedioxythiophene):polystyrene sulfonate (PEDOT:PSS) conducting polymer as the active material for monitoring the plant sap electrolyte content [115]. Such transistor containing device used in plant physiology and biology is termed as "bioristor." The bio-FET reflectS periodic variation in the drain current as the ion concentration varies during the circadian cycle of the plants. Sometimes, there is a reduction in the drain current with an increase in the gate-source voltage due to de-doping of PEDOT:PSS by plant sap. Owing to the biocompatibility of bioristors, they can be inserted, without affecting plant morphology, to monitor the immune response and variation in the ionic composition of the plant xylem sap in real time. The efficiency of the device can be enhanced using a PEDOT:PSS conducting polymer with piezoelectric materials. PEDOT:PSS enhances the performance of piezoelectric materials (e.g., PVDF), which blend with it [116]. Kim *et al.* fabricated a bottom-gate FET sensor with CNTs as semiconducting materials to monitor Hg^{2+} in aqueous samples [117]. The presence of Hg^{2+} produces a redox reaction at the CNT surface, resulting in an enhancement of the drain current, reflecting an LOD of 10 nM with a detection range of up to 1 mM. This drain current enhancement did not reflect the presence of other ions such as Cu^{2+} and Co^{2+}, reflecting its selectivity for Hg^{2+} ions [117]. Takemoto *et al.* developed a process for the fabrication of a highly transparent (up to 90% transmittance within the 400–800 nm wavelength region) electrode for a transistor device (OECT) with an organic crystal (C8-BTBT) as the active material [118]. This device is used to monitor plant's electric potential on leaves without compromising the photosynthetic process [118]. In place of this electrode, a transparent ferroelectric electrode may enhance the measurement accuracy and performance. Berto *et al.* designed an EG-FET device from quartz (piezoelectric), an organic active layer of pentacene, and a functionalized gate with anti-PPV antibodies. It has been used to sense plant pathogens such as plum pox viruses (PPVs), which significantly affect the productivity and quality of stone fruits [119, 120]. As the virus binds with antibody at the gate, the drain current starts decreasing, and its transconductance behavior reflects that the drain current is a function of virus concentration. This device was utilized for the *in vitro* analysis of *Nicotiana benthamiana* leaf extraction [119]. Such devices reflect results comparable to those of traditional sensitive methods (such as the ELISA method). Izumi *et al.* proposed a MEMS technology-based biosensor device that comprises an ISFET with an ion-sensitive membrane layer on the Si substrate, an electrical conductivity sensor, and a p-n diode sensor for temperature sensing to monitor the pH of phloem sap in cucumber stems [121]. The pH monitored range with this device was 4.01–9.18 evaluated based on the drain current variation with different pH control solutions. This revealed a highly sensitivity of 40.2 mV/pH. Therefore, this device can monitor parameters related to soil and fertilizer content.

10.12 Perspective

This study confirms the ferroelectric as a potential candidate for multiple biosensing because of its specific characteristic feature tenability to alter polarization concentration with the interaction of several types of stimuli such as heat, force, photon, electric field, and magnetic field. These ferroelectrics have applications in memory storage, transducers, and negative capacitance in multiple areas, apart from sensing. Thus, ferroelectric materials are used in the fabrication of e-skin, e-nose, and p-jets.

There are several types of materials; however, materials that can be used in sensing devices with lower power consumption for monitoring cells or tissues without increasing the cell temperature to avoid cell death are lacking. In addition, implanted sensing systems are needed to monitor the real-time analysis of cell activities, disease diagnosis, tissue conditions, etc., for compact, flexible, biocompatible, and biochemically stable materials with sensitivity towards multiple stimuli, which are capable of holding memory. This requirement will be fulfilled by ferroelectrics if researchers collaborate to fabricate FET-based devices on flexible materials in real-world conditions, apart from the experimental measurements of only laboratory conditions. Thus, this device can be exploited for real-time monitoring of agriculture and plants to avoid contamination of food products.

There are very few reports on ferroelectric biosensors in FET configurations for specific analyte sensing. Therefore, more studies are needed to improve biosensor utilization for health, environmental and agricultural monitoring.

10.13 Conclusions

Advancements in biosensor fabrication with ferroelectrics have been studied with explanations of biosensor components, working principle, interaction mechanism of ferroelectric with physical stimulus, mechanism of physical stimulus assisted changes in the ferroelectric in the physiological microenvironment, ferroelectric biosensor, and ferroelectric-based FETs applications for monitoring of health, environment, and agriculture. This reflects the potential utilization of ferroelectric polarization at low energy for sensing lower temperature entities such as biomolecules, cells, and tissues; however, very few reports are available. Therefore, nanostructured ferroelectric interaction behavior is not completely understood. However, some reports reflect the potential performances of ferroelectric biosensors with enhancement in their sensitivity, limit of detection, and lowering of response time along with high through output due to the generation of electron holes inside the ferroelectric with a very low amount of energy, even with energy associated with the biomolecules themselves. This capability of ferroelectrics makes them a futuristic material in the area of biosensors for monitoring health as well as environmental and agricultural monitoring.

References

1. Gebbers, R. and Adamchuk, V.I., Precision agriculture and food security. *Science*, 327, 5967, 828–831, 2010.
2. Ning, Y., Liu, W., Wang, G.-L., Balancing immunity and yield in crop plants. *Trends Plant Sci.*, 22, 12, 1069–1079, 2017.
3. Steinhorst, L. and Kudla, J., How plants perceive salt. *Nature*, 572, 318–320, 2019.
4. Savary, S., Willocquet, L., Pethybridge, S.J., Esker, P., McRoberts, N., Nelson, A., The global burden of pathogens and pests on major food crops. *Nat. Ecol. Evol.*, 3, 3, 430–439, 2019.
5. Li, X., Chen, D., Carrión, V.J., Revillini, D., Yin, S., Dong, Y., Zhang, T., Wang, X., Delgado-Baquerizo, M., Acidification suppresses the natural capacity of

soil microbiome to fight pathogenic Fusarium infections. *Nat. Commun.*, 14, 1, 5090, 2023.
6. Liu, K., Harrison, M.T., Yan, H., Liu, D.L., Meinke, H., Hoogenboom, G., Wang, B., Peng, B., Guan, K., *et al.*, Silver lining to a climate crisis in multiple prospects for alleviating crop waterlogging under future climates. *Nat. Commun.*, 14, 1, 765, 2023.
7. Boyer, J.S., Plant productivity and environment. *Science*, 218, 4571, 443–448, 1982.
8. Sekitani, T., Yokota, T., Zschieschang, U., Klauk, H., Bauer, S., Takeuchi, K., Takamiya, M., Sakurai, T., Someya, T., Organic nonvolatile memory transistors for flexible sensor arrays. *Science*, 326, 5959, 1516–1519, 2009.
9. Park, J., Kim, M., Lee, Y., Lee, H.S., Ko, H., Fingertip skin–inspired microstructured ferroelectric skins discriminate static/dynamic pressure and temperature stimuli. *Sci. Adv.*, 1, 9, e1500661, 2015.
10. Mannsfeld, S.CB., Tee, B.CK., Stoltenberg, R.M., Chen, C.V.HH., Barman, S., Muir, B.VO., Sokolov, A.N., Reese, C., Bao, Z., Highly sensitive flexible pressure sensors with microstructured rubber dielectric layers. *Nat. Mater.*, 10, 859–864, 2010.
11. Kim, D.-H., Lu, N., Ma, R., Kim, Y.-S., Kim, R.-H., Wang, S., Wu, J., Won, M., Tao, H., Islam, A., Yu, K.J., Kim, T., Chowdhury, R., Ying, M., Xu, L.Z., Li, M., Chung, H.-J., Keum, H., McCormick, M., Liu, P., Zhang, Y.-W., Omenetto, F.G., Huang, Y., Coleman, T., Rogers, J.A., Epidermal electronics. *Science*, 333, 6044, 838–843, 2011.
12. Pang, C., Lee, G.-Y., Kim, T.-I., Kim, S.M., Kim, H.N., Ahn, S.-H., Suh, K.-Y., A flexible and highly sensitive strain-gauge sensor using reversible interlocking of nanofibres. *Nat. Mater.*, 11, 9, 795–801, 2012.
13. Fan, F.-R., Lin, L., Zhu, G., Wu, W., Zhang, R., Wang, Z.L., Transparent triboelectric nanogenerators and self-powered pressure sensors based on micropatterned plastic films. *Nano Lett.*, 12, 6, 3109–3114, 2012.
14. Wang, C., Hwang, D., Yu, Z., Takei, K., Park, J., Chen, T., Ma, B., Javey, A., User-interactive electronic skin for instantaneous pressure visualization. *Nat. Mater.*, 12, 10, 899–904, 2013.
15. Wu, W., Wen, X., Wang, Z.L., Taxel-addressable matrix of vertical-nanowire piezotronic transistors for active and adaptive tactile imaging. *Science*, 340, 6135, 952–957, 2013.
16. Webb, R.C., Bonifas, A.P., Behnaz, A., Zhang, Y., Yu, K.J., Cheng, H., Shi, M., *et al.*, Ultrathin conformal devices for precise and continuous thermal characterization of human skin. *Nat. Mater.*, 12, 10, 938–944, 2013.
17. Gong, S., Schwalb, W., Wang, Y., Chen, Y., Tang, Y., Si, J., Shirinzadeh, B., Cheng, W., A wearable and highly sensitive pressure sensor with ultrathin gold nanowires. *Nat. Commun.*, 5, 1, 3132, 2014.
18. Zhang, A. and Lieber, C.M., Nano-bioelectronics. *Chem. Rev.*, 116, 1, 215–257, 2016.

19. Chen, S., Weitemier, A.Z., Zeng, X., He, L., Wang, X., Tao, Y., Huang, A.JY., *et al.*, Near-infrared deep brain stimulation via upconversion nanoparticle–mediated optogenetics. *Science*, 359, 6376, 679–684, 2018.
20. Yuan, X., Shi, J., Kang, Y., Dong, J., Pei, Z., Ji, X., Piezoelectricity, pyroelectricity, and ferroelectricity in biomaterials and biomedical applications. *Adv. Mater.*, 36, 3, 2308726, 2024.
21. Yang, F., Ng, H.K., Ju, X., Cai, W., Cao, J., Chi, D., Suwardi, A., Hu, G., Ni, Z., Wang, X.R., Lu, J., Wu, J., Emerging Opportunities for Ferroelectric Field-Effect Transistors: Integration of 2D Materials. *Adv. Funct. Mater.*, 34, 2310438, 2024.
22. He, Y., Xu, X., Xiao, S., Wu, J., Zhou, P., Chen, L., Liu, H., Research Progress and Application of Multimodal Flexible Sensors for Electronic Skin. *ACS Sens.*, 9, 5, 2275–2293, 2024.
23. Qu, J., Cui, G., Li, Z., Fang, S., Zhang, X., Liu, A., Han, M., Liu, H., Wang, X., Wang, X., Advanced Flexible Sensing Technologies for Soft Robots. *Adv. Funct. Mater.*, 34, 2401311, 2024.
24. Li, X.-H., Li, M.-Z., Li, J.-Y., Gao, Y.-Y., Liu, C.-R., Hao, G.-F., Wearable sensor supports *in-situ* and continuous monitoring of plant health in precision agriculture era. *Plant Biotechnol. J.*, 22, 6, 1516–1535, 2024.
25. Ali, M.M., Hashim, N., Aziz, S.A., Lasekan, O., Principles and recent advances in electronic nose for quality inspection of agricultural and food products. *Trends Food Sci. Technol.*, 99, 1–10, 2020.
26. Son, S.Y., Lee, G., Wang, H., Samson, S., Wei, Q., Zhu, Y., You, W., Integrating charge mobility, stability and stretchability within conjugated polymer films for stretchable multifunctional sensors. *Nat. Commun.*, 13, 1, 2739, 2022.
27. Valasek, J., Piezo-electric and allied phenomena in Rochelle salt. *Phys. Rev.*, 17, 4, 475, 1921.
28. Liu, Y., Hong-Ling, C., Zelisko, M., Wang, Y., Sun, J., Yan, F., Ma, F., *et al.*, Ferroelectric switching of elastin. *Proc. Natl. Acad. Sci.*, 111, 27, E2780–E2786, 2014.
29. Lang, S.B., Piezoelectricity, pyroelectricity and ferroelectricity in biomaterials: Speculation on their biological significance. *IEEE Trans. Dielectr. Electr. Insul.*, 7, 4, 466–473, 2000.
30. Ramesh, R. (Ed.), *Thin film ferroelectric materials and devices*, vol. 3, Springer Science & Business Media, New York, NY, USA, 2013.
31. Rega, R., Mugnano, M., Oleandro, E., Tkachenko, V., Giudice, D.D., Bagnato, G., Ferraro, P., Grilli, S., Gangemi, S., Detecting collagen molecules at picogram level through electric field-induced accumulation. *Sensors*, 20, 12, 3567, 2020.
32. Thévenot, D.R., Klara, T., Durst, R.A., Wilson, G.S., Electrochemical biosensors: recommended definitions and classification. *Biosens. Bioelectron.*, 16, 1-2, 121–131, 2001.
33. Chaubey, A. and Malhotra, B.D., Mediated biosensors. *Biosens. Bioelectron.*, 17, 6-7, 441–456, 2002.

34. Paddle, B.M., Biosensors for chemical and biological agents of defence interest. *Biosens. Bioelectron.*, 11, 11, 1079–1113, 1996.
35. Lowe, C.R., Biosensors. *Trends Biotechnol.*, 2, 3, 59–65, 1984.
36. Arnold, M.A. and Meyerhoff, M.E., Recent advances in the development and analytical applications of biosensing probes. *Crit. Rev. Anal. Chem.*, 20, 3, 149–196, 1988.
37. Pearson, J.E., Gill, A., Vadgama, P., Analytical aspects of biosensors. *Ann. Clin. Biochem.*, 37, 2, 119–145, 2000.
38. Sande, M.G., Rodrigues, J.L., Ferreira, D., Silva, C.J., Rodrigues, L.R., Novel biorecognition elements against pathogens in the design of state-of-the-art diagnostics. *Biosensors*, 11, 11, 418, 2021.
39. Rhoades, R. and Pflanzer, R.G., *Human Physiology*, Saunders College Publishing, New York, NY, USA, 1992.
40. So, H.-M., Park, D.-W., Jeon, E.-K., Kim, Y.-H., Kim, B.S., Lee, C.-K., Choi, S.Y., Kim, S.C., Chang, H., Lee, J.-O., Detection and titer estimation of Escherichia coli using aptamer-functionalized single-walled carbon-nanotube field-effect transistors. *Small (Weinheim an der Bergstrasse Germany)*, 4, 2, 197–201, 2008.
41. Douaki, A., Demelash Abera, B., Cantarella, G., Shkodra, B., Mushtaq, A., Ibba, P., Sarwar Inam, A.K.M., Petti, L., Lugli, P., Flexible screen printed aptasensor for rapid detection of furaneol: A comparison of CNTs and AgNPs effect on aptasensor performance. *Nanomaterials*, 10, 6, 1167, 2020.
42. Lee, G.J., Lim, J.E., Park, J.H., Choi, S.K., Hong, S., Park, H.K., Neurotransmitter detection by enzyme-immobilized CNT-FET. *Curr. Appl Phys.*, 9, 1, S25–S28, 2009.
43. Arkhypova, V.N., Dzyadevych, S.V., Soldatkin, A.P., Anna, V., Martelet, C., Jaffrezic-Renault, N., Development and optimisation of biosensors based on pH-sensitive field effect transistors and cholinesterases for sensitive detection of solanaceous glycoalkaloids. *Biosens. Bioelectron.*, 18, 8, 1047–1053, 2003.
44. Bhalla, P. and Singh, N., Generalized Drude scattering rate from the memory function formalism: an independent verification of the Sharapov-Carbotte result. *Eur. Phys. J. B*, 89, 1–8, 2016.
45. Li, L., Tan, J., Dun, M., Huang, X., Porous ZnFe2O4 nanorods with networked nanostructure for highly sensor response and fast response acetone gas sensor. *Sens. Actuators, B*, 248, 85–91, 2017.
46. Wilkins, E. and Atanasov, P., Glucose monitoring: state of the art and future possibilities. *Med. Eng. Phys.*, 18, 4, 273–288, 1996.
47. Shur, V.Y., Akhmatkhanov, A.R., Baturin, I.S., Nebogatikov, M.S., Dolbilov, M.A., Complex study of bulk screening processes in single crystals of lithium niobate and lithium tantalate family. *Phys. Solid State*, 52, 2147–21535, 2010.
48. Ievlev, A.V., Jesse, S., Morozovska, A.N., Strelcov, E., Eliseev, E.A., PershinKumar, Y.V.A., YaShur, V., Kalinin, S.V., Intermittency, quasiperiodicity

and chaos in probe-induced ferroelectric domain switching. *Nat. Phys.*, 10, 1, 59–66, 2014.
49. Tian, Y., Wei, L., Zhang, Q., Huang, H., Zhang, Y., Zhou, H., Ma, F., *et al.*, Water printing of ferroelectric polarization. *Nat. Commun.*, 9, 1, 3809, 2018.
50. Setvin, M., Reticcioli, M., Poelzleitner, F., Hulva, J., Schmid, M., Boatner, L.A., Franchini, C., Diebold, U., Polarity compensation mechanisms on the perovskite surface KTaO3 (001). *Science*, 359, 6375, 572–575, 2018.
51. Athenstaedt, H., Claussen, H., Schaper, D., Epidermis of human skin: pyroelectric and piezoelectric sensor layer. *Science*, 216, 4549, 1018–1020, 1982.
52. Li, J., Yang, M., Sun, X., Yang, X., Xue, J.J., Zhu, C., Liu, H., Xia, Y., Micropatterning of the ferroelectric phase in a poly (vinylidene difluoride) film by plasmonic heating with gold nanocages. *Angew. Chem. Int. Ed.*, 55, 44, 13828–13832, 2016.
53. Shamos, M.H. and Lavine, L.S., Piezoelectricity as a fundamental property of biological tissues. *Nature*, 213, 5073, 267–269, 1967.
54. Duarte, L.R., The stimulation of bone growth by ultrasound. *Arch. Orthop. traumatic Surg.*, 101, 153–159, 1983.
55. Safari, A. and Akdogan, E.K. (Eds.), *Piezoelectric and acoustic materials for transducer applications*, Springer Science & Business Media, New York, NY, USA, 2008.
56. Singh, S., Remeika, J.P., Potopowicz, J.R., Nonlinear optical properties of ferroelectric lead titanate. *Appl. Phys. Lett.*, 20, 3, 135–137, 1972.
57. Lenzo, P.V., Spencer, E.G., Ballman, A.A., Electro-Optic Coefficients of Ferroelectric Strontium Barium Niobate. *Appl. Phys. Lett.*, 11, 1, 23–24, 1967.
58. Scott, J.F., Applications of modern ferroelectrics. *Science*, 315, 5814, 954–959, 2007.
59. Kreisel, J., Alexe, M., Thomas, P.A., A photoferroelectric material is more than the sum of its parts. *Nat. Mater.*, 11, 4, 260–260, 2012.
60. Matsuo, H., Noguchi, Y., Miyayama, M., Gap-state engineering of visible-light-active ferroelectrics for photovoltaic applications. *Nat. Commun.*, 8, 1, 207, 2017.
61. Ciofani, G. and Menciassi, A. (Eds.), *Piezoelectric nanomaterials for biomedical applications*, Springer, Berlin, 2012.
62. Lang, S.B., Review of ferroelectric hydroxyapatite and its application to biomedicine. *Phase Transitions*, 89, 7-8, 678–694, 2016.
63. Ali, J., Najeeb, J., Ali, M.A., Aslam, M.F., Raza, A. J. J. B. B., Biosensors: their fundamentals, designs, types and most recent impactful applications: a review. *J. Biosens. Bioelectron.*, 8, 1, 1–9, 2017.
64. Bodhak, S., Bose, S., Bandyopadhyay, A., Role of surface charge and wettability on early stage mineralization and bone cell–materials interactions of polarized hydroxyapatite. *Acta Biomater.*, 5, 6, 2178–2188, 2009.
65. Orrego, S., Chen, Z., Krekora, U., Hou, D., Jeon, S.-Y., Pittman, M., Montoya, C., Chen, Y., Kang, S.H., Bioinspired materials with self-adaptable mechanical properties. *Adv. Mater.*, 32, 21, 1906970, 2020.

66. Tang, B., Zhang, B., Zhuang, J., Wang, Q., Dong, L., Cheng, K., Weng, W., Surface potential-governed cellular osteogenic differentiation on ferroelectric polyvinylidene fluoride trifluoroethylene films. *Acta Biomater.*, 74, 291–301, 2018.
67. Blázquez-Castro, A., García-Cabañes, A., Carrascosa, M., Biological applications of ferroelectric materials. *Appl. Phys. Rev.*, 5, 4, 041101, 2018.
68. Wang, Y., Barhoumi, A., Tong, R., Wang, W., Ji, T., Deng, X., Li, L., *et al.*, $BaTiO_3$-core Au-shell nanoparticles for photothermal therapy and bimodal imaging. *Acta Biomater.*, 72, 287–2945, 2018.
69. Malkinson, G., Mahou, P., Chaudan, E., Gacoin, T., Sonay, A.Y., Pantazis, P., Beaurepaire, E., Supatto, W., Fast *in vivo* imaging of SHG nanoprobes with multiphoton light-sheet microscopy. *ACS Photonics*, 7, 4, 1036–1049, 2020.
70. Heredia, A., Bdikin, I., Kopyl, S., Mishina, E., Semin, S., Sigov, A., German, K., Bystrov, V., Gracio, J., Kholkin, A.L., Temperature-driven phase transformation in self-assembled diphenylalanine peptide nanotubes. *J. Phys. D: Appl. Phys.*, 43, 46, 462001, 2010.
71. Tayi, A.S., Shveyd, A.K., Sue, A.C., Szarko, J.M., Rolczynski, B.S., Cao, D., Kennedy, T.J., Sarjeant, A.A., Stern, C.L., Paxton, W.F., Wu, W., Dey, S.K., Fahrenbach, A.C., Guest, J.R., Mohseni, H., Chen, L.X., Wang, K.L., Stoddart, J.F., Stupp, S.I., Room-temperature ferroelectricity in supramolecular networks of charge-transfer complexes. *Nature*, 488, 7412, 485–489, 2012.
72. Dong, G., Li, S., Yao, M., Zhou, Z., Zhang, Y.Q., Han, X., Luo, Z., Yao, J., Peng, B., Hu, Z., Huang, H., Jia, T., Li, J., Ren, W., Ye, Z.G., Ding, X., Sun, J., Nan, C.W., Chen, L.Q., Li, J., Liu, M., Super-elastic ferroelectric single-crystal membrane with continuous electric dipole rotation. *Science*, 366, 6464, 366–475, 2019.
73. Li, J.Y., The effective pyroelectric and thermal expansion coefficients of ferroelectric ceramics. *Mech. Mater.*, 36, 10, 949–958, 2004.
74. Tien, N.T., Seol, Y.G., Dao, L.H.A., Noh, H.Y., Lee, N.-E., Utilizing highly crystalline pyroelectric material as functional gate dielectric in organic thin-film transistors. *Adv. Mater.*, 21, 8, 910–915, 2009.
75. Sassi, U., Parret, R., Nanot, S., Bruna, M., Borini, S., De Fazio, D., Zhao, Z., Lidorikis, E., Koppens, F.H.L., Ferrari, A.C., Colli, A., Graphene-based mid-infrared room-temperature pyroelectric bolometers with ultrahigh temperature coefficient of resistance. *Nat. Commun.*, 8, 1, 14311, 2017.
76. Sun, J.-G., Yang, T.-N., Wang, C.-Y., Chen, L.-J., A flexible transparent one-structure tribo-piezo-pyroelectric hybrid energy generator based on bio-inspired silver nanowires network for biomechanical energy harvesting and physiological monitoring. *Nano Energy*, 48, 383–390, 2018.
77. Niu, H., Gao, S., Yue, W., Li, Y., Zhou, W., Liu, H., Highly morphology-controllable and highly sensitive capacitive tactile sensor based on epidermis-dermis-inspired interlocked asymmetric-nanocone arrays for detection of tiny pressure. *Small*, 16, 4, 1904774, 2020.

78. Nguyen, T.D., Deshmukh, N., Nagarah, J.M., Kramer, T., Purohit, P.K., Berry, M.J., McAlpine, M.C., Piezoelectric nanoribbons for monitoring cellular deformations. *Nat. Nanotechnol.*, 7, 9, 587–593, 2012.
79. Li, T., Feng, Z.-Q., Qu, M., Yan, K., Yuan, T., Gao, B., Wang, T., Dong, W., Zheng, J., Core/shell piezoelectric nanofibers with spatial self-orientated β-phase nanocrystals for real-time micropressure monitoring of cardiovascular walls. *ACS Nano*, 13, 9, 10062–10073, 2019.
80. Yihao, C,., Lu, S., Zhang, S., Li, Y., Qu, Z., Chen, Y., Lu, B., Wang, X., Feng, X., Skin-like biosensor system via electrochemical channels for noninvasive blood glucose monitoring. *Sci. Adv.*, 312, e1701629, 2017.
81. Boutet-Dieye, A. and Sarré-Charrier, M., Quand la peau fait corps avec l'écran. Rétrospective et perspective du tatouagetraditionnel au tatouageconnecté, in: *IMPEC: Interactions Multimodales par Ecran*, vol. 87, pp. 394–398, 2018.
82. Lee, H., Song, C., Hong, Y.S., Kim, M., Cho, H.R., Kang, T., Shin, K., Choi, S.H., Hyeon, T., Kim, D.-H., Wearable/disposable sweat-based glucose monitoring device with multistage transdermal drug delivery module. *Sci. Adv.*, 3, 3, e1601314, 2017.
83. Talary, M.S., Dewarrat, F., Huber, D., Caduff, A., *In vivo* life sign application of dielectric spectroscopy and non-invasive glucose monitoring. *J. Non-Cryst. Solids*, 353, 47-51, 4515–4517, 2007.
84. Huang, X., Liu, Y., Chen, K., Shin, W.-J., Lu, C.-J., Kong, G.-W., Patnaik, D., Lee, S.-H., Cortes, J.F., Rogers, J.A., Stretchable, wireless sensors and functional substrates for epidermal characterization of sweat. *Small*, 1015, 3083–3090, 2014.
85. Rose, D.P., Ratterman, M.E., Griffin, D.K., Hou, L., Kelley-Loughnane, N., Naik, R.R., Hagen, J.A., Papautsky, I., Heikenfeld, J.C., Adhesive RFID sensor patch for monitoring of sweat electrolytes. *IEEE Trans. Biomed. Eng.*, 62, 6, 1457–1465, 2014.
86. Gao, W., Emaminejad, S., Nyein, H.Y.Y., Challa, S., Chen, K., Peck, A., Fahad, H.M., *et al.*, Fully integrated wearable sensor arrays for multiplexed *in situ* perspiration analysis. *Nature*, 529, 7587, 509–514, 2016.
87. Selvarajan, S., Alluri, N.R., Chandrasekhar, A., Kim, S.-J., Unconventional active biosensor made of piezoelectric BaTiO3 nanoparticles for biomolecule detection. *Sens. Actuators, B*, 253, 1180–1187, 2017.
88. Pullano, S.A., Greco, M., Corigliano, D.M., Foti, D.P., Brunetti, A., Fiorillo, A.S., Cell-line characterization by infrared-induced pyroelectric effect. *Biosens. Bioelectron.*, 140, 111338, 2019.
89. Dagdeviren, C., Shi, Y., Joe, P., Ghaffari, R., Balooch, G., Usgaonkar, K., Gur, O., *et al.*, Conformal piezoelectric systems for clinical and experimental characterization of soft tissue biomechanics. *Nat. Mater.*, 14, 7, 728–736, 2015.

90. Andres, R.T. and Narayanaswamy, R., Fibre-optic pesticide biosensor based on covalently immobilized acetylcholinesterase and thymol blue. *Talanta*, 44, 8, 1335–1352, 1997.
91. Ivanov, A.N., Evtugyn, G.A., Gyurcsányi, R.E., Toth, K., Budnikov, H.C., Comparative investigation of electrochemical cholinesterase biosensors for pesticide determination. *Anal. Chim. Acta*, 404, 1, 55–65, 2000.
92. Yilmaz, E., Majidi, D., Ozgur, E., Denizli, A., Whole cell imprinting based Escherichia coli sensors: A study for SPR and QCM. *Sens. Actuators, B*, 209, 714–721, 2015.
93. Idil, N., Hedström, M., Denizli, A., Mattiasson, B., Whole cell based microcontact imprinted capacitive biosensor for the detection of Escherichia coli. *Biosens. Bioelectron.*, 87, 807–815, 2017.
94. de Cezaro, A., Ballen, S.C., Steffens, J., Steffens, C., Tyrosinase-based nanobiosensor for environmental monitoring of hormones in river water. *Environ. Res.*, 250, 118501, 2024.
95. Li, S., Simonian, A., Chin, B.A., Sensors for agriculture and the food industry. *Electrochem. Soc. Interface*, 19, 4, 41, 2010.
96. Wilson, A.D., Lester, D.G., Oberle, C.S., Development of conductive polymer analysis for the rapid detection and identification of phytopathogenic microbes. *Phytopathology*, 94, 5, 419–431, 2004.
97. Akmal, MH.M., Ahmad, F.B., Hisham, F., Hazmi, A.T., Biopolymer-based waste for biomaterials thin film in piezoelectric application, in: *Advanced Technology for the Conversion of Waste into Fuels and Chemicals*, pp. 355–381, Woodhead Publishing, United Kingdom (UK), 2021.
98. Ganeshkumar, R., Sopiha, K.V., Wu, P., Cheah, C.W., Zhao, R., Ferroelectric KNbO3 nanofibers: synthesis, characterization and their application as a humidity nanosensor. *Nanotechnology*, 2739, 395607, 2016.
99. Si, R., Xie, X., Li, T., Zheng, J., Cheng, C., Huang, S., Wang, C., TiO2/(K, Na) NbO3 nanocomposite for boosting humidity-sensing performances. *ACS Sens.*, 5, 5, 1345–1353, 2020.
100. Si, R., Xie, X., Li, T., Zheng, J., Cheng, C., Wang, C., TiO2/NaNbO3 heterojunction for boosted humidity sensing ability. *Sens. Actuators, B*, 309, 127803, 2020.
101. Wang, J., Zhao, B., Wang, C., TiO2/KNbO3 nanocomposite for enhanced humidity sensing performance. *Sens. Actuators, A*, 349, 114057, 2023.
102. Shkodra, B., Petrelli, M., Costa Angeli, M.A., Garoli, D., Nakatsuka, N., Lugli, P., Petti, L., Electrolyte-gated carbon nanotube field-effect transistor-based biosensors: Principles and applications. *Appl. Phys. Rev.*, 8, 4, 041325, 2021.
103. Wang, D., Noël, V., Piro, B., Electrolytic gated organic field-effect transistors for application in biosensors—A Review. *Electronics*, 5, 1, 9, 2016.
104. Golio, M. and Golio, J., *RF and Microwave Passive and Active Technologies*, CRC Press, Boca Raton, FL, USA, 2018.

105. Lee, M.L., Fitzgerald, E.A., Bulsara, M.T., Currie, M.T., Lochtefeld, A., Strained Si, SiGe, and Ge channels for high-mobility metal-oxide-semiconductor field-effect transistors. *J. Appl. Phys.*, 97, 1, 011101, 2005.
106. Lüssem, B., Riede, M., Leo, K., Doping of organic semiconductors. *Phys. Status Solidi (A)*, 210, 1, 9–43, 2013.
107. Zhao, Y., Guo, Y., Liu, Y., 25th anniversary article: recent advances in n-type and ambipolar organic field-effect transistors. *Adv. Mater.*, 25, 38, 5372–5391, 2013.
108. Zhang, S., Review of modern field effect transistor technologies for scaling. *J. Phys. Conf. Ser.*, 1617, 1, 012054, IOP Publishing, 2020.
109. Rivnay, J., Inal, S., Salleo, A., Owens, R.M., Berggren, M., Malliaras, G.G., Organic electrochemical transistors. *Nat. Rev. Mater.*, 3, 2, 1–14, 2018.
110. Bergveld, P., Thirty years of ISFETOLOGY: What happened in the past 30 years and what may happen in the next 30 years. *Sens. Actuators, B*, 88, 1, 1–20, 2003.
111. Polk, B.J., ChemFET arrays for chemical sensing microsystems, in: *Proceedings of the SENSORS, 2002 IEEE*, Orlando, FL, USA, vol. 1, pp. 732–735.
112. Enderby, J.E. and Neilson, G.W., The structure of electrolyte solutions. *Rep. Prog. Phys.*, 44, 6, 593, 1981.
113. Lee, K., Jang, S., Kim, K.L., Koo, M., Park, C., Lee, S., Lee, J., Wang, G., Park, C., Artificially intelligent tactile ferroelectric skin. *Adv. Sci.*, 7, 22, 2001662, 2020.
114. Le Gall, J., Vasilijević, S., Battaglini, N., Mattana, G., Noël, V., Brayner, R., Piro, B., Algae-functionalized hydrogel-gated organic field-effect transistor. Application to the detection of herbicides. *Electrochim. Acta*, 372, 137881, 2021.
115. Coppedè, N., Janni, M., Bettelli, M., Maida, C.L., Gentile, F., Villani, M., Ruotolo, R., *et al.*, An *in vivo* biosensing, biomimetic electrochemical transistor with applications in plant science and precision farming. *Sci. Rep.*, 7, 1, 16195, 2017.
116. Lim, J., Choi, S., Kim, H.S., Electrospinning behavior and piezoelectric property of PVDF/PEDOT: PSS composite web by PEDOT: PSS addition. *Polym. Adv. Technol.*, 34, 6, 1959–1966, 2023.
117. Kim, T.H., Lee, J., Hong, S., Highly selective environmental nanosensors based on anomalous response of carbon nanotube conductance to mercury ions. *J. Phys. Chem. C*, 113, 45, 19393–19396, 2009.
118. Takemoto, A., Teppei, A., Uemura, T., Noda, Y., Yoshimoto, S., Izumi, S., Tsuruta, S., Sekitani, T., Printable transparent microelectrodes toward mechanically and visually imperceptible electronics. *Adv. Intell. Syst.*, 2, 11, 2000093, 2020.
119. Berto, M., Vecchi, E., Baiamonte, L., Condò, C., Sensi, M., Di Lauro, M., De Stradis, M.S.A., Biscarini, F., Minafra, A., Bortolotti, C.A., Label free detection of plant viruses with organic transistor biosensors. *Sens. Actuators, B*, 281, 150–1565, 2019.

120. García, J.A., Glasa, M., Cambra, M., Candresse, T., P lum pox virus and sharka: a model potyvirus and a major disease. *Mol. Plant Pathol*, 15, 3, 226–241, 2014.
121. Ryosuke, I.,., Ono, A., Ishizuka, H., Terao, K., Takao, H., Kobayashi, T., Kataoka, I., Shimokawa, F., Biological information (pH/EC) sensor device for quantitatively monitoring plant health conditions, in: *2017 IEEE SENSORS*, IEEE, pp. 1–3, 2017.

11

Ferroelectric Application in Recent Nanoscale Device with ITRS Roadmap

Shiromani Balmukund Rahi[1] and Young Suh Song[2]*

[1]*University School of Information and Communication Technology, Gautam Buddha University, Uttar Pradesh, India*
[2]*Department of Computer Science, Korea Military Academy, Seoul, Republic of Korea*

Abstract
Recently, ferroelectric materials have garnered considerable attention in the field of material and electronic engineering owing to their potential use in nanoscale devices, spurred by the ongoing scaling of semiconductor technology as outlined in the International Technology Roadmap for Semiconductors (ITRS) and other engineering organizations. This chapter introduces ferroelectric applications, elucidates the characteristics of ferroelectric materials, and discusses the fundamental scaling principles outlined in the ITRS Roadmap. This chapter then explores the realm of nanoscale devices, examining various configurations, such as ultra-thin body MOSFETs, gate-all-around MOSFETs, and advanced engineering techniques, such as local high doping. Furthermore, this chapter investigates the incorporation of ferroelectric materials into various modern nanoscale devices, and emphasizes their possible advantages and potential applications. Additionally, this chapter examines the placement of ferroelectric technologies within the ITRS Roadmap framework and provides insights into the future prospects of ferroelectric applications in the semiconductor industry. Overall, it seems that with the help of the guideline provided by the ITRS roadmap and advancement of ferroelectric materials, semiconductor devices (including memory devices and memory devices) would be able to further able to achieve more improvement in terms of low power consumption and fast switching operation.

***Keywords*:** Ferroelectric, material engineering, semiconductor, nanodevice

**Corresponding author*: songyeongseo84@gmail.com

Balwinder Raj, Shiromani Balmukund Rahi and Nandakishor Yadav (eds.) FeFET Devices, Trends, Technology and Applications, (275–296) © 2025 Scrivener Publishing LLC

11.1 Introduction to Ferroelectric Application

Ferroelectric materials, distinguished by their inherent electric polarization, which can be reversed by an external electric field, have emerged as promising contenders for various technological endeavors [1, 2]. Recently, there has been burgeoning interest in leveraging the distinctive attributes of ferroelectric materials for application in nanoscale devices, fueled by the relentless evolution and downsizing of semiconductor technology [3].

The utility of ferroelectric materials extends across a broad spectrum of domains, including memory devices, sensors, actuators, and energy-harvesting systems [4–6]. Their capacity to preserve polarization states relies on an external power source, rendering them particularly appealing for non-volatile memory applications [7]. Moreover, ferroelectric materials exhibit piezoelectric characteristics, facilitating their deployment in sensors and actuators for the conversion of mechanical energy and sensing tasks [8, 9].

Within the realm of nanoscale devices, ferroelectric materials offer unparalleled advantages such as rapid switching speeds, minimal power consumption, and compatibility with complementary metal-oxide-semiconductor (CMOS) technology [10–15]. These attributes render them highly conducive for incorporation into sophisticated transistor architectures, including ultra-thin-body MOSFETs and gate-all-around MOSFETs, where precise manipulation of electrical properties is imperative for optimal device performance [16]. It seems like the combination of novel emerging materials with conventional CMOS technology could enable various pioneering technologies [17–22].

Furthermore, the amalgamation of ferroelectric materials into nanoscale devices has led to novel prospects for advanced device functionalities such as non-volatile logic and neuromorphic computing. By capitalizing on the polarization switching behavior inherent in ferroelectric materials, the realization of energy-efficient computing paradigms akin to synaptic plasticity observed in biological systems is attainable [23–28].

11.2 Ferroelectric Materials and Properties

Ferroelectric materials represent a class of substances that are notable for their distinct electrical characteristics, primarily marked by spontaneous electric polarization, which can be altered between two or more stable states by an external electric field. This reversible polarization stems from structural irregularity within the crystal lattice of ferroelectric materials, leading to the creation of electric dipoles [29–37]. When exposed to an external electric field, these dipoles align in the direction of the field, resulting in overall polarization of the material.

The defining feature of ferroelectricity lies in its hysteresis behavior, wherein the relationship between the polarization and the applied electric field forms a loop-like curve. This hysteresis loop signifies the capacity of ferroelectric materials to maintain their polarization state even after the removal of the external electric field, rendering them well suited for non-volatile memory applications [38–42].

Another notable property of ferroelectric materials is their piezoelectricity, which refers to their ability to produce an electric charge when subjected to mechanical stress, or vice versa. This characteristic is useful in sensors, actuators, and energy-harvesting systems, where mechanical energy can be converted into electrical energy and vice versa.

The ferroelectric phase transition temperature, commonly known as the Curie temperature (Tc), is a crucial parameter that influences the behavior of ferroelectric materials. Above the Curie temperature, these materials lose their spontaneous polarization and transition into a paraelectric phase, where the dipoles become randomly oriented.

Numerous materials exhibit ferroelectric properties, including perovskite oxides such as lead zirconate titanate (PZT), barium titanate (BaTiO3), and lithium niobate (LiNbO3), as well as organic ferroelectrics and polymers [43–49]. The selection of a ferroelectric material depends on factors such as the desired operating temperature, switching speed, and compatibility with other components in the device.

Specifically, perovskite oxides, including lead zirconate titanate (PZT), barium titanate (BaTiO3), and lithium niobate (LiNbO3), are a class of materials with crystalline structures characterized by the perovskite arrangement of atoms (Figure 11.1) and (Table 11.1). These materials exhibit ferroelectric properties owing to their specific crystal structure, which allows for the spontaneous polarization mentioned earlier.

Lead zirconate titanate (PZT) is one of the most widely studied ferroelectric materials. It possesses excellent piezoelectric properties, making it suitable for applications, such as actuators, sensors, and ultrasound transducers. PZT-based materials are also commonly used in non-volatile memory devices because of their highly remanent polarization and retention characteristics [50–55].

Barium titanate (BaTiO3) is another prominent ferroelectric material known for its high Curie temperature and large spontaneous polarization. It has applications in capacitors, sensors, and electromechanical devices. BaTiO3-based ceramics are utilized in multilayer ceramic capacitors

Figure 11.1 Three representative materials with ferroelectric properties.

Table 11.1 Electrical and thermal properties of novel materials.

Novel material	Pyroelectric coefficient	Curie temperature [K]
BaTiO3	1.9	408
PbTiO3	0.4	1,483
LiNbO3	2.7	765
LiTaO3	2.3	891
PZT	4.7	638

(MLCCs) for various electronic applications owing to their high dielectric constants and stabilities [56–61].

Lithium niobate (LiNbO3) is valued for its strong piezoelectric and electro-optic properties, which make it suitable for applications in telecommunications, optical modulators, and surface acoustic wave devices. LiNbO3-based devices are widely used in the telecommunication industry for signal processing and frequency conversion.

In addition to perovskite oxides, organic ferroelectrics and specific polymers also exhibit ferroelectric properties. Organic ferroelectrics, such as poly(vinylidene fluoride) (PVDF) and its copolymers, offer flexibility and compatibility with flexible electronic devices. They are utilized in sensors, energy harvesters, and memory devices [62–68]. Specific polymers, such as polyurethane and polyvinyl alcohol, have been engineered to exhibit ferroelectric behavior and have applications in sensors, actuators, and biomedical devices because of their biocompatibility and tunable properties.

11.3 Basic Scaling and ITRS Roadmap

As semiconductor technology progresses, scaling remains pivotal for propelling innovation and advancement. Scaling is the process of reducing the size of semiconductor components, such as transistors, to enhance performance, expand functionality, and lower costs [69–74]. The International Technology Roadmap for Semiconductors (ITRS, recently renamed as IRDS) serves as a guiding framework for the semiconductor sector by delineating key trends, hurdles, and objectives concerning the scaling of semiconductor technology. As shown in Table 11.2, the IRDS usually suggests a parameter roadmap that includes various parameter goals.

Table 11.2 Parameter roadmap suggested by IRDS.

	2014	2017	2020	2023	2026
Metal Half Pitch (nm)	27	16.9	11.9	8.4	6.0
Equivalent Oxide Thickness (EOT)	0.73–0.8	0.56–0.72	0.54–0.62	0.53	0.45
Physical Gate Length (nm)	18	14	10.6	8.1	5.9
Ring Oscillator Delay (RO delay) (ps)	4.11–7.12	3.30–7.18	2.60–2.66	2.13	1.80
Power Supply Voltage (V)	0.82	0.75	0.68	0.62	0.57

The foundational principles of scaling involve the diminution of semiconductor device dimensions, including gate length, gate oxide thickness, and interconnect pitch, to bolster device performance and density [23, 24]. This downsizing facilitates heightened transistor density on a chip, consequently augmenting the computational power and energy efficiency [25–29].

The ITRS Roadmap provides an extensive overview of the challenges and prospects inherent to scaling semiconductor technology. It delineates pivotal areas for research and development, such as material innovation, device architecture, and manufacturing methodologies, to foster ongoing scaling and progression [75–82].

The critical elements of the ITRS Roadmap encompass technology nodes, which signify specific milestones in semiconductor scaling, alongside forecasts for future advancements predicated on industry trends and technological capabilities. These prognostications aid in steering research and investment endeavors within the semiconductor domain, ensuring alignment with the objectives outlined on the roadmap.

Essentially, fundamental scaling principles and the ITRS Roadmap are integral drivers of innovation and advancements in semiconductor technology. By adhering to these principles and addressing the identified challenges and opportunities, the semiconductor industry can persist in pushing technological boundaries and delivering increasingly sophisticated and capable electronic devices [83–88].

11.4 Nanoscale Devices: Ultra-Thin-Body MOSFET, Gate-All-Around MOSFET, Gate, Channel, Source/Drain Engineering, Local High Doping for Better Subthreshold Swing

Nanoscale devices have marked a notable advancement in semiconductor technology, delivering superior performance and functionality compared to their larger counterparts. Within the realm of nanoscale devices, several key breakthroughs have emerged, such as ultra-thin-body MOSFETs, gate-all-around MOSFETs, and sophisticated engineering methods targeting the gate, channel, and source/drain regions [89–95].

Ultra-thin-body MOSFETs (also known as UTB MOSFET) feature a markedly reduced body thickness, facilitating enhanced control over the device's electrostatics. This thinning of the body enables superior gate control and mitigates short-channel effects (SCE), thereby bolstering the device performance and enabling greater integration densities on a chip.

Gate-all-around (GAA) MOSFETs represent a revolutionary shift in transistor architecture, with the gate electrode enveloping the channel from all sides [96–102]. This configuration affords superior electrostatic control and alleviates short-channel effects, resulting in an enhanced transistor performance and diminished leakage currents.

Additionally, advancements in gate, channel, and source/drain engineering have played a pivotal role in improving device performance. Engineers have devised innovative methods to optimize the properties of gate dielectrics, channel materials, and source/drain regions, thereby achieving lower power consumption, increased speed, and enhanced reliability.

Moreover, the adoption of local high-doping techniques has been instrumental in improving the subthreshold swing of nanoscale devices [103–108]. By selectively doping specific regions of the transistor with higher concentrations of dopants, engineers can reduce the energy required to switch the device on and off, thereby enhancing the energy efficiency and overall performance.

11.5 Nanoscale Devices with Ferroelectric Applications

Nanoscale devices incorporating ferroelectric applications represent an exciting frontier in semiconductor technology, capitalizing on the unique attributes of ferroelectric materials to introduce innovative functionalities and enhance overall performance. These materials, distinguished by their innate electric polarization, which can be altered by an external electric field, confer distinct advantages when integrated into nanoscale devices.

A notable application of ferroelectric materials in nanoscale devices pertains to nonvolatile memory technologies. Ferroelectric random-access memory (FeRAM) exploits the polarization switching phenomenon of ferroelectric materials to store data without requiring a continuous power supply, offering rapid read/write speeds and robust endurance [109–114]. FeRAM holds considerable potential for applications demanding swift, dependable, and energy-efficient memory solutions, such as mobile devices and Internet of Things (IoT) devices.

Another compelling area of interest is ferroelectric field-effect transistors (FeFETs), in which ferroelectric materials are integrated into the gate dielectric of the transistor. By leveraging the polarization-dependent electrical characteristics of ferroelectric materials, FeFETs enable the incorporation of nonvolatile memory capabilities within transistor devices, thereby unlocking opportunities for unified logic and memory functionalities in a

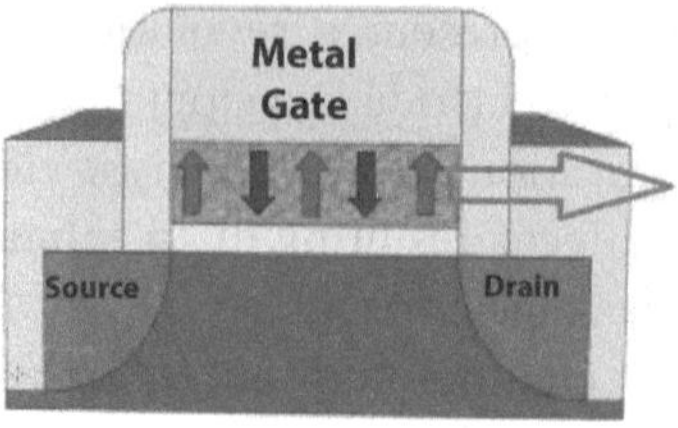

Figure 11.2 Illustration describing the principle and advantages of FeFET.

single device. FeFETs exhibit promise for applications in energy-efficient computing, neuromorphic computing, and integrated circuits (Figure 11.2).

Moreover, ferroelectric materials are useful in nanoelectromechanical systems (NEMS) and sensors, leveraging their piezoelectric properties for mechanical sensing and actuation. By integrating ferroelectric materials into nanoscale sensors, engineers can develop highly sensitive and energy-efficient devices tailored for diverse sensing applications, including pressure, strain, and biomedical sensing [115–119].

Furthermore, the integration of ferroelectric materials into nanoscale devices facilitates the exploration of emerging technologies, such as spintronics and quantum computing. Ferroelectric materials present intriguing opportunities to manipulate spin states and governing quantum states, thereby paving the way for groundbreaking advancements in next-generation computing and information processing.

11.6 Advantages and Potential Applications of Ferroelectric Materials

Ferroelectric materials have numerous advantages and demonstrate a diverse array of potential applications in multiple technological domains. These exceptional materials, defined by their inherent electric polarization that can be reversed by an external electric field, possess several key characteristics that render them highly attractive for various purposes (Figure 11.3).

As described in the previous section, one of the most notable benefit of ferroelectric materials is their 'non-volatile' nature, allowing them to preserve their polarization state even in the absence of an external power supply. Specifically, the term 'non-volatile' means that the information stored

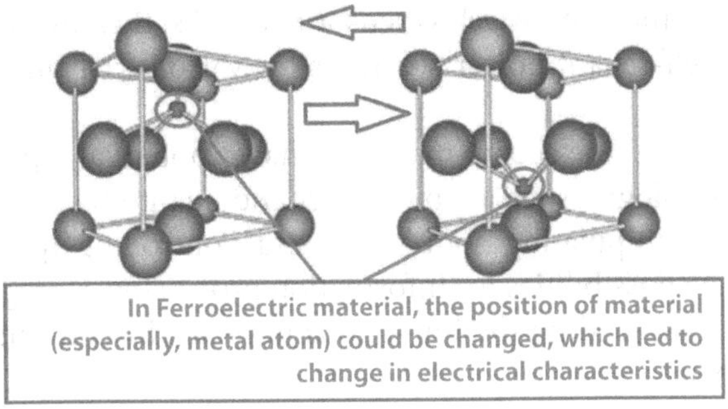

Figure 11.3 Basic operation principle of ferroelectric materials.

in the memory device can hold the information for a long time. This feature positions ferroelectric materials as prime candidates for non-volatile memory applications such as ferroelectric random-access memory (FeRAM). FeRAM offers swift read/write speeds, robust endurance, and minimal power consumption, making it well suited for applications requiring dependable and energy-efficient data storage solutions.

Moreover, ferroelectric materials demonstrate piezoelectric properties, enabling them to generate an electric charge when subjected to mechanical stress or conversely, undergo mechanical deformation when exposed to an electric field. This attribute is useful in sensors, actuators, and energy-harvesting systems. Ferroelectric-based sensors exhibit remarkable sensitivity and precision for detecting various physical parameters, including pressure, strain, temperature, and acceleration. In addition, ferroelectric actuators efficiently convert electrical energy into mechanical motion, facilitating precise control and manipulation in micro- and nanoscale systems.

Furthermore, the distinctive polarization-switching behavior of ferroelectric materials offers avenues for innovative device functionalities and applications. Ferroelectric field-effect transistors (FeFETs) capitalize on this polarization switching to introduce nonvolatile memory capabilities within transistor devices, thereby enabling the integration of logic and memory functionalities within a single device. FeFETs are promising for applications in low-power computing, neuromorphic computing, and integrated circuits.

Ferroelectric materials also play a crucial role in energy-harvesting systems, where they transform mechanical energy from vibrations or thermal differentials into electrical energy. This capability renders ferroelectric materials suitable for powering autonomous sensor nodes, wearable electronics, and wireless sensor networks, thereby providing sustainable and self-reliant power solutions for various IoT and monitoring applications.

Additionally, ferroelectric materials have potential applications in emerging technologies such as spintronics and quantum computing. Their adeptness in manipulating spin states and controlling quantum states introduces new possibilities for information processing, storage, and communication, transcending the limitations imposed by conventional semiconductor devices.

11.7 Positioning of Ferroelectric Technologies in the ITRS Roadmap

Ferroelectric technologies are positioned to significantly influence the trajectory of semiconductor technology, as reflected by their strategic placement within the International Technology Roadmap for Semiconductors (ITRS). This roadmap serves as a comprehensive framework that delineates the course of semiconductor advancements, and the integration of ferroelectric technologies underscores their pivotal role in driving innovation and progress within the industry.

Embedded within the ITRS Roadmap, ferroelectric technologies are strategically positioned to address critical challenges and opportunities identified by industry experts. These technologies are expected to make substantial contributions to meeting the evolving demands across diverse applications, including memory devices, logic circuits, and sensors. By harnessing the distinctive attributes of ferroelectric materials, such as their nonvolatile nature and rapid switching speeds, semiconductor manufacturers can explore novel pathways to enhance device performance, energy efficiency, and functionality.

Furthermore, the incorporation of ferroelectric materials into field-effect transistors (FeFETs) facilitates the realization of non-volatile memory functionalities within transistor devices. It seems that incorporating ferroelectric techniques into FeFETs could achieve the performance goal suggested by IRDS. This paves the way for the development of integrated logical and memory solutions. FeFETs are expected to play a pivotal role in advancing low-power computing, neuromorphic computing, and integrated circuits, aligning with the industry's pursuit of energy-efficient and high-performance computing architectures.

11.8 Possible Challenge in Future Ferroelectric Applications

As ferroelectric materials have garnered increasing attention across various technological sectors, it is imperative to recognize and confront the potential challenges that may arise in their future applications. Despite their numerous advantages, ferroelectric materials encounter several obstacles that hinder their widespread adoption and implementation. Therefore, comprehending and mitigating these challenges is vital to unlocking the potential of ferroelectric technologies. In this context, we delved into several potential challenges, as presented in Table 11.3.

Table 11.3 Possible future challenges in ferroelectric application.

Type of challenge	Description
Stability and Endurance	The primary hurdle associated with ferroelectric materials is their stability and endurance over prolonged periods. Due to repeated polarization switching, these materials may undergo fatigue or degradation, resulting in a decline in the device performance. It is imperative to develop strategies to improve the stability and endurance of ferroelectric materials. This requires optimizing the material composition and device architecture to ensure long-term reliability.
Scalability	Another significant obstacle pertains to the scalability of ferroelectric devices, particularly their adaptation to smaller dimensions in nanoscale technologies. As the device size decreases, maintaining the favorable properties of ferroelectric materials becomes increasingly challenging. Factors such as domain-wall pinning and size-dependent polarization behavior may emerge, impacting device reliability and performance. Overcoming scalability hurdles necessitates the exploration of innovative fabrication techniques and material-engineering approaches tailored to nanoscale dimensions.
Integration Compatibility	The integration of ferroelectric materials into existing semiconductor processes and device architectures is another challenge. Seamless integration into mainstream semiconductor manufacturing processes requires compatibility with established fabrication techniques and materials. The development of compatible interfaces and interface engineering strategies is crucial for achieving robust device performance and reliability.

(*Continued*)

Table 11.3 Possible future challenges in ferroelectric application. (*Continued*)

Type of challenge	Description
Material Selection and Synthesis	The selection of suitable ferroelectric materials is pivotal for determining device performance and functionality. However, identifying materials with the requisite properties for specific applications poses a significant challenge. Additionally, the synthesis of high-quality ferroelectric thin films or nanostructures with precise control over the composition, crystal structure, and defect density is a non-trivial task. Advancements in material synthesis techniques and the exploration of novel ferroelectric materials are essential for overcoming these challenges.
Power Consumption	Despite offering low-power operation in certain scenarios, minimizing the power consumption remains a formidable challenge, particularly in high-density memory arrays and logic circuits. Optimizing the power efficiency of ferroelectric devices requires addressing issues such as leakage currents, coercive voltage requirements, and energy dissipation during polarization switching.
Reliability and Variability	Ensuring reliability and reducing variability in ferroelectric devices pose significant obstacles. Variations in material properties, fabrication processes, and operating conditions can result in device-to-device variability and reliability concerns. Enhancing device reliability and reducing variability necessitate the development of robust device designs, the incorporation of error correction techniques, and the implementation of effective reliability testing methodologies.

11.9 Conclusion

This chapter provides an in-depth explanation of the concept and utilization of ferroelectric applications. Despite the diverse challenges of employing ferroelectric materials, each material exhibits unique strengths and weaknesses. Through the integration of the various electrical properties of ferroelectric materials with conventional CMOS materials, scientists and engineers anticipate the design and development of next-generation semiconductor devices with markedly enhanced electrical and device characteristics. The significance of ferroelectric and novel materials in the design and development of next-generation semiconductors and electronic devices cannot be overstated.

References

1. Nautiyal, A., Dabra, N., Hundal, J.S., Pathak, N.P., Nath, R., Investigation on structural and ferroelectric properties of spray deposited Cs1-xKxNO3: PVA : composite films. *IEEE Trans. Dielectr. Electr. Insul.*, 22, 1, 251–256, Feb. 2015, doi: 10.1109/TDEI.2014.004099.
2. Kämpfe, T., *et al.*, Heavy Ion Irradiation Effects on Structural and Ferroelectric Properties of HfO2 Films. *2020 Joint Conference of the IEEE International Frequency Control Symposium and International Symposium on Applications of Ferroelectrics (IFCS-ISAF)*, Keystone, CO, USA, pp. 1–3, 2020, doi: 10.1109/IFCS-ISAF41089.2020.9234942.
3. Esseni, D., Driussi, F., Lizzit, D., Massarotto, M., Segatto, M., Modelling and Simulations of Ferroelectric Materials and Ferroelectric-Based Nanoelectronic Devices : (Invited Paper). *2023 International Conference on Simulation of Semiconductor Processes and Devices (SISPAD)*, Kobe, Japan, pp. 9–12, 2023, doi: 10.23919/SISPAD57422.2023.10319574.
4. Takada, Y., *et al.*, Comparative Study of Hydrogen- and Deuterium-Induced Degradation of Ferroelectric (Pb,La)(Zr,Ti)O3 Capacitors Using Time-of-Flight Secondary Ion Measurement. *IEEE Trans. Ultrason. Ferroelectr. Freq. Control*, 63, 10, 1668–1673, Oct. 2016, doi: 10.1109/TUFFC.2016.2593585.
5. Wang, D., *et al.*, Ferroelectric Switching in Sub-20 nm Aluminum Scandium Nitride Thin Films. *IEEE Electron Device Lett.*, 41, 12, 1774–1777, Dec. 2020, doi: 10.1109/LED.2020.3034576.
6. Ye, Z.-G., Tan, X., Bokov, A.A., Introduction to the IEEE International Symposium on Applications of Ferroelectrics and International Symposium on Piezoresponse Force Microscopy and Nanoscale Phenomena in Polar Materials. *IEEE Trans. Ultrason. Ferroelectr. Freq. Control*, 59, 9, 1853–1854, September 2012, doi: 10.1109/TUFFC.2012.2395.
7. Mantri, S. and Daniels, J.E., Ferroelectric Domain Continuity Over Grain Boundaries for Tetragonal, Orthorhombic, and Rhombohedral Crystal Symmetries. *IEEE Trans. Ultrason. Ferroelectr. Freq. Control*, 65, 9, 1517–1524, Sept. 2018. doi: 10.1109/TUFFC.2018.2827406.
8. Yang, E., Kim, K.R., Chang, J., Theoretical Evaluation of Two-Dimensional Ferroelectric Material CuInP2S6 for Ferroelectric Tunnel Junction Device. *IEEE Electron Device Lett.*, 42, 10, 1472–1475, Oct. 2021, doi: 10.1109/LED.2021.3103518.
9. Kim, I.-J. and Lee, J.-S., Ferroelectric Transistor With Grooved Structure for Reliable Multi-Level Characteristics. *IEEE Electron Device Lett.*, 45, 3, 352–355, March 2024, doi: 10.1109/LED.2023.3347630.
10. Mikolajick, T., Schroeder, U., Slesazeck, S., Unleashing the Potential of Integrated Ferroelectric Devices with Hafnium Oxide. *2022 Device Research Conference (DRC)*, Columbus, OH, USA, pp. 1–2, 2022, doi: 10.1109/DRC55272.2022.9855802.

11. Manente, D., Padovan, F., Seebacher, D., Bassi, M., Bevilacqua, A., A 28-GHz Stacked Power Amplifier with 20.7-dBm Output P1dB in 28-nm Bulk CMOS. *IEEE Solid-State Circuits Lett.*, 3, 170–173, 2020, doi: 10.1109/LSSC.2020.3009973.
12. Jung, S.-G., Jang, D., Min, S.-J., Park, E., Yu, H.-Y., Performance Analysis on Complementary FET (CFET) Relative to Standard CMOS With Nanosheet FET. *IEEE J. Electron Devices Soc.*, 10, 78–82, 2022, doi: 10.1109/JEDS.2021.3136605.
13. Wang, Y., Chiu, T.-Y., Chien, C.-C., Tsai, W.-H., Wang, H., An E-Band High-Performance Variable Gain Low Noise Amplifier for Wireless Communications in 90-nm CMOS Process. *IEEE Microwave Wireless Compon. Lett.*, 32, 9, 1095–1098, Sept. 2022, doi: 10.1109/LMWC.2022.3161295.
14. Rathore, S., *et al.*, Self-Heating Aware Threshold Voltage Modulation Conforming to Process and Ambient Temperature Variation for Reliable Nanosheet FET. *2023 IEEE International Reliability Physics Symposium (IRPS)*, Monterey, CA, USA, pp. 1–5, 2023, doi: 10.1109/IRPS48203.2023.10117918.
15. Tayal, S., *et al.*, Incorporating Bottom-Up Approach Into Device/Circuit Co-Design for SRAM-Based Cache Memory Applications. *IEEE Trans. Electron Devices*, 69, 11, 6127–6132, Nov. 2022, doi: 10.1109/TED.2022.3210070.
16. Jaisawal, R.K., *et al.*, Self-Heating and Interface Traps Assisted Early Aging Revelation and Reliability Analysis of Negative Capacitance FinFET. *2023 7th IEEE Electron Devices Technology & Manufacturing Conference (EDTM)*, Seoul, Korea, Republic of, pp. 1–3, 2023, doi: 10.1109/EDTM55494.2023.10103127.
17. Song, Y.S., Kim, S., Kim, J.H., Kim, G., Lee, J.-H., Choi, W.Y., Enhancement of Thermal Characteristics and On-Current in GAA MOSFET by Utilizing Al2O3-Based Dual-κ Spacer Structure. *IEEE Trans. Electron Devices*, 70, 1, 343–348, Jan. 2023, doi: 10.1109/TED.2022.3223321.
18. Park, M.-K., *et al.*, Cointegration of the TFT-Type AND Flash Synaptic Array and CMOS Circuits for a Hardware-Based Neural Network. *IEEE Trans. Electron Devices*, 70, 1, 93–98, Jan. 2023, doi: 10.1109/TED.2022.3220726.
19. Song, Y.S., Tayal, S., Rahi, S.B., Kim, J.H., Upadhyay, A.K., Park, B.-G., Thermal-Aware IC Chip Design by Combining High Thermal Conductivity Materials and GAA MOSFET. *2022 5th International Conference on Circuits, Systems and Simulation (ICCSS)*, Nanjing, China, pp. 135–140, 2022, doi: 10.1109/ICCSS55260.2022.9802341.
20. Che, C., Lei, K.-M., Martins, R.P., Mak, P.-I., A 0.4-V 8400-μm2 Voltage Reference in 65-nm CMOS Exploiting Well-Proximity Effect. *IEEE Trans. Circuits Syst. II Express Briefs*, 70, 10, 3822–3826, Oct. 2023, doi: 10.1109/TCSII.2023.3289500.
21. Kang, S.J., Kim, J.H., Song, Y.S., Go, S., Kim, S., Investigation of Self-Heating Effects in Vertically Stacked GAA MOSFET With Wrap-Around Contact.

IEEE Trans. Electron Devices, 69, 3, 910–914, March 2022, doi: 10.1109/TED.2022.3140283.

22. Lu, X., Law, M.-K., Jiang, Y., Zhao, X., Mak, P.-I., Martins, R.P., A 4-μm Diameter SPAD Using Less-Doped N-Well Guard Ring in Baseline 65-nm CMOS. *IEEE Trans. Electron Devices*, 67, 5, 2223–2225, May 2020, doi: 10.1109/TED.2020.2982701.
23. Song, Y.S., Kim, J.H., Kim, G., Kim, H.-M., Kim, S., Park, B.-G., Improvement in Self-Heating Characteristic by Incorporating Hetero-Gate-Dielectric in Gate-All-Around MOSFETs. *IEEE J. Electron Devices Soc.*, 9, 36–41, 2021, doi: 10.1109/JEDS.2020.3038391.
24. Rahi, S.B., Asthana, P., Gupta, S., Heterogate junctionless tunnel field-effect transistor: future of low-power devices. *J. Comput. Electron.*, 16, 30–38, 2017, https://link.springer.com/article/10.1007/s10825-016-0936-9.
25. Rahi, S.B., Ghosh, B., Bishnoi, B., Temperature effect on hetero structure junctionless tunnel FET. *J. Semicond.*, 36, 3, 034002, Mar. 2015, https://iopscience.iop.org/article/10.1088/1674-4926/36/3/034002/meta.
26. Upadhyay, A.K., Rahi, S.B., Tayal, S., Song, Y.S., Recent progress on negative capacitance tunnel FET for low-power applications: Device perspective. *Microelectron. J.*, 129, 105583, Nov. 2022, https://www.sciencedirect.com/science/article/abs/pii/S0026269222002129.
27. Koričić, M., Žilak, J., Suligoj, T., Double-Emitter Reduced-Surface-Field Horizontal Current Bipolar Transistor With 36 V Breakdown Integrated in BiCMOS at Zero Cost. *IEEE Electron Device Lett.*, 36, 2, 90–92, Feb. 2015, doi: 10.1109/LED.2014.2385107.
28. Song, Y.S., Thoutam, L.R., Tayal, S., Rahi, S.B., Samuel, T.S.A., *Handbook of Emerging Materials for Semiconductor Industry*, Springer Nature, Singapore, 2024.
29. Chen, X., *et al.*, Printed Organic Synaptic Transistor Array for One-to-Many Neural Response. *IEEE Electron Device Lett.*, 43, 3, 394–397, March 2022, doi: 10.1109/LED.2022.3144662.
30. Bhardwaj, A., Kumar, P., Raj, B., Anand, S., Design and Performance Optimization of doping-less Vertical Nanowire TFET using Gate Stack Technique. *J. Electron. Mater. (JEMS)*, Springer, 41, 7, 4005–4013, 2022.
31. Singh, J. and Raj, B., Tunnel Current Model of Asymmetric MIM Structure Levying Various Image Forces to Analyze the Characteristics of Filamentary Memristor. *Appl. Phys. A*, Springer, 125, 3, 203.1 to 203.11, Feb 2019.
32. Goyal, C., Ubhi, J.S., Raj, B., Low Leakage Zero Ground Noise Nanoscale Full Adder using Source Biasing Technique. *J. Nanoelectron. Optoelectron.*, American Scientific Publishers, 14, 360–370, March 2019.
33. Singh, G., Sarin, R.K., Raj, B., A Novel Robust Exclusive-OR Function Implementation in QCA Nanotechnology with Energy Dissipation Analysis. *J. Comput. Electron.*, Springer, 15, 2, 455–465, June 2016.

34. Wadhwa, G., Kamboj, P., Raj, B., Design optimisation of junctionless TFET biosensor for high sensitivity. *Adv. Nat. Sci.: Nanosci. Nanotechnol.*, 10, 7, 045001, 2019.
35. Bansal, P. and Raj, B., MEMRISTOR MODELING AND ANALYSIS FOR LINEAR DOPANT DRIFT KINETICS. *J. Nanoeng. Nanomanuf.*, American Scientific Publishers, 6, 1–7, 2016.
36. Ashima, Vaithiyanathan, D., Raj, B., Performance Analysis of Charge Plasma induced Graded Channel Si nanotube. *J. Eng. Res. (JER)*, EMSME Special Issue, 15, 146–154, Aug 2021.
37. Singh Tomar, A., Kumar Magraiya, V., Raj, B., Scaling of Access and Data Transistor for High Performance DRAM Cell Design. *Quantum Matter*, 2, 412–416, Oct. 2013.
38. Jain, N. and Raj, B., Parasitic Capacitance and Resistance Model Development and Optimization of Raised Source/Drain SOI FinFET Structure for Analog Circuit Applications. *J. Nanoelectron. Optoelectronins*, ASP, USA, 13, 531–539, Apl 2018.
39. Singh, S., Vishvakarma, S.K., Raj, B., Analytical Modeling of Split-Gate Junction-Less Transistor for a Biosensor Application. *Sens. Bio-Sens.*, Elsevier, 18, 31–36, April 2018.
40. Gopal, M. and Raj, B., Low Power 8T SRAM Cell Design for High Stability Video Applications. *ITSI Trans. Electr. Electron. Eng.*, 1, 5, 91–97, 2013.
41. Raj, B., Mitra, J., Bihani, D.K., Rangharajan, V., Saxena, A.K., Dasgupta, S., Analysis of Noise Margin, Power and Process Variation for 32 nm FinFET Based 6T SRAM Cell. *J. Comput. (JCP)*, Academy Publisher, FINLAND, 5, 6, 1–8, 2010.
42. Sharma, D., Mehra, R., Raj, B., Comparative Analysis of Photovoltaic Technologies for High Efficiency Solar Cell Design. *Superlattices Microstruct.*, Elsevier, 153, 106861, May 2021.
43. Kaur, P., Buttar, A.S., Raj, B., A comprehensive Analysis of Nanoscale Transistor based Biosensor: A Review. *Indian J. Pure Appl. Phys.*, 59, 304–318, April 2021.
44. Yadav, D., Raj, B., Raj, B., Design and Simulation of Low Power Microcontroller for IoT Applications. *J. Sens. Lett.*, ASP, 18, 401–409, May 2020.
45. Singh, S. and Raj, B., A 2-D Analytical Surface Potential and Drain current Modeling of Double-Gate Vertical t-shaped Tunnel FET. *J. Comput. Electron.*, Springer, 19, 1154–1163, Apl 2020.
46. Singh, J. and Raj, B., An Accurate and Generic Window function for Nonlinear Memristor Model. *J. Comput. Electron.*, Springer, 18, 2, 640–647, June 2019.
47. Kaur, M., Gupta, N., Kumar, S., Raj, B., Singh, A.K., Comparative RF and Crosstalk Analysis of Carbon Based Nano Interconnects. *IET Circuits Devices Syst.*, 15, 6, 493–503, Feb 2021.

48. Kandasamy, N., Ahmad, F., Ajitha, D., Raj, B., Telagam, N., Quantum Dot Cellular Automata based Scan Flip Flop and Boundary Scan Register. *IETE J. Res.*, 66, 535–548, 2020.
49. Singh, J. and Raj, B., Design and Investigation of 7T2M NVSARM with Enhanced Stability and Temperature Impact on Store/Restore Energy. *IEEE Trans. Very Large Scale Integr. Syst.*, 27, 6, 1322–1328, June 2019.
50. Bhardwaj, A.K., Gupta, S., Raj, B., Singh, A., Impact of Double Gate Geometry on the Performance of Carbon Nanotube Field Effect Transistor Structures for Low Power Digital Design. *Comput. Theor. Nanosci.*, ASP, 16, 1813–1820, 2019.
51. Jain, N. and Raj, B., Thermal Stability Analysis and Performance Exploration of Asymmetrical Dual-k underlap Spacer (ADKUS) SOI FinFET for Security and Privacy Applications. *Indian J. Pure Appl. Phys. (IJPAP)*, 57, 352–360, May 2019.
52. Jain, N. and Raj, B., Dual-k Spacer Region Variation at the Drain Side of Asymmetric SOI FinFET Structure: Performance Analysis towards the Analog/RF Design Applications. *J. Nanoelectron. Optoelectron.*, American Scientific Publishers, 14, 349–359, March 2019.
53. Singh, J., Sharma, S., Raj, B., Khosla, M., Analysis of barrier layer thickness on performance of In1-xGaxAs based Gate Stack Cylindrical Gate Nanowire MOSFET. *JNO*, ASP, 13, 1473–1477, Oct 2018.
54. Jain, N. and Raj, B., Analysis and Performance Exploration of High-k SOI FinFETs Over the Conventional Low-k SOI FinFET toward Analog/RF Design. *J. Semicond. (JoS)*, IOP Science, 39, 12, 124002-1-7, Dec 2018.
55. Goyal, C., Ubhi, J.S., Raj, B., A reliable leakage reduction technique for approximate full adder with reduced ground bounce noise. *J. Math. Probl. Eng.*, Hindawi, 2018, Article ID 3501041, 16, 15 Oct 2018.
56. Anuradha, Singh, J., Raj, B., Khosla, M., Design and Performance Analysis of Nano-scale Memristor-based Nonvolatile SRAM. *J. Sens. Lett.*, American Scientific Publishers, 16, 798–805, Oct 2018.
57. Wadhwa, G. and Raj, B., Parametric Variation Analysis of Charge-Plasma-based Dielectric Modulated JLTFET for Biosensor Application. *IEEE Sens. J.*, 18, 15, 6070–6077, AUGUST 1 2018, (SCI).
58. Singh, J. and Raj, B., Comparative Analysis of Memristor Models for Memories Design. *JoS*, IoP, 39, 7, 074006-1-12, July 2018.
59. Yadav, D., Chouhan, S.S., Vishvakarma, S.K., Raj, B., Application Specific Microcontroller Design for IoT based WSN. *Sens. Lett.*, ASP, 16, 374–385, May 2018.
60. Singh, G., Sarin, R.K., Raj, B., Fault-Tolerant Design and Analysis of Quantum-Dot Cellular Automata Based Circuits. *IEEE/IET Circuits Devices Syst.*, 12, 638–64, 2018.
61. Singh, J. and Raj, B., Modeling of Mean Barrier Height Levying Various Image Forces of Metal Insulator Metal Structure to Enhance the Performance

of Conductive Filament Based Memristor Model. *IEEE Nanotechnol.*, 17, 2, 268–267, March 2018.

62. Jain, A., Sharma, S., Raj, B., Analysis of Triple Metal Surrounding Gate (TM-SG) III-V Nanowire MOSFET for Photosensing Application. *Opto-electronics J.*, Elsevier, 26, 2, 141–148, May 2018.
63. Jain, A., Sharma, S., Raj, B., Design and Analysis of High Sensitivity Photosensor Using Cylindrical Surrounding Gate MOSFET For Low Power Sensor Applications. *Eng. Sci. Technol. Int. J.*, Elsevier's, 19, 4, 1864–1870, December 2016.
64. Saiphani Kumar, G., Singh, A., Raj, B., Design and Analysis of Gate All Around CNTFET based SRAM cell Design. *J. Comput. Electron.*, Springer, 17, 1, 138–145, March 2018.
65. Singh, G., Sohi, B.S., Raj, B., Material Properties Analysis of Graphene Base Transistor (GBT) for VLSI Analog Circuits. *Indian J. Pure Appl. Phys. (IJPAP)*, 55, 896–902, Dec-2017.
66. Kumar, S. and Raj, B., Estimation of Stability and Performance metric for Inward Access Transistor based 6T SRAM Cell Design using n-type/p-type DMDG-GDOV TFET. *IEEE VLSI Circuits Syst. Lett.*, 3, 2, 25–39, June 2017.
67. Sharma, S., Kumar, A., Pattanaik, M., Raj, B., Forward Body Biased Multimode Multi-Threshold CMOS Technique for Ground Bounce Noise Reduction in Static CMOS Adders. *Int. J. Inf. Electron. Eng.*, 3, 3, 567–572, 2013.
68. Singh, H., Kumar, P., Raj, B., Performance Analysis of Majority Gate SET Based 1-bit Full Adder. *Int. J. Comput. Commun. Eng. (IJCCE)*, 2, 4, 624–626, IACSIT Press Singapore, ISSN: 2010-3743.
69. Bhardwaj, A.K., Gupta, S., Raj, B., Investigation of Parameters for Schottky Barrier (SB) Height for Schottky Barrier Based Carbon Nanotube Field Effect Transistor Device. *J. Nanoelectron. Optoelectron.*, ASP, 15, 783–791, July 2020.
70. Bansal, P. and Raj, B., MEMRISTOR: A VERSATILE NONLINEAR MODEL FOR DOPANT DRIFT AND BOUNDARY ISSUES. *JCTN*, American Scientific Publishers, 14, 5, 2319–2325, May 2017.
71. Jain, N. and Raj, B., An Analog and Digital Design Perspective Comprehensive Approach on Fin-FET (Fin-Field Effect transistor) Technology - A Review. *Rev. Adv. Sci. Eng. (RASE)*, ASP, 5, 1–14, 2016.
72. Raj, B., Saxena, A.K., Dasgupta, S., Analytical Modeling for the Estimation of Leakage Current and Subthreshold Swing Factor of Nanoscale Double Gate FinFET Device. *Microelectron. Int.*, 26, 53–63, 2009.
73. Soniya, S.S., Wadhwa, G., Raj, B., An Analytical Modeling for Dual Source Vertical Tunnel Field Effect Transistor" submitted to. *Int. J. Recent Technol. Eng. (IJRTE)*, UK, 8, 2, 603–608, July 2019.
74. Singh, S. and Raj, B., Design and analysis of hetrojunction Vertical T-shaped Tunnel Field Effect Transistor. *J. Electron. Mater. Springer*, 48, 10, 6253–6260, October 2019.

75. Goyal, C., Ubhi, J.S., Raj, B., A Low Leakage CNTFET based Inexact Full Adder for Low Power Image Processing Applications. *Int. J. Circuit Theory Appl.*, Wiley, 47, 9, 1446–1458, September 2019.
76. Raj, B., Saxena, A.K., Dasgupta, S., A Compact Drain Current and Threshold Voltage Quantum Mechanical Analytical Modeling for FinFETs. *J. Nanoelectron. Optoelectron. (JNO)*, USA, 3, 2, 163–170, 2008.
77. Wadhwa, G. and Raj, B., An Analytical Modeling of Charge Plasma based Tunnel Field Effect Transistor with Impacts of Gate underlap Region. *Superlattices Microstruct.*, Elsevier, 142, 106512, June 2020.
78. Singh, S. and Raj, B., Modeling and Simulation analysis of SiGe hetrojunction Double GateVertical t-shaped Tunnel FET. *Superlattices Microstruct.*, Elsevier, 142, 106496, June 2020.
79. Jain, N. and Raj, B., Device and Circuit Co-Design Perspective Comprehensive Approach on FinFET Technology - A Review. *J. Electron Devices*, 23, 1, 1890–1901, 2016.
80. Khandelwal, S., Raj, B., Gupta, R.D., FinFET based 6T SRAM Cell Design: Analysis of Performance Metric, Process Variation and Temperature Effect. *J. Comput. Theor. Nanosci.*, ASP, USA, 12, 2500–2506, 2015.
81. Singh, S., Yadav, S., Rahul, J., Srivastava, A., Raj, B., Impact of HfO_2 in Graded Channel Dual Insulator Double Gate MOSFET. *J. Comput. Theor. Nanosci.*, American Scientific Publishers, 12, 6, 950–953, April 2015.
82. Sharma, V.K., Pattanaik, M., Raj, B., PVT variations aware low leakage INDEP approach for nanoscale CMOS Circuits. *Microelectron. Reliab.*, Elsevier, 54, 90–99, 2014.
83. Raj, B., Saxena, A.K., Dasgupta, S., Quantum Mechanical Analytical Modeling of Nanoscale DG FinFET: Evaluation of Potential, Threshold Voltage and Source/Drain Resistance. *Elsevier's J. Mater. Sci. Semicond. Process.*, Elsevier, 16, 4, 1131– 1137, 2013.
84. Gopal, M., Prasad, S.S.D., Raj, B., 8T SRAM Cell Design for Dynamic and Leakage Power Reduction. *Int. J. Comput. Appl.*, Foundation of Computer Science, New York, USA, 71, 9, 43–48, June 2013.
85. Pattanaik, M., Raj, B., Sharma, S., Kumar, A., Diode Based Trimode Multi-Threshold CMOS Technique for Ground Bounce Noise Reduction in Static CMOS Adders. *Adv. Mater. Res.*, Trans Tech Publications, Switzerland, 548, 885–889, 2012.
86. Raj, B., Saxena, A.K., Dasgupta, S., Nanoscale FinFET Based SRAM Cell Design: Analysis of Performance metric, Process variation, Underlapped FinFET and Temperature effect. *IEEE Circuits Syst. Mag.*, 11, 2, 38–50, 2011.
87. Sharma, V.K., Pattanaik, M., Raj, B., Leakage Current ONOFIC Approach for Deep Submicron VLSI Circuit Design International Journal of Electrical, Computer, Electronics and Communication Engineering. *World Acad. Sci. Eng. Technol.*, 7, 4, 239–244, 2013.
88. Kaur, P., Gill, S.S., Raj, B., Comparative Analysis of OFETs Materials and Devices for Sensor Applications. *J. Silicon*, Springer, 14, 4463–4471, 2022.

89. Sharma, S.K., Kumar, P., Raj, B., Raj, B., $In_{1-x}Ga_xAs$ Double Metal Gate-Stacking Cylindrical Nanowire MOSFET for highly sensitive Photo detector. *J. Silicon*, Springer, 14, 3535–3541, 2022.
90. Raj, B., Saxena, A.K., Dasgupta, S., Analytical Modeling of Quasi Planar Nanoscale Double Gate FinFET with Source/Drain Resistance and field dependent carrier mobility: A quantum mechanical study. *J. Comput. (JCP)*, Academy Publisher, FINLAND, 4, 9, 1–8, 2009.
91. Bhushan, S., Khandelwal, S., Raj, B., Analyzing Different Mode FinFET based Memory Cell at Different Power Supply for Leakage Reduction. *Seventh International Conference on Bio-Inspired Computing: Theories and Application, (BIC-TA 2012)*, Advances in Intelligent Systems and Computing, vol. 202, pp. 89–100, 2012, 2013.
92. Singh, J. and Raj, B., Temperature Dependent Analytical Modeling and Simulations of Nanoscale Memristor. *J. Eng. Sci. Technol. Int. J.*, Elsevier's, 21, 862–868, Oct 2018.
93. Singh, S., Bala, S., Raj, B., Raj, B., Improved Sensitivity of Dielectric Modulated Junctionless Transistor for Nanoscale Biosensor Design. *Sens. Lett.*, ASP, 18, 328–333, Apl 2020.
94. Kumar, V., Vishvakarma, S.K., Raj, B., Design and Performance Analysis of ASIC for IoT Applications. *Sens. Lett.*, ASP, 18, 31–38, Jan 2020.
95. Jaiswal, A., Sarin, R.K., Raj, B., Sukhija, S., A Novel Circular Slotted Microstrip-fed Patch Antenna with Three Triangle Shape Defected Ground Structure for Multiband Applications. *Adv. Electromagn. (AEM)*, 7, 3, 56–63, Aug 2018.
96. Wadhwa, G. and Raj, B., Label Free Detection of Biomolecules using Charge-Plasma-Based Gate Underlap Dielectric Modulated Junctionless TFET. *J. Electron. Mater. (JEMS)*, Springer, 47, 8, 4683–4693, August 2018.
97. Singh, G., Sarin, R.K., Raj, B., Design and Performance Analysis of a New Efficient Coplanar Quantum-Dot Cellular Automata Adder. *Indian J. Pure Appl. Phys. (IJPAP)*, 55, 97–103, February 2017.
98. Kaur, P., Pandey, V., Raj, B., Comparative Study of Efficient Design, Control and Monitoring of Solar Power using IoT. *Sens. Lett.*, ASP, 18, 419–426, May 2020.
99. Bhardwaj, A.K., Gupta, S., Raj, B., Development & Analysis of compact model for Double Gate Schottky Barrier CNTFET. *J. Nanoelectron. Optoelectron.*, ASP, 15, 1199–1208, Aug 2020.
100. Wadhwa, G., Kamboj, P., Singh, J., Raj, B., Design and Investigation of Junctionless DGTFET for Biological Molecule Recognition. *Trans. Electr. Electron. Mater.*, Springer, 22, 282–289, 2021.
101. Verma, S.K., Singh, S., Wadhwa, G., Raj, B., Detection of Biomolecules using Charge-Plasma based Gate Underlap Dielectric Modulated Dopingless TFET. *Trans. Electr. Electron. Mater. (TEEM)*, Springer, 21, 528–535, June 2020.

102. Jain, N. and Raj, B., Impact of underlap spacer region variation on Electrostatic and analog/RF performance of symmetrical high-k SOI FinFET at 20 nm channel length. *J. Semicond. (JoS)*, IOP Science, 38, 12, 122002, Dec 2017.
103. Singh, S. and Raj, B., Analytical Modeling and Simulation analysis of T-shaped III-V heterojunction Vertical T-FET. *Superlattices Microstruct.*, Elsevier, 147, 106717, Nov 2020.
104. Singh, G., Sarin, R.K., Raj, B., Design and Analysis of Area Efficient QCA Based Reversible Logic Gates. *J. Microprocess. Microsys.*, Elsevier, 52, 59–68, May 2017.
105. Singh, S. and Raj, B., Parametric variation analysis on hetero-junction Vertical t-shape TFET for Supressing ambipolar conduction. *Indian J. Pure Appl. Phys.*, 58, 478–485, June 2020.
106. Soniya, S.S., Wadhwa, G., Raj, B., Design and Analysis of Dual Source Vertical Tunnel Field Effect Transistor for high performance. *Trans. Electr. Electron. Mater.*, Springer, 21, 74–82, Oct 2019.
107. Kaur, M., Gupta, N., Kumar, S., Raj, B., Arun Kumar Singh, RF Performance Analysis of Intercalated Graphene Nanoribbon Based Global Level Interconnects. *J. Comput. Electron.*, Springer, 19, 1002–1013, June 2020.
108. Wadhwa, G. and Raj, B., Design and Performance Analysis of Junctionless TFET Biosensor for high sensitivity. *IEEE Nanotechnol.*, 18, 567–574, 2019.
109. Singh, J. and Raj, B., Enhanced Nonlinear Memristor Model Encapsulating Stochastic Dopant Drift. *JNO*, ASP, 14, 958–963, 2019.
110. Kumar, S. and Raj, B., Analysis of I_{ON} and Ambipolar Current for Dual-Material Gate-drain Overlapped DG-TFET. *J. Nanoelectron. Optoelectron.*, American Scientific Publishers, USA, 11, 323–333, June 2016.
111. Anjum, N., Bali, T., Raj, B., Design And Simulation Of Handwritten Multiscript Character Recognition. *Int. J. Adv. Res. Comput. Commun. Eng.*, 2, 7, 2544–2549, July 2013.
112. Singh, K. and Raj, B., Performance and analysis of temperature dependent Multi-walled Carbon Nanotubes as Global Interconnects at different technology nodes. *J. Comput. Electron.*, Springer, 14, 2, 469–476, June 2015.
113. Kumar, S. and Raj, B., Compact channel potential analytical Modeling of DG-TFET based on Evanescent-mode Approach. *J. Comput. Electron.*, Springer, 14, 2, 820–827, July 2015.
114. Singh, K. and Raj, B., Temperature Dependent Modeling and Performance Evaluation of Multi-Walled CNT and Single-Walled CNT as Global Interconnects. *J. Electron. Mater.*, Springer, 44, 12, 4825–4835, Dec 2015.
115. Sharma, V.K., Pattanaik, M., Raj, B., INDEP approach for leakage reduction in nanoscale CMOS circuits. *Int. J. Electron.*, Taylor & Francis, 102, 2, 200–215, 2014.
116. Singh, K. and Raj, B., Influence of Temperature on MWCNT bundle, SWCNT bundle and Copper interconnects for Nanoscaled Technology nodes. *J. Mater. Sci.: Mater. Electron.*, Springer, 26, 8, 6134–6142, 2015.

117. Anjum, N., Bali, T., Raj, B., Design and Simulation of Handwritten Gurumukhi and Devanagri Numerical Recognition. *Int. J. Comput. Appl.*, Foundation of Computer Science, New York, USA, 73, 12, 16–21, 2013.
118. Khandelwal, S., Gupta, V., Raj, B., Gupta, R.D., Process Variability Aware Low Leakage Reliable Nano Scale DG-FinFET SRAM Cell Design Technique. *J. Nanoelectron. Optoelectron.*, 10, 6, 810–817, December 2015.
119. Sharma, V.K., Pattanaik, M., Raj, B., ONOFIC Approach: Low Power High Speed Nanoscale VLSI Circuits Design. *Int. J. Electron.*, Taylor & Francis, 101, 1, 61–73, 2014.

12

Recent Electron Mobility Models for FeFET

Shiromani Balmukund Rahi[1] and Young Suh Song[2*]

[1]*University School of Information and Communication Technology, Gautam Buddha University, Uttar Pradesh, India*
[2]*Department of Computer Science, Korea Military Academy, Seoul, Republic of Korea*

Abstract
The electron mobility is one of the most important parameters in electronic applications and transistor design. With the advent of ferroelectric field-effect transistors (FeFETs), understanding electron mobility has become increasingly important. This chapter addresses recent electron mobility models for FeFETs. From classical electron mobility models to quantum mechanical electron mobility models, various electron models have been explained so that readers can understand the concept of electron mobility in modern semiconductor transistors. Specifically, as the dimensions of the device have decreased repeatedly, various complex quantum mechanisms make it difficult to interpret the movement of electrons and calculate electron mobility. In this chapter, we would like to look into the recent electron mobility models that have been developed for modern transistors and aimed to achieve high accuracy, compared to conventional electron mobility models.

Keywords: Ferroelectric, material engineering, semiconductor, nanodevice

12.1 Introduction to Electron Mobility and FeFET

Recently, with the advent of modern semiconductor chips, the electron mobility (a fundamental parameter in semiconductor physics) has

**Corresponding author*: songyeongseo84@gmail.com

Balwinder Raj, Shiromani Balmukund Rahi and Nandakishor Yadav (eds.) FeFET Devices, Trends, Technology and Applications, (297–318) © 2025 Scrivener Publishing LLC

delineated the capacity of electrons to traverse a material under the influence of an electric field influence [1–4]. It holds pivotal significance in determining electronic devices' electrical conductivity and performance [5]. Within field-effect transistors (FETs), including ferroelectric field-effect transistors (FeFETs), electron mobility profoundly impacts device functionality, speed, and power consumption [6–8].

FeFETs constitute a transistor subclass that integrates a ferroelectric material into the gate stack, facilitating nonvolatile memory operations within the device [9]. The ferroelectric layer's polarization switching modulates the channel conductivity, enabling programmable and nondestructive read-and-write functionalities [10]. As FeFET technology progresses, comprehending and precisely modeling the electron mobility within these devices has become increasingly crucial for optimizing performance and reliability [11–15].

FeFETs introduce additional complexities to electron mobility modeling compared to conventional FETs due to the ferroelectric layer's presence [16]. The ferroelectric material's polarization state influences the electrostatics of the device, thereby affecting the electron transport properties in the channel region. Consequently, accurate electron mobility models for FeFETs require accounting for the interplay between the ferroelectric polarization, material characteristics, and device geometry.

12.2 Classical Electron Mobility Models

Classical models of electron mobility serve as the cornerstone for understanding the electron transport within semiconductor materials. These models are rooted in semiclassical transport theory, which blends classical mechanics with quantum mechanics to elucidate electron motion within crystalline solids [17–22].

One of the earliest and most prevalent classical mobility models is the Drude model, which was devised by Paul Drude in the late 19th century. This model conceptualizes electrons as charged entities that navigate through a lattice of positively charged ions. It attributes electron mobility to scattering events arising from interactions between lattice vibrations (phonons) and impurities. Although the Drude model offers valuable insights into electron transport phenomena, its accuracy diminishes in modern semiconductor devices, where quantum effects play a significant role [23–28].

Another widely employed classical mobility model in semiconductor physics is Matthiessen's rule, which determines the collective impact of various scattering mechanisms on electron mobility. According to Matthiessen's rule, the total electron mobility is the sum of the individual mobilities influenced by diverse scattering mechanisms such as acoustic phonon scattering, ionized impurity scattering, and alloy scattering.

Moreover, the Kane model, proposed by John Kane, expands classical mobility theory to encompass high-field and low-temperature effects [29–35]. By incorporating parameters to accommodate band non-parabolicity and temperature sensitivity of electron mobility, the Kane model offers a more precise depiction of electron transport in semiconductors across a broader range of conditions.

The electron mobility models can be explained first by Equations (12.1) and (12.2).

$$\mu_{eff} = \frac{\mu_0}{1+\left(\frac{E_{eff}}{E_0}\right)^V} \tag{12.1}$$

$$\mu_{eff} = \frac{\mu_0}{1+\varphi(V_G - V_{th})} \tag{12.2}$$

In Equations (12.1) and (12.2), the effective electron mobility (μ_{eff}) shows the actual electron mobility under certain conditions, such as an electric field and temperature. The effective electron mobility can be calculated using Equations (12.1) and (12.2) by utilizing the intrinsic electron mobility (μ_0). Intrinsic electron mobility is the natural electron mobility of the material [36–43]. The intrinsic electron mobilities of various materials are listed in Table 12.1.

Table 12.1 Intrinsic electron mobility and bandgap of various novel materials.

Material	Ge	Si	AlP	AlSb	GaAs	GaP	InP
Intrinsic electron mobility (μ_0) (m^2V^{-1} sec^{-1})	0.39	0.135	0.006	0.02	0.85	0.011	0.46
Bandgap (eV)	0.67	1.1	2.4	1.6	1.4	2.3	1.3

However, classical electron mobility models have certain limitations. For example, it is difficult for classical electron mobility models to consider various quantum-mechanical phenomena. As the device size and material thickness become smaller, it would be better to consider a novel quantum mechanical phenomenon. For this reason, quantum-mechanical models for electrons will be described in the next section [44–51].

12.3 Quantum Mechanical Models for Electron Mobility

In contrast to conventional electron mobility models rooted in semiclassical transport theory, quantum mechanical models offer a deeper understanding of the electron transport phenomena in semiconductor materials. These models account for the wave-like nature of electrons and discrete energy levels inherent in the semiconductor band structure.

A fundamental quantum mechanical model utilized for describing electron mobility is the Kubo–Greenwood formula. This formula employs quantum mechanics principles to compute the electrical conductivity of a material based on electron wavefunctions and scattering mechanisms occurring within it. By considering various scattering processes such as phonon scattering, impurity scattering, and electron–electron scattering, the Kubo–Greenwood formula provides valuable insights into electron mobility behavior in semiconductors [52–57].

The Boltzmann transport equation (BTE) is another notable approach to electron mobility modeling. This equation is a robust tool for characterizing the electron transport in semiconductor materials subjected to an electric field. By incorporating different scattering mechanisms, the BTE facilitates the calculation of electron mobility as a function of the carrier concentration, temperature, and material properties. It offers a systematic framework for analyzing the electron transport in both bulk and nanostructured semiconductors.

Density functional theory (DFT) is another prominent quantum mechanical method for investigating the electron mobility in semiconductors. Based on quantum mechanics principles, DFT provides a computational strategy for determining the electronic structures and transport properties of materials. By solving the Schrödinger equation for electrons within the crystal lattice, DFT enables accurate predictions of electron mobility and other electronic characteristics.

Moreover, in nanostructured semiconductor materials, where quantum confinement effects become pronounced, quantum mechanical models such as tight-binding models and effective mass approximations are employed [58–66]. These models account for the discrete energy levels and confinement effects on electron mobility as the dimensions approach the quantum confinement regime, offering insights into the electron behavior in such systems.

Recently, more complicated mobility models have incorporated various quantum mechanics. Specifically, modern mobility models are presented in Table 12.2.

Table 12.2 Modern electron mobility models with various quantum mechanics.

Model name	Explanation
Yamaguchi Model	Considers doping concentration and local electric field effects. However, usually used for temperature 300 K.
Lombardi Model	Compact model including doping concentration, local electric field, and temperature. Good for non-planar devices, and also good for temperature-related research.
Modified Watt	More sophisticated model, compared to watt model. This is extended version of watt model especially with non-surface nodes. This model is known for best model for planar MOS devices
Shirahata Model	This model includes doping concentration and electric field effect. This could be an alternative surface mobility model that could be combined with the following Klaasen model.
Klaasen Model	Includes local doping concentration dependence. This models could utilize individual mobility to both majority and minority careers. This models are usually recommended for bipolar devices.
Watt Model	This is known as transverse field model usually applied to surface nodes only. It is good to utilize this models for calculation of surface electron mobility.
Tasch Model	Incorporates transverse field dependence. However, this model usually requires very fine grid.

12.4 Density Functional Theory (DFT) Approaches for Electron Mobility

Density Functional Theory (DFT) stands as a powerful computational tool used to investigate electron mobility within semiconductor materials [67–74].

Table 12.3 The domain of electron mobility research with DFT.

Domain	Description
Band Structure Calculations	DFT accurately calculates the electronic band structure of semiconductor materials, outlining the allowable energy states for electrons within the crystal lattice. This analysis aids in understanding the effective mass of charge carriers and their mobility behavior.
Scattering Mechanisms	DFT analyses can be used to identify and quantify various scattering mechanisms that affect electron mobility, such as phonon scattering, impurity scattering, and defects. By examining the interactions between electrons and scattering centers, DFT facilitates the evaluation of scattering rates and their impact on electron mobility.
Lattice Dynamics	Through DFT-based simulations, researchers have explored the lattice dynamics of semiconductors, including the vibrational properties of atoms and phonon dispersion relation. Understanding lattice dynamics is crucial for predicting electron-phonon interactions, which significantly influence electron mobility through phonon scattering processes.
Charge Transport Properties	DFT computations provide insights into the charge transport characteristics, such as carrier mobility, conductivity, and diffusivity. By solving the Schrödinger equation for electrons within an external electric field, DFT incorporates electron-electron interactions and the influence of the crystal structure, offering precise predictions of the electron mobility.
Defect Engineering	DFT simulations were used to investigate the effects of defects, dopants, and interfaces on electron mobility within semiconductors. By analyzing the electronic structure and energetics of defects, DFT assists in identifying strategies to optimize the material quality and improve the electron transport properties.

Based on quantum mechanics, DFT offers a theoretical framework for predicting electronic properties by examining the quantum behavior of electrons within a crystalline lattice.

In the field of electron mobility research, DFT presents various methodologies for probing electron transport in semiconductors, as shown in Table 12.3.

12.5 Empirical Electron Mobility Models and Parameter Extraction Techniques

Empirical models for electron mobility and associated parameter extraction techniques are pivotal tools for comprehending and forecasting the electron transport dynamics in semiconductor materials. In contrast to theoretical models grounded in quantum mechanics, empirical approaches rely on experimental data to establish correlations between the electron mobility and diverse material attributes [75–82]. These models are often coupled with parameter extraction methods to delineate the principal factors influencing the electron mobility in semiconductors.

One prevalent empirical strategy is the empirical pseudopotential method (EPM), which delves into the scattering phenomena encountered by electrons in semiconductors. The EPM utilizes a set of empirical parameters to depict the impact of scattering events arising from phonons, impurities, and defects on electron mobility [83–91]. By fitting the experimental mobility data to the EPM framework, researchers can deduce the values of these parameters and gain insight into the predominant scattering mechanisms within the material.

Another widely embraced empirical model is the Kane dispersion relation, which elucidates the relationship between the energy and momentum of charge carriers in semiconductors [92–98]. By incorporating material-specific parameters, such as effective mass and bandgap energy, the Kane model forecasts electron mobility as a function of temperature and carrier concentration. Parameter extraction methods entail fitting the experimental mobility data to the Kane model and extracting the material parameter values.

In addition to these empirical models, machine learning techniques have emerged as potent tools for predicting the electron mobility in semiconductor materials. Machine learning algorithms trained on extensive datasets of experimental mobility data can discern intricate relationships between material properties and electron mobility [99–105]. By leveraging these associations, machine learning models can accurately forecast the electron mobility in novel materials or under diverse conditions.

Parameter extraction methods for empirical models typically involve fitting experimental data to theoretical formulations or numerical simulations [106–112]. Techniques such as least squares fitting, Bayesian inference, and genetic algorithms are employed to optimize the model parameters and diminish the disparity between the experimental and simulated mobility data. These methodologies enable researchers to derive precise values for material parameters and corroborate the predictive capabilities of empirical models.

12.6 Challenges and Limitations in Modeling FeFET Electron Mobility

Modeling electron mobility in ferroelectric field-effect transistors (FeFETs) poses numerous challenges and limitations owing to the distinctive properties of ferroelectric materials and their interaction with semiconductor structures [113–118]. It is crucial to understand and address these obstacles to ensure accurate prediction and optimization of FeFET device performance. The following are some significant challenges and limitations encountered in modeling the electron mobility of FeFETs. A detailed explanation is presented in Table 12.4.

Table 12.4 The possible challenges in modeling FeFET electron mobility.

Challenge type	Explanation
Ferroelectric Polarization Dynamics	The switching behavior of ferroelectric materials directly affects the electrostatics of FeFET devices and, consequently, electron mobility. However, modeling the dynamics of ferroelectric polarization, including polarization reversal and domain-wall motion, is intricate and requires sophisticated theoretical frameworks. Incorporating these dynamics into device simulations poses a significant challenge for accurately predicting the electron mobility in FeFETs.

(Continued)

Table 12.4 The possible challenges in modeling FeFET electron mobility. (*Continued*)

Challenge type	Explanation
Nonlinear Effects	FeFETs exhibit nonlinear behavior because of the polarization-dependent capacitance of the ferroelectric layer. As the polarization state changes, the capacitance of the gate dielectric fluctuates, resulting in nonlinear device characteristics. Modeling these nonlinear effects accurately requires advanced simulation techniques capable of capturing the dynamic interaction between ferroelectric polarization and charge transport.
Multiscale Modeling	FeFET devices operate at the nanoscale, where quantum mechanical effects become increasingly significant. However, conventional device simulation tools based on classical transport models may not fully capture the quantum behavior of electrons in FeFETs. Multiscale modeling approaches that integrate quantum mechanical simulations with device-level simulations are imperative to account for quantum effects and accurately predict electron mobility in FeFETs.
Interface Effects	The interface between the ferroelectric layer and semiconductor channel plays a pivotal role in determining device performance. Interface defects, roughness, and charge-trapping phenomena can significantly influence electron mobility in FeFETs. Modeling these interface effects requires comprehensive knowledge of the electronic structure and interface properties, posing a challenge in accurately incorporating them into device simulations.
Material Variability	Variations in the material properties, including ferroelectric polarization, semiconductor doping, and interface quality, introduce device-to-device variability in FeFETs. Modeling the impact of material variability on electron mobility requires statistical approaches that consider the distribution of the material parameters and their influence on device performance. Addressing material variability is essential to ensure the reliability and manufacturability of FeFET devices.

Overcoming these challenges and limitations in modeling FeFET electron mobility necessitates interdisciplinary collaboration among material scientists, device engineers, and computational physicists. Advanced simulation techniques, experimental validation, and innovative device designs are essential for developing accurate predictive models of FeFET devices and unlocking their full potential for next-generation electronics.

12.7 Future Directions and Emerging Trends in FeFET Electron Mobility Modeling

In contrast to the previous section, this section explains the future direction and possibilities of utilizing FeFET electron mobility modeling in conjunction with classical CMOS electron mobility modeling [17–21]. As the development of ferroelectric field-effect transistor (FeFET) progresses, researchers are exploring novel approaches and emerging trends in modeling electron mobility to overcome current challenges and fully exploit the potential of FeFET devices including semiconductor applications [22–30]. Table 12.5 shows some future directions and emerging trends in FeFET electron mobility modeling [119, 120].

Table 12.5 Future direction and possibilities in electron mobility modeling with FeFET.

Type	Description
Advancements in Simulation Techniques	Future research endeavors are likely to concentrate on refining simulation techniques capable of accurately capturing the intricate interplay between ferroelectric polarization dynamics, charge transport phenomena, and device characteristics. Multiscale modeling strategies that integrate quantum mechanical simulations with device-level simulations will increase the significance of modeling electron mobility in FeFETs operating at the nanoscale level.

(*Continued*)

Table 12.5 Future direction and possibilities in electron mobility modeling with FeFET. (*Continued*)

Type	Description
Incorporation of Quantum Mechanical Effects	As FeFET devices shrink to smaller dimensions, quantum mechanical effects become more prominent. Future models will need to incorporate quantum mechanical phenomena, such as tunneling and quantum confinement, to provide accurate predictions of the electron mobility in FeFETs. Quantum mechanical simulations based on density functional theory (DFT) and tight-binding methods are pivotal in elucidating these effects. The DFT is explained in the previous section in this chapter.
Utilization of Machine Learning Techniques	Machine learning (ML) algorithms offer promising avenues for predicting electron mobility in FeFETs by uncovering complex relationships between material properties, device parameters, and electron transport behavior. Future studies may leverage machine learning techniques to develop data-driven models capable of accurately predicting electron mobility in FeFET devices with high precision and efficiency.
Exploration of Interface Engineering Strategies	The interface effects between the ferroelectric layer and the semiconductor channel exert a significant influence on the electron mobility in FeFETs. Future modeling endeavors will concentrate on comprehending and optimizing the interface properties to mitigate scattering effects and enhance charge transport. Atomistic simulations and interface engineering strategies play a crucial role in unraveling the underlying mechanisms and designing FeFET devices with enhanced performance.
Advancements in Material Design and Synthesis	Progress in material design and synthesis will continue to propel innovation in the FeFET technology. Future research will explore novel ferroelectric materials tailored to optimize the electron mobility and device performance. Computational screening techniques and high-throughput simulations will facilitate the discovery and characterization of new ferroelectric materials that are suitable for FeFET applications.

(*Continued*)

Table 12.5 Future direction and possibilities in electron mobility modeling with FeFET. (*Continued*)

Type	Description
Focus on Reliability and Variability Analysis	Enhancing device reliability and reducing device-to-device variability are imperative for the widespread adoption of the FeFET technology. Future modeling efforts will focus on quantifying the impact of material variability, interface defects, and operational conditions on the electron mobility and device reliability. Probabilistic modeling techniques and stochastic simulations will enable researchers to evaluate device performance under realistic conditions and to develop strategies for enhancing reliability.

12.8 Conclusion

This chapter provides a detailed explanation of the FeFET and electron mobility models. Considering the advancements in semiconductor technologies and novel emerging materials (including ferroelectric materials), it is important to understand the fundamental properties of electron movement and electron mobility. With the advancement of recent electron models, it is expected that it will be possible to understand the behavior of the electrons properly and specifically in FeFETs. Although there are several challenges, there is still rooms for the utilization of novel FeFET materials with the advancement of recent electron mobility models.

References

1. Zhang, X.-W., *et al.*, MoS2 Field-Effect Transistors With Lead Zirconate-Titanate Ferroelectric Gating. *IEEE Electron Device Lett.*, 36, 8, 784–786, Aug. 2015, doi: 10.1109/LED.2015.2440249.
2. Liu, W.-D., Huang, Z.-Y., Ma, J., Zheng, Z.-W., Cheng, C.-H., Impact of Series-Connected Ferroelectric Capacitor in HfO_2-Based Ferroelectric Field-Effect Transistors for Memory Application. *IEEE J. Electron Devices Soc.*, 8, 1076–1081, 2020, doi: 10.1109/JEDS.2020.3029268.
3. Yurchuk, E., *et al.*, Impact of Scaling on the Performance of HfO2-Based Ferroelectric Field Effect Transistors. *IEEE Trans. Electron Devices*, 61, 11, 3699–3706, Nov. 2014, doi: 10.1109/TED.2014.2354833.

4. Peng, L., Jia, Y., Qi, S., Lixuan, K., Yexi, R., Guanyu, W., Performance Measurement and Analysis of Hafnium-Oxide Ferroelectric Field Effect Transistors on Silicon Substrate. *2023 International Conference on Communication, Circuits, and Systems (IC3S)*, BHUBANESWAR, India, pp. 1–5, 2023, doi: 10.1109/IC3S57698.2023.10169839.
5. Huang, K., *et al.*, $Hf_{0.5}Zr_{0.5}O_2$ Ferroelectric Embedded Dual-Gate MoS_2 Field Effect Transistors for Memory Merged Logic Applications. *IEEE Electron Device Lett.*, 41, 10, 1600–1603, Oct. 2020, doi: 10.1109/LED.2020.3019681.
6. Yoon, S.-J., Min, D.-H., Moon, S.-E., Park, K.S., Won, J.I., Yoon, S.-M., Improvement in Long-Term and High-Temperature Retention Stability of Ferroelectric Field-Effect Memory Transistors With Metal–Ferroelectric–Metal–Insulator–Semiconductor Gate-Stacks Using Al-Doped HfO2 Thin Films. *IEEE Trans. Electron Devices*, 67, 2, 499–504, Feb. 2020, doi: 10.1109/TED.2019.2961117.
7. Zhou, H., Ocker, J., Müller, S., Pesic, M., Mikolajick, T., Polarization Switching and Charge Trapping in HfO2-Based Ferroelectric Transistors. *IEEE Electron Device Lett.*, 44, 11, 1903–1906, Nov. 2023, doi: 10.1109/LED.2023.3318294.
8. Choe, G. and Yu, S., Multigate Ferroelectric Transistor Design Toward 3-nm Technology Node. *IEEE Trans. Electron Devices*, 68, 11, 5908–5911, Nov. 2021, doi: 10.1109/TED.2021.3108477.
9. Xi, F., Han, Y., Tiedemann, A., Grützmacher, D., Zhao, Q.-T., 4-Terminal Ferroelectric Schottky Barrier Field Effect Transistors as Artificial Synapses. *ESSDERC 2021 - IEEE 51st European Solid-State Device Research Conference (ESSDERC)*, Grenoble, France, pp. 291–294, 2021, doi: 10.1109/ESSDERC53440.2021.9631818.
10. Hattori, J., Ikegami, T., Fukuda, K., Ota, H., Migita, S., Asai, H., Device Simulation of Negative-Capacitance Field-Effect Transistors With a Ferroelectric Gate Insulator. *2018 International Conference on Simulation of Semiconductor Processes and Devices (SISPAD)*, Austin, TX, USA, pp. 214–219, 2018, doi: 10.1109/SISPAD.2018.8551737.
11. Tan, Q., *et al.*, Impact of ZnO Polarization on the Characteristics of Metal–Ferroelectric–ZnO Field Effect Transistor. *IEEE Trans. Electron Devices*, 58, 8, 2738–2742, Aug. 2011, doi: 10.1109/TED.2011.2157348.
12. Pandey, P. and Kaur, H., Drain Current Model of Reconfigurable Ferroelectric Field Effect Transistor (R-Fe-FET). *2018 4th International Conference for Convergence in Technology (I2CT)*, Mangalore, India, pp. 1–5, 2018, doi: 10.1109/I2CT42659.2018.9057805.
13. Mondal, S., *et al.*, ScAlN-Based ITO Channel Ferroelectric Field-Effect Transistors With Large Memory Window. *IEEE Trans. Electron Devices*, 70, 9, 4618–4621, Sept. 2023, doi: 10.1109/TED.2023.3295792.
14. Kwak, B., Lee, K., Park, N.-H., Jeon, S.J., Kim, H., Kwon, D., Recessed Channel Ferroelectric-Gate Field-Effect Transistor Memory With Ferroelectric Layer Between Dual Metal Gates. *IEEE Trans. Electron Devices*, 69, 3, 1054–1057, March 2022, doi: 10.1109/TED.2022.3144621.

15. Gomez, J., Dutta, S., Ni, K., Joshi, S., Datta, S., Steep Slope Ferroelectric Field Effect Transistor. *2019 Electron Devices Technology and Manufacturing Conference (EDTM)*, Singapore, pp. 59–61, 2019, doi: 10.1109/EDTM.2019.8731115.
16. Gomez, J., *et al.*, Significance of Multi and Few Domain Ferroelectric Switching Dynamics for Steep-Slope Non-Hysteretic Ferroelectric Field Effect Transistor. *2019 Device Research Conference (DRC)*, Ann Arbor, MI, USA, pp. 247–248, 2019, doi: 10.1109/DRC46940.2019.9046482.
17. Schenk, T. and Mueller, S., A New Generation of Memory Devices Enabled by Ferroelectric Hafnia and Zirconia. *2021 IEEE International Symposium on Applications of Ferroelectrics (ISAF)*, Sydney, Australia, pp. 1–11, 2021, doi: 10.1109/ISAF51943.2021.9477377.
18. Song, Y.S., Kim, S., Kim, J.H., Kim, G., Lee, J.-H., Choi, W.Y., Enhancement of Thermal Characteristics and On-Current in GAA MOSFET by Utilizing Al2O3-Based Dual-κ Spacer Structure. *IEEE Trans. Electron Devices*, 70, 1, 343–348, Jan. 2023, doi: 10.1109/TED.2022.3223321.
19. Weinreich, W., Lederer, M., Czernohorsky, M., Enablement of CMOS integrated sensor, harvesting and storage applications by ferroelectric HfO2. *2022 International Conference on IC Design and Technology (ICICDT)*, Hanoi, Vietnam, pp. 84–87, 2022, doi: 10.1109/ICICDT56182.2022.9933079.
20. Tayal, S., *et al.*, Incorporating Bottom-Up Approach Into Device/Circuit Co-Design for SRAM-Based Cache Memory Applications. *IEEE Trans. Electron Devices*, 69, 11, 6127–6132, Nov. 2022, doi: 10.1109/TED.2022.3210070.
21. Song, Y.S., Tayal, S., Rahi, S.B., Kim, J.H., Upadhyay, A.K., Park, B.-G., Thermal-Aware IC Chip Design by Combining High Thermal Conductivity Materials and GAA MOSFET. *2022 5th International Conference on Circuits, Systems and Simulation (ICCSS)*, Nanjing, China, pp. 135–140, 2022, doi: 10.1109/ICCSS55260.2022.9802341.
22. Udayakumar, K.R., *et al.*, Low-power ferroelectric random access memory embedded in 180nm analog friendly CMOS technology. *2013 5th IEEE International Memory Workshop*, Monterey, CA, USA, pp. 128–131, 2013, doi: 10.1109/IMW.2013.6582115.
23. Rathore, S., *et al.*, Self-Heating Aware Threshold Voltage Modulation Conforming to Process and Ambient Temperature Variation for Reliable Nanosheet FET. *2023 IEEE International Reliability Physics Symposium (IRPS)*, Monterey, CA, USA, pp. 1–5, 2023, doi: 10.1109/IRPS48203.2023.10117918.
24. Rahi, S.B., Asthana, P., Gupta, S., Heterogate junctionless tunnel field-effect transistor: future of low-power devices. *J. Comput. Electron.*, 16, 30–38, 2017, https://link.springer.com/article/10.1007/s10825-016-0936-9.
25. Jaisawal, R.K., *et al.*, Self-Heating and Interface Traps Assisted Early Aging Revelation and Reliability Analysis of Negative Capacitance FinFET. *2023 7th IEEE Electron Devices Technology & Manufacturing

Conference (EDTM), Seoul, Korea, Republic of, pp. 1–3, 2023, doi: 10.1109/EDTM55494.2023.10103127.

26. Fichtner, S., Kaden, D., Lofink, F., Wagner, B., A Generic CMOS Compatible Piezoelectric Multilayer Actuator Approach Based on Permanent Ferroelectric Polarization Inversion In Al1-X ScxN. *2019 20th International Conference on Solid-State Sensors, Actuators and Microsystems & Eurosensors III (TRANSDUCERS & EUROSENSORS XXXIII)*, Berlin, Germany, pp. 289–292, 2019, doi: 10.1109/TRANSDUCERS.2019.8808624.
27. Upadhyay, A.K., Rahi, S.B., Tayal, S., Song, Y.S., Recent progress on negative capacitance tunnel FET for low-power applications: Device perspective. *Microelectron. J.*, 129, 105583, Nov. 2022, https://www.sciencedirect.com/science/article/abs/pii/S0026269222002129.
28. Rahi, S.B., Ghosh, B., Bishnoi, B., Temperature effect on hetero structure junctionless tunnel FET. *J. Semicond.*, 36, 3, 034002, Mar. 2015, https://iopscience.iop.org/article/10.1088/1674-4926/36/3/034002/meta.
29. Song, Y.S., Thoutam, L.R., Tayal, S., Rahi, S.B., Arun Samuel, T.S., *Handbook of Emerging Materials for Semiconductor Industry*, Springer Nature, Singapore, 2024.
30. Chen, X., *et al.*, Printed Organic Synaptic Transistor Array for One-to-Many Neural Response. *IEEE Electron Device Lett.*, 43, 3, 394–397, March 2022, doi: 10.1109/LED.2022.3144662.
31. Bhardwaj, A., Kumar, P., Raj, B., Anand, S., Design and Performance Optimization of doping-less Vertical Nanowire TFET using Gate Stack Technique. *J. Electron. Mater. (JEMS)*, Springer, 41, 7, 4005–4013, 2022.
32. Singh, J. and Raj, B., Tunnel Current Model of Asymmetric MIM Structure Levying Various Image Forces to Analyze the Characteristics of Filamentary Memristor. *Appl. Phys. A*, Springer, 125, 3, 203.1 to 203.11, Feb 2019.
33. Goyal, C., Ubhi, J.S., Raj, B., Low Leakage Zero Ground Noise Nanoscale Full Adder using Source Biasing Technique. *J. Nanoelectron. Optoelectron.*, American Scientific Publishers, 14, 360–370, March 2019.
34. Singh, G., Sarin, R.K., Raj, B., A Novel Robust Exclusive-OR Function Implementation in QCA Nanotechnology with Energy Dissipation Analysis. *J. Comput. Electron.*, Springer, 15, 2, 455–465, June 2016.
35. Wadhwa, G., Kamboj, P., Raj, B., Design optimisation of junctionless TFET biosensor for high sensitivity. *Adv. Nat. Sci.: Nanosci. Nanotechnol.*, 10, 7, 045001, 2019.
36. Bansal, P. and Raj, B., Memristor Modeling and Analysis for Linear Dopant Drift Kinetics. *J. Nanoeng. Nanomanuf.*, American Scientific Publishers, 6, 1–7, 2016.
37. Ashima, Vaithiyanathan, D., Raj, B., Performance Analysis of Charge Plasma induced Graded Channel Si nanotube. *J. Eng. Res. (JER)*, EMSME Special Issue, 146–154, Aug 2021.

38. Tomar, A.S., Magraiya, V.K., Raj, B., Scaling of Access and Data Transistor for High Performance DRAM Cell Design. *Quantum Matter*, 2, 412–416, Oct. 2013.
39. Jain, N. and Raj, B., Parasitic Capacitance and Resistance Model Development and Optimization of Raised Source/Drain SOI FinFET Structure for Analog Circuit Applications. *J. Nanoelectron. Optoelectron.*, ASP USA, 13, 531–539, Apl 2018.
40. Singh, S., Vishvakarma, S.K., Raj, B., Analytical Modeling of Split-Gate Junction-Less Transistor for a Biosensor Application. *Sens. Bio-sensing*, Elsevier, 18, 31–36, April 2018.
41. Gopal, M. and Raj, B., Low Power 8T SRAM Cell Design for High Stability Video Applications. *ITSI Transaction Electr. Electron. Eng.*, 1, 5, 91–97, 2013.
42. Raj, B., Mitra, J., Bihani, D.K., Rangharajan, V., Saxena, A.K., Dasgupta, S., Analysis of Noise Margin, Power and Process Variation for 32 nm FinFET Based 6T SRAM Cell. *J. Comput. (JCP)*, Academy Publisher Finland, 5, 6, 1–8, 2010.
43. Sharma, D., Mehra, R., Raj, B., Comparative Analysis of Photovoltaic Technologies for High Efficiency Solar Cell Design. *Superlattices Microstruct.*, Elsevier, 153, 106861, May 2021.
44. Kaur, P., Buttar, A.S., Raj, B., A comprehensive Analysis of Nanoscale Transistor based Biosensor: A Review. *Indian J. Pure Appl. Phys.*, 59, 304–318, April 2021.
45. Yadav, D., Raj, B., Raj, B., Design and Simulation of Low Power Microcontroller for IoT Applications. *J. Sens. Lett.*, ASP, 18, 401–409, May 2020.
46. Singh, S. and Raj, B., A 2-D Analytical Surface Potential and Drain current Modeling of Double-Gate Vertical t-shaped Tunnel FET. *J. Comput. Electr.*, Springer, 19, 1154–1163, Apl 2020.
47. Singh, J. and Raj, B., An Accurate and Generic Window function for Non-linear Memristor Model. *J. Comput. Electr.*, Springer, 18, 2, 640–647, June 2019.
48. Kaur, M., Gupta, N., Kumar, S., Raj, B., Singh, A.K., Comparative RF and Crosstalk Analysis of Carbon Based Nano Interconnects. *IET Circuit Devices Syst.*, 15, 6, 493–503, Feb 2021.
49. Kandasamy, N., Ahmad, F., Ajitha, D., Raj, B., Telagam, N., Quantum Dot Cellular Automata based Scan Flip Flop and Boundary Scan Register. *IETE J. Res.*, 66, 535–548, 2020.
50. Singh, J. and Raj, B., Design and Investigation of 7T2M NVSARM with Enhanced Stability and Temperature Impact on Store/Restore Energy. *IEEE Trans. Very Large Scale Integr. Syst.*, 27, 6, 1322–1328, June 2019.
51. Bhardwaj, A.K., Gupta, S., Raj, B., Singh, A., Impact of Double Gate Geometry on the Performance of Carbon Nanotube Field Effect Transistor Structures for Low Power Digital Design. *Comput. Theor. Nanosci.*, ASP, 16, 1813–1820, 2019.

52. Jain, N. and Raj, B., Thermal Stability Analysis and Performance Exploration of Asymmetrical Dual-k underlap Spacer (ADKUS) SOI FinFET for Security and Privacy Applications. *Indian J. Pure Appl. Phys. (IJPAP)*, 57, 352–360, May 2019.
53. Jain, N. and Raj, B., Dual-k Spacer Region Variation at the Drain Side of Asymmetric SOI FinFET Structure: Performance Analysis towards the Analog/RF Design Applications. *J. Nanoelectron. Optoelectron.*, American Scientific Publishers, 14, 349–359, March 2019.
54. Singh, J., Sharma, S., Raj, B., Khosla, M., Analysis of barrier layer thickness on performance of In1-xGaxAs based Gate Stack Cylindrical Gate Nanowire MOSFET. *JNO*, ASP, 13, 1473–1477, Oct 2018.
55. Jain, N. and Raj, B., Analysis and Performance Exploration of High-k SOI FinFETs Over the Conventional Low-k SOI FinFET toward Analog/RF Design. *J. Semicond. (JoS)*, IOP Science, 39, 12, 124002-1-7, Dec 2018.
56. Goyal, C., Ubhi, J.S., Raj, B., A reliable leakage reduction technique for approximate full adder with reduced ground bounce noise. *J. Math. Probl. Eng.*, Hindawi, 2018, Article ID 3501041, 16, 15 Oct 2018.
57. Anuradha, J.S., Raj, B., Khosla, M., Design and Performance Analysis of Nano-scale Memristor-based Nonvolatile SRAM. *J. Sens. Lett.*, American Scientific Publishers, 16, 798–805, Oct 2018.
58. Wadhwa, G. and Raj, B., Parametric Variation Analysis of Charge-Plasma-based Dielectric Modulated JLTFET for Biosensor Application. *IEEE Sens. J.*, 18, 15, 6070–6077, AUGUST 1 2018, SCI.
59. Singh, J. and Raj, B., Comparative Analysis of Memristor Models for Memories Design. *JoS*, IoP, 39, 7, 074006-1-12, July 2018.
60. Yadav, D., Chouhan, S.S., Vishvakarma, S.K., Raj, B., Application Specific Microcontroller Design for IoT based WSN. *Sens. Lett.*, ASP, 16, 374–385, May 2018.
61. Singh, G., Sarin, R.K., Raj, B., Fault-Tolerant Design and Analysis of Quantum-Dot Cellular Automata Based Circuits. *IEEE/IET Circuits Devices Syst.*, 12, 638–64, 2018.
62. Singh, J. and Raj, B., Modeling of Mean Barrier Height Levying Various Image Forces of Metal Insulator Metal Structure to Enhance the Performance of Conductive Filament Based Memristor Model. *IEEE Nanotechnol.*, 17, 2, 268–267, March 2018.
63. Jain, A., Sharma, S., Raj, B., Analysis of Triple Metal Surrounding Gate (TM-SG) III-V Nanowire MOSFET for Photosensing Application. *Opto-electron. J.*, Elsevier, 26, 2, 141–148, May 2018.
64. Jain, A., Sharma, S., Raj, B., Design and Analysis of High Sensitivity Photosensor Using Cylindrical Surrounding Gate MOSFET For Low Power Sensor Applications. *Eng. Sci. Technol. Int. J.*, Elsevier's, 19, 4, 1864–1870, December 2016.

65. Saiphani Kumar, G., Singh, A., Raj, B., Design and Analysis of Gate All Around CNTFET based SRAM cell Design. *J. Comput. Electron.*, Springer, 17, 1, 138–145, March 2018.
66. Goyal, C., Ubhi, J.S., Raj, B., A Low Leakage CNTFET based Inexact Full Adder for Low Power Image Processing Applications. *Int. J. Circuit Theory Appl.*, Wiley, 47, 9, 1446–1458, September 2019.
67. Raj, B., Saxena, A.K., Dasgupta, S., A Compact Drain Current and Threshold Voltage Quantum Mechanical Analytical Modeling for FinFETs. *J. Nanoelectron. Optoelectron. (JNO)*, USA, 3, 2, 163–170, 2008.
68. Wadhwa, G. and Raj, B., An Analytical Modeling of Charge Plasma based Tunnel Field Effect Transistor with Impacts of Gate underlap Region. *Superlattices Microstruct.*, Elsevier, 142, 106512, June 2020.
69. Singh, S. and Raj, B., Modeling and Simulation analysis of SiGe hetrojunction Double GateVertical t-shaped Tunnel FET. *Superlattices Microstruct.*, Elsevier, 142, 106496, June 2020.
70. Jain, N. and Raj, B., Device and Circuit Co-Design Perspective Comprehensive Approach on FinFET Technology - A Review. *J. Electron Devices*, 23, 1, 1890–1901, 2016.
71. Kumar, S. and Raj, B., Analysis of I_{ON} and Ambipolar Current for Dual-Material Gate-drain Overlapped DG-TFET. *J. Nanoelectron. Optoelectron.*, American Scientific Publishers, USA, 11, 323–333, June 2016.
72. Anjum, N., Bali, T., Raj, B., Design And Simulation Of Handwritten Multiscript Character Recognition. *Int. J. Adv. Res. Comput. Commun. Eng.*, 2, 7, 2544–2549, July 2013.
73. Singh, K. and Raj, B., Performance and analysis of temperature dependent Multi-walled Carbon Nanotubes as Global Interconnects at different technology nodes. *J. Comput. Electron.*, Springer, 14, 2, 469–476, June 2015.
74. Kumar, S. and Raj, B., Compact channel potential analytical Modeling of DG-TFET based on Evanescent–mode Approach. *J. Comput. Electron.*, Springer, 14, 2, 820–827, July 2015.
75. Singh, K. and Raj, B., Temperature Dependent Modeling and Performance Evaluation of Multi-Walled CNT and Single-Walled CNT as Global Interconnects. *J. Electron. Mater.*, Springer, 44, 12, 4825–4835, Dec 2015.
76. Sharma, V.K., Pattanaik, M., Raj, B., INDEP approach for leakage reduction in nanoscale CMOS circuits. *Int. J. Electron.*, Taylor & Francis, 102, 2, 200–215, 2014.
77. Singh, K. and Raj, B., Influence of Temperature on MWCNT bundle, SWCNT bundle and Copper interconnects for Nanoscaled Technology nodes. *J. Mater. Sci. Mater. Electron.*, Springer, 26, 8, 6134–6142, 2015.
78. Anjum, N., Bali, T., Raj, B., Design and Simulation of Handwritten Gurumukhi and Devanagri Numerical Recognition. *Int. J. Comput. Appl.*, Foundation of Computer Science, New York, USA, 73, 12, 16–21, 2013.

79. Khandelwal, S., Gupta, V., Raj, B., Gupta, R.D., Process Variability Aware Low Leakage Reliable Nano Scale DG-FinFET SRAM Cell Design Technique. *J. Nanoelectron. Optoelectron.*, 10, 6, 810–817, December 2015.
80. Sharma, V.K., Pattanaik, M., Raj, B., ONOFIC Approach: Low Power High Speed Nanoscale VLSI Circuits Design. *Int. J. Electron.*, Taylor & Francis, 101, 1, 61–73, 2014.
81. Khandelwal, S., Raj, B., Gupta, R.D., FinFET based 6T SRAM Cell Design: Analysis of Performance Metric, Process Variation and Temperature Effect. *J. Comput. Theor. Nanosci.*, ASP, USA, 12, 2500–2506, 2015.
82. S.S., Yadav, S., Rahul, J., Srivastava, A., Raj, B., Impact of HfO_2 in Graded Channel Dual Insulator Double Gate MOSFET. *J. Comput. Theor. Nanosci.*, American Scientific Publishers, 12, 6, 950–953, April 2015.
83. Sharma, V.K., Pattanaik, M., Raj, B., PVT variations aware low leakage INDEP approach for nanoscale CMOS Circuits. *Microelectron. Reliab.*, Elsevier, 54, 90–99, 2014.
84. Raj, B., Saxena, A.K., Dasgupta, S., Quantum Mechanical Analytical Modeling of Nanoscale DG FinFET: Evaluation of Potential, Threshold Voltage and Source/Drain Resistance. *Elsevier's J. Mater. Sci. Semicond. Process.*, Elsevier, 16, 4, 1131–1137, 2013.
85. Gopal, M., Prasad, S.S.D., Raj, B., 8T SRAM Cell Design for Dynamic and Leakage Power Reduction, International Journal of Computer Applications. *Found. Comput. Sci.*, New York, USA, 71, 9, 43–48, June 2013.
86. Pattanaik, M., Raj, B., Sharma, S., Kumar, A., Diode Based Trimode Multi-Threshold CMOS Technique for Ground Bounce Noise Reduction in Static CMOS Adders. *Adv. Mater. Res.*, Trans Tech Publications, Switzerland, 548, 885–889, 2012.
87. Raj, B., Saxena, A.K., Dasgupta, S., Nanoscale FinFET Based SRAM Cell Design: Analysis of Performance metric, Process variation, Underlapped FinFET and Temperature effect. *IEEE Circuits Syst. Mag.*, 11, 2, 38–50, 2011.
88. Sharma, V.K., Pattanaik, M., Raj, B., Leakage Current ONOFIC Approach for Deep Submicron VLSI Circuit Design. *Int. J. Electr. Comput. Electron. Commun. Eng.*, World Academy Sciences, Engineering Technology, 7, 4, 239–244, 2013.
89. Kaur, P., Gill, S.S., Raj, B., Comparative Analysis of OFETs Materials and Devices for Sensor Applications. *J. Silicon*, Springer, 14, 4463–4471, 2022.
90. Sharma, S.K., Kumar, P., Raj, B., Raj, B., $In_{1-x}Ga_xAs$ Double Metal Gate-Stacking Cylindrical Nanowire MOSFET for highly sensitive Photo detector. *J. Silicon*, Springer, 14, 3535–3541, 2022.
91. Raj, B., Saxena, A.K., Dasgupta, S., Analytical Modeling of Quasi Planar Nanoscale Double Gate FinFET with Source/Drain Resistance and field dependent carrier mobility: A quantum mechanical study. *J. Comput. (JCP)*, Academy Publisher, FINLAND, 4, 9, 1–8, 2009.
92. Bhushan, S., Khandelwal, S., Raj, B., Analyzing Different Mode FinFET based Memory Cell at Different Power Supply for Leakage Reduction.

Seventh International Conference on Bio-Inspired Computing: Theories and Application, (BIC-TA 2012), Advances in Intelligent Systems and Computing, vol. 202, pp. 89–100, 2012, 2013.

93. Singh, J. and Raj, B., Temperature Dependent Analytical Modeling and Simulations of Nanoscale Memristor. *J. Eng. Sci. Technol. Int. J.*, Elsevier's, 21, 862–868, Oct 2018.
94. Singh, S., Bala, S., Raj, B., Raj, B., Improved Sensitivity of Dielectric Modulated Junctionless Transistor for Nanoscale Biosensor Design. *Sens. Lett.*, ASP, 18, 328–333, Apl 2020.
95. Kumar, V., Vishvakarma, S.K., Raj, B., Design and Performance Analysis of ASIC for IoT Applications. *Sens. Lett.*, ASP, 18, 31–38, Jan 2020.
96. Jaiswal, A., Sarin, R.K., Raj, B., Sukhija, S., A Novel Circular Slotted Microstrip-fed Patch Antenna with Three Triangle Shape Defected Ground Structure for Multiband Applications. *Adv. Electromagn. (AEM)*, 7, 3, 56–63, Aug 2018.
97. Wadhwa, G. and Raj, B., Label Free Detection of Biomolecules using Charge-Plasma-Based Gate Underlap Dielectric Modulated Junctionless TFET. *J. Electron. Mater. (JEMS)*, Springer, 47, 8, 4683–4693, August 2018.
98. Singh, G., Sarin, R.K., Raj, B., Design and Performance Analysis of a New Efficient Coplanar Quantum-Dot Cellular Automata Adder. *Indian J. Pure Appl. Phys. (IJPAP)*, 55, 97–103, February 2017.
99. Kaur, P., Pandey, V., Raj, B., Comparative Study of Efficient Design, Control and Monitoring of Solar Power using IoT. *Sens. Lett.*, ASP, 18, 419–426, May 2020.
100. Bhardwaj, A.K., Gupta, S., Raj, B., Development & Analysis of compact model for Double Gate Schottky Barrier CNTFET. *J. Nanoelectron. Optoelectron.*, ASP, 15, 1199–1208, Aug 2020.
101. Wadhwa, G., Kamboj, P., Singh, J., Raj, B., Design and Investigation of Junctionless DGTFET for Biological Molecule Recognition. *Trans. Electr. Electron. Mater.*, Springer, 22, 282–289, 2021.
102. Verma, S.K., Singh, S., Wadhwa, G., Raj, B., Detection of Biomolecules using Charge-Plasma based Gate Underlap Dielectric Modulated Dopingless TFET. *Trans. Electr. Electron. Mater. (TEEM)*, Springer, 21, 528–535, June 2020.
103. Jain, N. and Raj, B., Impact of underlap spacer region variation on Electrostatic and analog/RF performance of symmetrical high-k SOI FinFET at 20 nm channel length. *J. Semicond. (JoS)*, IOP Sciences, 38, 12, 122002, Dec 2017.
104. Singh, S. and Raj, B., Analytical Modeling and Simulation analysis of T-shaped III-V heterojunction Vertical T-FET. *Superlattices Microstruct.*, Elsevier, 147, 106717, Nov 2020.
105. Singh, G., Sarin, R.K., Raj, B., Design and Analysis of Area Efficient QCA Based Reversible Logic Gates. *J. Microprocess. Microsys.*, Elsevier, 52, 59–68, May 2017.

106. Singh, S. and Raj, B., Parametric variation analysis on hetero-junction Vertical t-shape TFET for Supressing ambipolar conduction. *Indian J. Pure Appl. Phys.*, 58, 478–485, June 2020.
107. Soniya, S.S., Wadhwa, G., Raj, B., Design and Analysis of Dual Source Vertical Tunnel Field Effect Transistor for high performance. *Trans. Electr. Electron. Mater.*, Springer, 21, 74–82, Oct 2019.
108. Kaur, M., Gupta, N., Kumar, S., Raj, B., Arun Kumar Singh, RF Performance Analysis of Intercalated Graphene Nanoribbon Based Global Level Interconnects. *J. Comput. Electron.*, Springer, 19, 1002–1013, June 2020.
109. Wadhwa, G. and Raj, B., Design and Performance Analysis of Junctionless TFET Biosensor for high sensitivity. *IEEE Nanotechnol.*, 18, 567–574, 2019.
110. Singh, J. and Raj, B., Enhanced Nonlinear Memristor Model Encapsulating Stochastic Dopant Drift. *JNO*, ASP, 14, 958–963, 2019.
111. Singh, G., Sohi, B.S., Raj, B., Material Properties Analysis of Graphene Base Transistor (GBT) for VLSI Analog Circuits. *Indian J. Pure Appl. Phys. (IJPAP)*, 55, 896–902, Dec-2017.
112. Kumar, S. and Raj, B., Estimation of Stability and Performance metric for Inward Access Transistor based 6T SRAM Cell Design using n-type/p-type DMDG-GDOV TFET. *IEEE VLSI Circuits Syst. Lett.*, 3, 2, 25–39, June 2017.
113. Sharma, S., Kumar, A., Pattanaik, M., Raj, B., Forward Body Biased Multimode Multi-Threshold CMOS Technique for Ground Bounce Noise Reduction in Static CMOS Adders. *Int. J. Inf. Electron. Eng.*, 3, 3, 567–572, 2013.
114. Singh, H., Kumar, P., Raj, B., Performance Analysis of Majority Gate SET Based 1-bit Full Adder. *Int. J. Comput. Commun. Eng. (IJCCE)*, IACSIT Press Singapore, 2, 4, 1–2, 2013, ISSN: 2010-3743.
115. Bhardwaj, A.K., Gupta, S., Raj, B., Investigation of Parameters for Schottky Barrier (SB) Height for Schottky Barrier Based Carbon Nanotube Field Effect Transistor Device. *J. Nanoelectron. Optoelectron.*, ASP, 15, 783–791, July 2020.
116. Bansal, P. and Raj, B., Memristor: A Versatile Nonlinear Model for Dopant Drift and Boundary Issues. *JCTN*, American Scientific Publishers, 14, 5, 2319–2325, May 2017.
117. Jain, N. and Raj, B., An Analog and Digital Design Perspective Comprehensive Approach on Fin-FET (Fin-Field Effect transistor) Technology - A Review. *Rev. Adv. Sci. Eng. (RASE)*, ASP, 5, 1–14, 2016.
118. Raj, B., Saxena, A.K., Dasgupta, S., Analytical Modeling for the Estimation of Leakage Current and Subthreshold Swing Factor of Nanoscale Double Gate FinFET Device. *Microelectron. Int.*, UK, 26, 53–63, 2009.
119. Soniya, S.S., Wadhwa, G., Raj, B., An Analytical Modeling for Dual Source Vertical Tunnel Field Effect Transistor. *Int. J. Recent Technol. Eng. (IJRTE)*, 8, 2, 603–608, July 2019.
120. Singh, S. and Raj, B., Design and analysis of hetrojunction Vertical T-shaped Tunnel Field Effect Transistor. *J. Electron. Mater.*, Springer, 4810, 6253–6260, October 2019.

About the Editors

Balwinder Raj, PhD is an associate professor at the National Institute of Technology, Jalandhar, India. He has authored and co-authored five books, 12 book chapters, and over 100 research papers in peer-reviewed national and international journals and conferences. His areas of interest include nanoscale semiconductor device modeling, nanoelectronics and their applications in hardware security, sensors, and circuit design, FinFET-based memory design, low-power very large-scale integrated design, and field programmable gate array implementation.

Shiromani Balmukund Rahi, PhD is an assistant professor in the School of Information and Communication Technology, Gautam Buddha University, Greater Noida, India. He has published 25 research articles, two conference proceedings, 25 book chapters for various projects, and seven books for reputed internal publishers. He also serves as a reviewer for various national and international journals, conferences, and workshops. His research interests include semiconductor device modeling and simulation, tunnel FETs, NCFETs, and Nanosheet FETs.

Nandakishor Yadav, PhD is a senior scientist with the Fraunhofer Institute for Photonic Microsystems, Dresden, Germany with over ten years of research and teaching experience. He has published over 50 research articles in peer-reviewed journals and conferences. His research interests include very large-scale interface design, Ferroelectric memory, and peripheral circuit design.

Index

Also of Interest

Books by the same editors

Field Effect Transistors, Edited by P. Suveetha Dhanaselvam, K. Srinivasa Rao, Shiromani Balmukund Rahi, and Dharmendra Singh Yadav, ISBN: 9781394248476. *Field Effect Transistors* is an essential read for anyone interested in the future of electronics, as it provides a comprehensive yet accessible exploration of innovative semiconductor devices and their applications, making it a perfect resource for both beginners and seasoned professionals in the field.

Integrated Devices for Artificial Intelligence and VLSI: VLSI Design, Simulation and Applications, Edited by Balwinder Raj, Suman Lata Tripathi, Tarun Chaudhary, K. Srinivasa Rao, and Mandeep Singh, ISBN: 9781394204359. With its in-depth exploration of the close connection between microelectronics, AI, and VLSI technology, this book offers valuable insights into the cutting-edge techniques and tools used in VLSI design automation, making it an essential resource for anyone seeking to stay ahead in the rapidly evolving field of VLSI design.

Advanced Ultra Low-Power Semiconductor Devices; Design and Applications, Edited by Shubham Tayal, Abhishek Kumar Upadhyay, Shiromani Balmukund Rahi, and Young Suh Song, ISBN: 9781394166411. Written and edited by a team of experts in the field, this important new volume broadly covers the design and applications of metal oxide semiconductor field effect transistors.

Check out these other related titles from Scrivener Publishing

Condition-Based Maintenance and Residual Life Prediction, Edited by Chandan Deep Singh, Davinder Singh. Kanwal Jit Singh, and Harleen Kaur, ISBN: 9781119933120. *Condition-Based Maintenance and Residual Life Prediction* is essential for those looking to effectively implement condition-based maintenance strategies and enhance fault detection through a comprehensive understanding of vibration data analysis and residual life prediction, addressing key challenges in the field.

Power Devices and Internet of Things for Intelligent System Design, Edited by Angsuman Sarkar and Arpan Deyasi, ISBN: 9781394305780. Unlock the potential of cutting-edge advancements in power electronics and IoT with "Power Devices and Internet of Things for Intelligent System Design," a vital resource that bridges the gap between industry and academia, inspiring innovative solutions across diverse fields like agriculture, healthcare, and security.

Intelligent Green Technologies for Smart Cities, Edited by Suman Lata Tripathi, Souvik Ganguli, Abhishek Kumar, and Tengiz Magradze, ISBN: 9781119816065. Presenting the concepts and fundamentals of smart cities and developing "green" technologies, this volume, written and edited by a global team of experts, also goes into the practical applications that can be utilized across multiple disciplines and industries, for both the engineer and the student.

Hybrid Intelligent Optimization Approaches for Smart Energy: A Practical Approach, Edited by Senthilkumar Mohan, A. John, Sanjeevikumar Padmanaban, and Yasir Hamid, ISBN: 9781119821243. Written and edited by a group of experts in the field, this is a comprehensive and up-to-date description of current energy optimization techniques, such as artificial intelligence techniques, machine learning, deep learning, and IoT techniques and their future trends.

NANODEVICES FOR INTEGRATED CIRCUIT DESIGN, Edited by Suman Lata Tripathi, Abhishek Kumar, K. Srinivasa Rao, and Prasantha R. Mudimela, ISBN: 9781394185788. Written and edited by a team of experts in the field, this important new volume broadly covers the design of nano-devices and their integrated applications in digital and analog integrated circuits (IC) design.

EXPLAINABLE MACHINE LEARNING MODELS AND ARCHITECTURES: Real-Time System Implementation, Edited by Suman Lata Tripathi and Mufti Mahmud, ISBN: 9781394185849. This cutting-edge new volume covers the hardware architecture implementation, the software implementation approach, and the efficient hardware of machine learning applications.

MACHINE LEARNING TECHNIQUES FOR VLSI CHIP DESIGN, Edited by Abhishek Kumar, Suman Lata Tripathi, and K. Srinivasa Rao, ISBN: 9781119910398. This cutting-edge new volume covers the hardware architecture implementation, the software implementation approach, and the efficient hardware of machine learning applications with FPGA or CMOS circuits, and many other aspects and applications of machine learning techniques for VLSI chip design.

INTELLIGENT GREEN TECHNOLOGIES FOR SMART CITIES, Edited by Suman Lata Tripathi, Souvik Ganguli, Abhishek Kumar, and Tengiz Magradze, ISBN: 9781119816065. Presenting the concepts and fundamentals of smart cities and developing "green" technologies, this volume, written and edited by a global team of experts, also goes into the practical applications that can be utilized across multiple disciplines and industries, for both the engineer and the student.

DESIGN AND DEVELOPMENT OF EFFICIENT ENERGY SYSTEMS, Edited by Suman Lata Tripathi, Dushyant Kumar Singh, Sanjeevikumar Padmanaban, and P. Raja, ISBN: 9781119761631. Covering the concepts and fundamentals of efficient energy systems, this volume, written and edited by a global team of experts, also goes into the practical applications that can be utilized across multiple industries, for both the engineer and the student.

Electrical and Electronic Devices, Circuits, and Materials: Technical Challenges and Solutions, Edited by Suman Lata Tripathi, Parvej Ahmad Alvi, and Umashankar Subramaniam, ISBN: 9781119750369. Covering every aspect of the design and improvement needed for solid-state electronic devices and circuit and their reliability issues, this new volume also includes overall system design for all kinds of analog and digital applications and developments in power systems.

MODELING AND OPTIMIZATION OF OPTICAL COMMUNICATION NETWORKS, Edited by Chandra Singh, Rathishchandra R Gatti, K.V.S.S.S.S. Sairam, and Ashish Singh, ISBN: 9781119839200. *Modeling and Optimization of Optical Communication Networks* is a comprehensive and authoritative book that delves into the optical networks, principles, technologies, and practical applications of optical networks, equipping readers with the knowledge needed to design, implement, and optimize optical networks for various applications, from telecommunications to scientific research. With its comprehensive coverage and up-to-date insights, this book serves as an essential reference in the field of optical networks.

WIRELESS COMMUNICATION SECURITY: Mobile and Network Security Protocols, Edited by Manju Khari, Manisha Bharti, and M. Niranjanamurthy, ISBN: 9781119777144. Presenting the concepts and advances of wireless communication security, this volume, written and edited by a global team of experts, also goes into the practical applications for the engineer, student, and other industry professionals.

ADVANCES IN DATA SCIENCE AND ANALYTICS, Edited by M. Niranjanamurthy, Hemant Kumar Gianey, and Amir H. Gandomi, ISBN: 9781119791881. Presenting the concepts and advances of data science and analytics, this volume, written and edited by a global team of experts, also goes into the practical applications that can be utilized across multiple disciplines and industries, for both the engineer and the student, focusing on machining learning, big data, business intelligence, and analytics.

ARTIFICIAL INTELLIGENCE AND DATA MINING IN SECURITY FRAMEWORKS, Edited by Neeraj Bhargava, Ritu Bhargava, Pramod Singh Rathore, and Rashmi Agrawal, ISBN: 9781119760405. Written and edited by a team of experts in the field, this outstanding new volume offers solutions to the problems of security, outlining the concepts behind allowing computers to learn from experience and understand the world in terms of a hierarchy of concepts.

MACHINE LEARNING AND DATA SCIENCE: Fundamentals and Applications, Edited by Prateek Agrawal, Charu Gupta, Anand Sharma, Vishu Madaan, and Nisheeth Joshi, ISBN: 9781119775614. Written and edited by a team of experts in the field, this collection of papers reflects the most up-to-date and comprehensive current state of machine learning and data science for industry, government, and academia.

MEDICAL IMAGING, Edited by H. S. Sanjay, and M. Niranjanamurthy, ISBN: 9781119785392. Written and edited by a team of experts in the field, this is the most comprehensive and up-to-date study of and reference for the practical applications of medical imaging for engineers, scientists, students, and medical professionals.

SECURITY ISSUES AND PRIVACY CONCERNS IN INDUSTRY 4.0 APPLICATIONS, Edited by Shibin David, R. S. Anand, V. Jeyakrishnan, and M. Niranjanamurthy, ISBN: 9781119775621. Written and edited by a team of international experts, this is the most comprehensive and up-to-date coverage of the security and privacy issues surrounding Industry 4.0 applications, a must-have for any library.

CYBER SECURITY AND DIGITAL FORENSICS: Challenges and Future Trends, Edited by Mangesh M. Ghonge, Sabyasachi Pramanik, Ramchandra Mangrulkar, and Dac-Nhuong Le, ISBN: 9781119795636. Written and edited by a team of world renowned experts in the field, this groundbreaking new volume covers key technical topics and gives readers a comprehensive understanding of the latest research findings in cyber security and digital forensics.

DEEP LEARNING APPROACHES TO CLOUD SECURITY, Edited by Pramod Singh Rathore, Vishal Dutt, Rashmi Agrawal, Satya Murthy Sasubilli, and Srinivasa Rao Swarna, ISBN: 9781119760528. Covering one of the most important subjects to our society today, this editorial team delves into solutions taken from evolving deep learning approaches, solutions allow computers to learn from experience and understand the world in terms of a hierarchy of concepts.

MACHINE LEARNING TECHNIQUES AND ANALYTICS FOR CLOUD SECURITY, Edited by Rajdeep Chakraborty, Anupam Ghosh and Jyotsna Kumar Mandal, ISBN: 9781119762256. This book covers new methods, surveys, case studies, and policy with almost all machine learning techniques and analytics for cloud security solutions.

SECURITY DESIGNS FOR THE CLOUD, IOT AND SOCIAL NETWORKING, Edited by Dac-Nhuong Le, Chintin Bhatt and Mani Madhukar, ISBN: 9781119592266. The book provides cutting-edge research that delivers insights into the tools, opportunities, novel strategies, techniques, and challenges for handling security issues in cloud computing, Internet of Things and social networking.

DESIGN AND ANALYSIS OF SECURITY PROTOCOLS FOR COMMUNICATION, Edited by Dinesh Goyal, S. Balamurugan, Sheng-Lung Peng and O.P. Verma, ISBN: 9781119555643. The book combines analysis and comparison of various security protocols such as HTTP, SMTP, RTP, RTCP, FTP, UDP for mobile or multimedia streaming security protocol.

SMART GRIDS FOR SMART CITIES VOLUME 2: Real-Time Applications in Smart Cities, Edited by O.V. Gnana Swathika, K. Karthikeyan, and Sanjeevikumar Padmanaban, ISBN: 9781394215874. Written and edited by a team of experts in the field, this second volume in a two-volume set focuses on an interdisciplinary perspective on the financial, environmental, and other benefits of smart grid technologies and solutions for smart cities.

SMART GRIDS AND INTERNET OF THINGS, Edited by Sanjeevikumar Padmanaban, Jens Bo Holm-Nielsen, Rajesh Kumar Dhanaraj, Malathy Sathyamoorthy, and Balamurugan Balusamy, ISBN: 9781119812449. Written and edited by a team of international professionals, this groundbreaking new volume covers the latest technologies in automation, tracking, energy distribution and consumption of Internet of Things (IoT) devices with smart grids.

Printed in the USA/Agawam, MA
September 16, 2026

910661.102